FRONTIERS IN THE ECONOMICS OF ENVIRONMENTAL REGULATION AND LIABILITY

Frontiers in the Economics of Environmental Regulation and Liability

Edited by

MARCEL BOYER
University of Montreal

YOLANDE HIRIART
Université des Sciences Sociales Toulouse

DAVID MARTIMORT
Université des Sciences Sociales Toulouse

ASHGATE

Published by
Ashgate Publishing Limited
Gower House
Croft Road
Aldershot
Hampshire GU11 3HR
England

Ashgate Publishing Company
Suite 420
101 Cherry Street
Burlington, VT 05401-4405
USA

Ashgate website: http://www.ashgate.com

British Library Cataloguing in Publication Data
Frontiers in the economics of environmental regulation and
    liability. - (Ashgate studies in environmental and natural
    resource economics)
    1. Environmental law - Economic aspects 2. Liability for
    environmental damages 3. Industrial management -
    Environmental aspects 4. Economic development -
    Environmental aspects 5. Industrial safety - Law and
    legislation
    I. Boyer, Marcel II. Hiriart, Yolande III. Martimort, David
    333.7

Library of Congress Control Number: 2005938625

ISBN-10: 0-7546-4208-9
ISBN-13: 978-0-7546-4208-4

Printed and bound in Great Britain by Antony Rowe Ltd, Chippenham, Wiltshire.

Acknowledgments

First of all, our gratitude goes to all the contributors who have kindly accepted to participate to this collective endeavour. We would also like to warmly thank our collaborators Marie-Pierre Boé and Florence Chauvet who helped us on this project. Florence Chauvet managed the earlier stages with great skill and patience. Marie-Pierre did a huge amount of editorial work on the final document with a rarely matched efficiency and kindness.

Contents

List of Figures

List of Tables

List of Contributors

Alessandra Arcuri, *Erasmus University Rotterdam.*

Dieter Balkenborg, *University of Exeter.*

Philippe Bontems, *Université de Toulouse and INRA.*

Marcel Boyer, *Université de Montréal.*

Pierre Dubois, *Université de Toulouse and INRA.*

Winand Emons, *University of Bern.*

Karine Gobert, *Université de Sherbrooke.*

Patrick González, *Université Laval.*

Anthony Heyes, *Royal Holloway College, University of London.*

Yolande Hiriart, *Université de Toulouse.*

Catherine Liston-Heyes, *Royal Holloway College, University of London.*

Philippe Mahenc, *Université de Perpignan.*

Nathalie De Marcellis-Warin, *École polytechnique de Montréal.*

David Martimort, *Université de Toulouse.*

Thomas J. Miceli, *University of Connecticut.*

Michel Moreaux, *Université de Toulouse.*

Ingrid Peignier, *CIRANO.*

Michel Poitevin, *Université de Montréal.*

Donatella Porrini, *Universita di Lecce.*

François Salanié, *Université de Toulouse and INRA.*

Kathleen Segerson, *University of Connecticut.*

Bernard Sinclair-Desgagné, *HEC Montréal and École Polytechnique Paris.*

Nicolas Treich, *Université de Toulouse and INRA.*

Tomislav Vukina, *North Carolina State University.*

Introduction

As noticed by the great philosopher Ulrich Beck in his path-breaking book *The Risky Society: Towards a New Modernity*, the major challenge faced by our modern societies is "the impossibility to impute threats to external causes. Contrary to all cultures and previous eras of social evolutions [...], the risky society is nowadays confronted to itself". That the major risks borne by our societies are by large endogenous, be they a threat for the environment, the health or the industry, is the main concern of this book and the common thread of all the contributions we have chosen to gather below.

Industrial and/or environmental accidents may indeed be understood as resulting from three sets of complex interactions: first, between different technologies which might be combined into the production processes which generate hazards; second, between different economic agents with their own objectives which sometimes are conflicting; and finally, between those agents and those technologies. In each case, beliefs and incentives play a key role in generating risks and this book precisely aims at understanding how beliefs on risk form and how this process affects incentives in both the public and the private management of risk that our societies must undertake.

Economic agents whose interactions affect the likelihood of hazards include lawmakers, regulatory agencies, firms sometimes arranged into networks or linked through contractual relationships, individuals taken either as consumers, as workers or as voters, and lastly courts. Those agents may have different beliefs on the likelihoods of a given hazard. Those beliefs may be based on objective scientific knowledge available through experts' reports and advices or they may sometimes be much more subjective in nature. Both ethical values and objective scientific knowledge contribute indeed in helping agents to build a conceptual framework which enables them to assess the amount of sustainable risks that a society should accept to reconcile growth and safety. In this search for such a sustainable economic development, one must properly understand that the main difficulty lies in the fact that the various agents involved in generating risks or in assessing their consequences for society may be endowed with different information sets and that the convergence of their expectations is a major challenge of risk management be it publicly or privately undertaken.

As already pointed out above, one of the main factors affecting the likelihood of industrial or environmental hazards is the interplay between different technologies (for instance, dangerous chemicals being transported through an urban network). Complexity in the multistage production processes generating hazards makes it in general quite difficult to understand the sources and consequences of those risks. The formation of the individuals' beliefs about these technological risks is not well understood even though those perceptions constitute a key driver of individuals' behavior. In particular, it is well known that individuals may sometimes share only very rough estimates of the true risk that a given activity or technology represents. Since their behavior is determined by their subjective beliefs rather than by the true characteristics of the technologies, it is crucial to ascertain those beliefs and characterize the optimal response that either the regulator or the individuals will have in front of systematic errors in risk perception.

The agents' incentives and contractual relationships are other key factors affecting the likelihood and size or severity of accidents. Information and beliefs play a significant role in assessing the consequences of incentives on risk management, in particular when agents do not share common knowledge of preferences and technologies. Information sets are in part endogenously determined by the willingness and efforts that agents are ready to undertake to become more informed. This is true of lawmakers, regulators, firms, individuals, and, last but not least, judges. A better understanding of the design of public and private policies towards reducing the probability of accident occurrences and their severity undoubtfully rests on a proper modeling of the agents preferences and information sets, taking possibly into account this endogeneity. Besides individual agents characteristics, the contractual relationships that agents entertain between themselves brings us to an even more complex level of analysis. As incentive theory has forcefully taught us over the last thirty years or so,[1] the imperfect and incomplete knowledge that different agents may have of each other objectives and information is a key factor limiting the effectiveness of any private or public management of risk. The links between those agents are often hierarchical in nature and can be viewed as a web of complex contractual relationships stroke in a given institutional and legal environment. With each contractual layer comes some agency cost that must explicitly be taken into account at the time of revising the standard cost-benefit analysis that must be made to assess if a given risk should be borne by society.

Imperfections in the contractual relationships of the various agents involved in risk management affect the distribution and size of those risks. Relationships between lawmakers in charge of enacting laws and regulators who implement them; between those regulators and the regulated firms who must comply with the existing standards and regulations; between banks and firms under joint liability rules; or between owners and managers of ventures involved in hazardous production processes are all examples of contracts which affect the standard cost-benefit approach. Along this complex hierarchical structure, incentives for safety care trickle-down in a non-trivial manner with transaction costs due to informational asymmetries and conflicting interests

---

[1]See Laffont and Martimort (2002).

having a bite at each layer. Informational asymmetries and conflicting interests compound to undermine the process by which the probability and/or severity of accidents are mitigated. Moreover, higher-up "constitutional" welfare analysis may impose limits on the powers of different agents, in particular of regulators, making the seemingly first-best optimal solutions unattainable. Those limits and constraints may be set up in order to satisfy other welfare conditions outside the consideration of the modeler or of the institutions at play.

The papers gathered in this book represent different viewpoints on the determining factors explaining the emergence and distribution of industrial or environmental accidents. We have chosen to arbitrarily classify those papers into three different sets corresponding to either their focus on beliefs formation vis-à-vis risk or misconduct, their concerns for a better design of legal rules imposed ex post on injurers or finally their common concerns on how private incentives are affected by risk regulation and liability rules in asymmetric information settings. Those three directions represent the state of the art in the field of regulation and liability and we hope that they will motivate other researchers to accompany contributors in the present volume in the search for a more effective picture of the institutional and incentive constraints faced by major players in public and private management of industrial/environmental risk.

Basic Principles of Risk Regulation, Beliefs and Awareness

In modern societies, risky activities are a significant consequence of technological development. Unfortunately, the hazards brought up by these activities are only partly unknown. That is why, according to Alessandra Arcuri, the analysis of decisions related to whether and how to regulate these activities has gained importance. This view is developed in her paper entitled *The Case for a procedural Version of the Precautionary Principle Erring on the Side of Environmental Preservation.*

The Precautionary Principle is certainly one of the main guides for public decision-making under uncertainty when it comes to environmental or health policies. Given that its application will necessarily affect activities by imposing potentially heavy costs on industries and society as a whole, it is essential that the Precautionary Principle as well as its multiple interpretations and implications be properly understood. Arcuri presents the available options for decision-makers and argues that, in order for the Principle to be operational and useful, its application should lead to decisions that err on the side of environmental preservation but nevertheless include procedural norms. This would allow a reduction of uncertainty when no outcome can readily be identified as optimal. Moreover, better information and formation of qualified viewpoints would come about, thanks to such procedural norms.

In the first part of her paper, Arcuri presents a brief overview of the origins of the Principle, of its adoption at the international level, and of its different and extreme versions. She proposes a version of the principle that combines different points of view. Arcuri claims that the Precautionary Principle is better complemented by cost effectiveness analysis than by cost-benefit analysis due to the uncertainty that characterizes the addressed settings. She insists on the democratization of scientific knowledge, warning against the risks of "expertocracy" and encouraging greater interaction between experts and lay men. Without a proper understanding of legal structures, refined economic analyses would likely turn out to be sterile. Arcuri proposes a procedural version of the Principle based on sound science and opened to democratic scrutiny in order to ensure that all scientific facts and moral values be taken into account before the decision-maker adopts any decision.

One of the key lesson of Arcuri's paper is that knowledge on risks and its management should be diffused as largely as possible within our societies. This is also an underlying concern of Nathalie De Marcellis-Warin, Ingrid Peignier and Bernard Sinclair-Desgagné in their article entitled *Informational Regulation of Industrial Safety - An examination of the U.S. Local Emergency Planning Committee*. These authors discuss the trend in "regulation by information". Regulation by information was earlier on characterized by Kleindorfer and Ordts (1998) as follows: "Information regulation is any regulation which provides to affected stakeholders information on the operations of regulated entities, usually with the expectations that such stakeholders will then exert pressure on these entities to comply with regulation in a manner which serves the interests of the stakeholders". Since the 1980s, the regulation of industrial risks towards human health and the environment has indeed moved towards a greater empowerment of risk bearers and various stakeholders by providing them with specific information concerning hazardous activities in their area.

In the U.S., organizations at the state and local levels were created in order to develop emergency plans in the event that toxic chemicals be disastrously released and to subsequently provide information regarding covered facilities. More than 4000 such local organizations (LEPCs) exist in the United States and are considered as essential institutions in developing cooperation between federal and state agencies, in linking the chemical industry and the general public, thereby potentially fostering the effectiveness of the "regulation by information" of industrial risks.

De Marcellis, Peignier and Sinclair-Desgagné portray the functioning of the LEPCs, present a summary of related literature, report on some interviews they made with local officials in Vermont and Maryland, and provide preliminary data from a recent survey they conducted on a subset of some 200

LEPCs. International comparisons are also offered in addition to an appraisal of the actual LEPCs.

The authors conclude their overall appraisal by qualifying the performance of LEPCs as favorable but mixed. On the one hand, a fair proportion of LEPCs seems to be working effectively and to have met expectations in terms of promoting public awareness and involvement. Furthermore, the fact that other countries are also seeking to implement similar devices at home may indicate that this approach to practicing informational regulation is right and doable. On the other hand, there is considerable heterogeneity in performance across LEPCs, as an equally significant proportion of them still faces minimal public interest. Clearly, the actual process is a learning one which is subject to trial and error. Further creativity and effort need to be deployed vis-à-vis the transfer of best practices across LEPCs and the design of effective two-way communication channels with and between stakeholders. Therefore, a superior communication system with and between stakeholders is required if best practices across LEPCs ought to be conveyed.

Risk regulation is the theme of the article by François Salanié and Nicolas Treich *Regulating an agent with different beliefs*. In most of the existing incentive regulation literature, the problem faced by the regulator is the design of incentives given to various economic actors to ensure the consistency of the development of risky activity with social welfare.[2] This paper focuses on a different source of inconsistency: biased perceptions of risks by consumers. It is well known that people may misperceive the risks they face, that is, their beliefs may diverge from observed frequencies and known scientific evidence. As a consequence, even in the absence of any conflict of interests between the regulator and the consumers, risk regulation may be warranted. A better informed regulator might indeed judge as faulty the consumers' actions when the latters have misspecified beliefs. This may trigger regulatory intervention. Salanié and Treich consider a paternalistic regulator who bases its action on the true interest of the people rather than a populist regulator who would base its actions on the opinion or perceptions of the people on the underlying risk even when this perception is known to be systematically wrong.

Salanié and Treich address the following question: Under which conditions do public risks misperceptions call for a more stringent regulatory intervention? The main traditional regulatory tool is taxation and its effect is immediate: the consumption of merit goods, that is, goods which are wrongly perceived as dangerous should be subsidized to increase consumption and de-merit goods, that is, goods which are wrongly perceived as safe should be taxed to discourage consumption. Nonetheless, there exist other instruments of public intervention, like direct risk-reduction safety programs (in health, food, location, and design characteristics, for instance), that the authors regroup under the concept of norms.

---

[2]See Laffont and Tirole (1993) and Laffont (1995).

Two main results are derived. First, the optimal norm level depends monotonically on the absolute distance between the true characteristics of risks (true beliefs) and their perception by the public, regardless of whether the representative agent is a pessimist or optimist, i.e., regardless of the sign of the difference in beliefs. The intuitive reason for this systematic distortion is that, if risks are underestimated, the regulator might raise the norms to make people "really" safer, while if risks are overestimated, the regulator might raise the norms to make people feel "really" happier. In such a case, norms are raised whether risks are overestimated or underestimated. This two-sided monotonicity makes the use of norms quite different from the use of taxes. When both taxes and safety norms are available instruments for the regulator, Salanié and Treich show that only taxes should be used to correct misbehavior although they cannot correct the beliefs of consumers. With the norms, the regulator also aims at correcting the impact of consumers' beliefs on their behavior but through changes in the consumers environment rather than through prices. Now, risk misperceptions do not automatically imply higher norms. In fact, the result depends on a particular condition related to the agents' preferences through the elasticity of demand. A case-by-case study is required in order to derive specific conclusions. Finally, Salanié and Treich conjecture that many compulsory liability norms may be imposed by the state on firms in order to circumvent problems posed by both consumers' misperception of risk and agency costs.

In modern societies, much legislations governing firms' risky environmental operations incorporate "whistleblower" clauses and several protections for insiders divulging information to regulators also exist under statute and common law. In their article *Using Information from Insiders to Target Environmental Enforcement*, Anthony Heyes and Catherine Liston-Heyes argue that, although regulatory agencies lack pertinent firm-specific information to target investigative and enforcement actions, it remains essential to analyze the motivations behind every "whistleblower" disclosure. Although the authors do not propose a formal model, they dwelve into the issues that influence how and when information from such sources is helpful to regulators, and how it should be managed. Different approaches used in the elaboration of economic models of the enforcement/compliance game between regulators and firms are discussed. In some circumstances, much of a regulatory agency's work will be complaints-driven, where "whistleblowers" report to the agency their suspicions of any illegal behavior. A major difficulty that is encountered in this kind of context is discerning the reasons why an individual complains when he is not an actual or potential victim and is not paid to do so. The policy response to whistleblowing depends crucially on those motives.

In *Environmental Protection, Consumer Awareness, Product Characteristics, and Market Power*, Marcel Boyer, Philippe Mahenc and Michel Moreaux model the behavior of a polluting monopolist facing environmentally aware consumers in a market à la Hotelling. The production of the firm causes global environmental harm touching consumers and non-consumers alike, while consumption produces a specific damage affecting consumers only.

The authors depart from the usual polluting monopolist model by assuming that the firm makes sequential decisions regarding first the product characteristics, variant or location (typically a long term decision), then the pollution intensity of the production technology (a medium term decision), and finally the price at which the product will be marketed (a short term decision). Those decisions jointly determine if the market is fully or partially covered, that is, if all or a subset of potential consumers do buy and consume the product. The authors show that an unregulated monopolist may indeed increase emissions while restricting output. Consequently, the monopolist could produce too little while at the same time pollute too much because the link between pollution and production depends on an endogenous decision regarding the technology employed. When consumers are environmentally aware, the demand for the monopolist's product may shift down as the intensity of pollution increases. Hence in trying to avoid such a negative demand effect, the monopolist may internalize the damage caused by the pollution she emits and incurred by the consumers. This means that the sole pressure of market forces invites the monopolist to reduce in part her pollutant emissions in order to sustain the consumption plans of her "conscious" consumers.

The paper thus shows that the emergence of environmental awareness is a decisive factor in the monopolist's internalization of the externality due to pollution. Two different contexts are compared. The first represents a standard unregulated monopolist choosing the profit maximizing product variant, pollution intensity and market coverage or price. In the second, the monopolist chooses first the product quality, then a regulator chooses the pollution intensity or production technology, and third the monopolist chooses the price or market coverage. At each decision node, the decision maker, either the monopolist or the regulator, observes the choices made previously and anticipates the future choices in a subgame perfect fashion. Boyer, Mahenc and Moreaux show that the same variant is chosen in both cases. Having the power to make consumers pay for pollution abatement, the unregulated monopolist will pollute less when confronted with environmentally concerned consumers. When the market is not fully covered, more or fewer consumers may be served but always at a higher price compared to the price that would prevail in the absence of consumption-specific harm. On the other hand, if the market is fully covered and if the chosen pollution intensity level is relatively elastic with respect to the consumption-specific damage factor, the monopolist will raise her price at the same time as the consumers become more environmentally aware. Hence, part of the externality associated with the consumption-specific damage is internalized by the unregulated monopolist. Moreover, the authors show that the unregulated monopolist, no matter the market coverage, will pollute more but never produce more than the regulated monopolist. Alternatively, a regulated monopolist will always raise her price without reducing production. The paper identifies two main reasons to explain such a result. Firstly, a stricter standard of pollution intensity raises the consumers' surplus when the latter are environmentally concerned, thus increasing the amount that can be captured by the monopolist. Secondly, the monopolist increases her price since production costs soar when stricter standards of pollution intensity are imposed. In summary, environmental regulation always leads to an increase in price but never to a reduction in production, a rather striking result.

Advances in Legal Design

When an ex ante risk regulation turns out to be difficult or inefficient, one has to rely *ex post* (after an accident has taken place) on the court to find out guilty parties involved in the process having generated risks and to compensate victims for the harm they suffer. A second set of contributions in this book addresses possible innovations in the legal design to improve incentives towards safety care and reduce the likelihood of an accident.

In *Subgame-Perfect Punishment for Repeat Offenders*, Winand Emons looks at the efficiency-based rationale for an optimal law enforcement with an increasing sanction scheme. Contrary to the widely accepted and embedded principle of escalating sanctions, the author states that only under rather special circumstances does this particular scheme prove to be optimal.

The model developed by Emons is one in which agents who may commit a crime twice must be efficiently deterred. Furthermore, the agents are wealth constrained, that is, increasing the fine for the first offense automatically leads to a reduction in the possible sanction for the second offense. The agents may follow history dependent strategies, i.e., commit the crime a second time depending on whether they have been apprehended or not the first time. As to the government, she tries to optimize the probability of apprehension.

To begin with, Emons assumes that the government can commit to sanction schemes, using any set of threats to discipline offenders. His main (surprising) result is that, because the agents have a fixed amount of wealth available for penalties, it is preferable for the government to increase the fine for the first offense that is detected at the expense of the fine for the second offense. Optimally, the sanction for the first offense should amount to the entire wealth of the agents while the sanction for the second offense should be zero. Hence, the optimal sanction scheme is decreasing rather than increasing in the number of offenses for the reason that shifting scarce wealth from the second to the first sanction will increase deterrence.

Next, Emons relaxes different assumptions under which the above result rests. First the Government may not be able to commit to any particular sanction scheme. This captures the large discretion that judges often possess when deciding the size of the penalty of a wrongdoer, but also the time consistency problem to apply a sanction at some cost for an offense that has already taken place and can no longer be changed. This inevitably modifies the optimal enforcement schemes. The author also presents an alternative to this option, where the punishment of the second crime is set at a level that deters the act. Once again, Emons finds that the government will stick to the decreasing sanction scheme only when the benefits of the second crime are low. In the event that the benefit and/or the harm of the second crime are significant, the government will prefer to deter the second crime should the first crime have occurred. Schemes where each crime is prevented by its corresponding sanction, entailing equal sanctions in both periods, will be time consistent.

The findings are quite robust and could also be applied to non-monetary sanctions. To sum up, if the government can commit to penalties, decreasing sanctions are always optimal; if she cannot commit to penalties, decreasing sanctions may still be optimal but so may be equal sanctions.

In his article *Optimal Liability Rule with many Judgment-Proof Tortfeasors*, Patrick González characterizes the optimal multiplayer liability for negligence rule with judgment-proof "injurers"/players, that is to say when players have a limited liability.[3] When life or environmental matters are concerned, the magnitude of damages can rapidly attain levels that go far-beyond the actual capacity of injurers to pay for damages. It is for this reason that a liability rule aimed at providing incentives to players to exert proper care must take into consideration the limited solvency constraint.

González allows an arbitrary number of players and considers the optimal rule as one providing the maximum possible incentives for the players to undertake effort given their liability constraints. Parallel to Bergstrom, Blume and Varian (1986) concept in which the voluntary provision to a public good is affected by a redistribution of wealth when the redistribution modifies the subset of net contributors, González's liability rule determines the subset of players upon which liability should be imposed in order to stimulate a proper level of care.

González finds that in the event that assets are insufficient, i.e., when it is impossible to implement the first-best allocation, it becomes strictly efficient to make all players strictly liable except one that may be, what he names, a "deep-pocket" or again, a "victim". The latter is characterized by his responsiveness to financial incentives under the strict liability rule. In fact, he is the most responsive when it comes to monetary enticements and, because of that, is subjected to the negligence rule, which states that he shall evade liability if he exerts proper care in accident prevention. The monetary resources that are collected from the strictly liable players are utilized as an additional pecuniary incentive to the deep-pocket. As a result, the total spending on effort is assumed by the later. Finally, the author explains that the second-best alter-

---

[3] See Shavell (1986).

native to this option is obtained when monetary incentives are concentrated in modifying the precautionary behavior of a single player.

In *A Tort for Risk and Endogenous Bankruptcy*, Thomas Miceli and Kathleen Segerson consider the potential effects of allowing people, exposed to toxic substances originating from industrial activities, to sue for expected damages at the time of exposure. This would treat the exposure itself as a tort, that is, define a "tort for risk". A tort for risk would reduce the possibility that victims who waited until the time of illness to file suit be properly compensated if the injurer is insolvent at the time the disease is contracted. Such a proposition would certainly influence the extent to which victims are compensated as well as the injurer's incentives to reduce the magnitude or likelihood of exposure, hence affecting social welfare.

Under the traditional tort law, where victims can only file damage claims once the symptoms of illness actually developed, bankruptcy is considered as an exogenous event, i.e., driven by factors unrelated to liability. Miceli and Sergerson study the relationship between a tort for risk and bankruptcy. A tort for risk could induce bankruptcy because victims may "race to file" at exposure for fear that the injurer will have insufficient assets later on to pay actual damage claims. The results show that a race to file can indeed arise in equilibrium under certain conditions, particularly for firms that have relatively low inter-temporal asset streams. This may be an undesirable effect since it could lead to inefficient situations (impediment of socially beneficial future activity) and because many victims could be left under-compensated. On the contrary, if the firm's assets level is high enough, the threat of bankruptcy dissipates, eliminating the incentives for victims to file damage claims early on. The tort for risk rule may even produce a desirable effect in staving off bankruptcy from future tort suits if it allows the firm to pay some of its liability in expected terms (an insurance effect), thereby leaving her enough assets to pay fully any future illness claims.

Finally, the authors show how allowing a tort for risk can stimulate the injurer to adopt preventive measures in order to avoid industrial accidents. To sum up, a tort for risk can have several welfare effects that sometimes work in opposite directions, thus triggering an ambiguous final outcome on social welfare.

Liability Regimes and their Consequences on Private Contracting

The previous papers have taken guilty parties in isolation from their economic environment. Other contributions to this volume precisely tackle this issue and investigate the consequences of various liability rules on contractual relationships between principals and agents involved in producing hazards. The alarming speed with which environmentally risky production activities spread around the world and the urgency to appropriately compensate the victims of accidents are just two of the reasons why this subject is of great importance.

In *Environmental Risk Regulation and Liability under Adverse Selection and Moral Hazard*, Yolande Hiriart and David Martimort consider imperative to grasp the different contracting possibilities available to firms involved in environmentally risky activities in order to properly direct public policies intended to correct environmental externalities and to comprehend the overall repercussions of risk regulation and liability rules on social welfare. Risk regulation not only impacts on safety care but it might also distort contractual relationships between principals and their judgment-proof agents. This is especially the case in contexts where private transactions are plagued with an adverse selection problem, for instance when the agent has private information on its cost function for the services he delivers to the principal.

The authors first consider the impact of an optimal ex ante risk regulation on private contracting between a buyer (principal) and a seller (agent) under adverse selection and moral hazard, i.e., when the agent has private knowledge of both his production cost and his effort towards safety care. In the absence of a liability problem, there exists a strict dichotomy between economic and risk regulation. The first-best effort can be implemented even under moral hazard although the standard trade-off between extraction of the adverse selection rent and efficiency calls for some output distortions and for giving up some socially costly adverse selection rent to the most efficient agent.

This is no longer true when the agent is protected by limited liability. There are now two sources of information rent which compound each other. One is related to the fact that only rewards can be used to induce effort, the other comes again from adverse selection. What Hiriart and Martimort show is that these two sources of rents are deeply intertwined. One cannot reduce the adverse selection rent of the least efficient agent without also undermining his incentives to exert effort. The optimal risk regulation exhibits thus an endogenous complementarity between production and safety care.

Once the optimal risk regulation is characterized, Hiriart and Martimort observe that in many cases, such ex ante regulation is not feasible. The paper then moves on to the analysis of a weaker form of control. A particular example is an extended liability regime under which liability for accident damages is imposed both on the seller (primary liable party) and the buyer (residual liable party). Had the principal (the buyer of the agent's services) owned sufficient financial resources, a fine equal to the total damage would generate a second-best optimal level of care. However, most of the time, the size of the damage caused surpasses the gains from trading with the agent and the value of the principal's own assets, preventing the efficient use of such a liability rule. When the judge can ask from the contractual partners to pay fines at most equal to the whole value of their contracts, Hiriart and Martimort show that output distortions are much more pronounced under the extended liability regime than under the optimal risk regulation, capturing the difference between the private incentives of the principal and the social incentives of the regulator to induce safety care.

In his article *Judgment-Proofness and Extended Liability in the presence of Adverse Selection*, Dieter Balkenborg characterizes the welfare maximizing extended liability rule for environmental accidents, that a firm may cause when she has private information on the accident probability. This is a useful complement to the existing literature which has mostly analyzed the case of moral hazard, i.e., the case where this probability is endogenously chosen by the firm.[4] The firm is assumed to have no equity. It operates under limited liability (judgment-proof) and needs financing from an uninformed lender in a bilateral relationship. The firm and its lender negotiate a financial contract under asymmetric information. The extended liability rule chosen by the social planner affects this financial contracting stage. To solve the bargaining problem, Balkenborg uses a cooperative game-theory approach where an interim efficient direct mechanism is agreed upon. Among the various possibilities for such mechanisms, the weighted neutral bargaining solution due to Myerson (1984), which generalizes the Nash bargaining solution to incomplete information contexts, is selected. This solution attributes weights to the parties which correspond to their relative bargaining powers.

Balkenborg characterizes the weighted neutral bargaining outcome for a given liability rule imposed by the social planner. The outcome of such a mechanism not only yields the probability that the project will be realized as a function of the probability of accident announced by the firm; it also describes the payments made to the lender by the firm according to her type depending on whether an accident occurs or not. Let us consider a three type example. If the level of the joint liability imposed by the social planner exceeds the value of the project and if the project is profitable for the types with low and medium probabilities of accident, but not for the high probability type, then the project is realized (with probability 1) only if the probability of accident is either low or medium. The firm receives a bonus (rent) in the no-accident state and gets zero otherwise. The firm with a high probability of accident receives an upfront payment equal to the payment she can expect by lying on the probability of accident. Finally, the bonus given to the firm in the no-accident state is the value for which the lender just breaks even. There is no way to avoid paying a bribe to the firm having a high probability of accident because this firm has no equity and operates under limited liability.

Given the characterization of the weighted neutral bargaining outcome in the different cases, one can determine the solution to the central planner problem. Balkenborg shows that there exists an optimal lender-firm joint and strict liability rule, which may be full or punitive, that is, equal to or exceeding the actual social value of the damage. In general, the liability is full. It will be punitive when the project is profitable only if the probability of an accident is

---

[4]See Pitchford (1995) and Boyer and Laffont (1997) among others.

low, the probability of a medium type (medium probability of an accident) is small, and the expected loss of running the firm under a medium probability of an accident is small. Increasing the liability and making it punitive induce a financial contract such that the firm is shut down under a medium probability of accident and the rent of the high type is reduced.

In their paper *Environmental Regulation of Livestock Production Contracts*, Philippe Bontems, Pierre Dubois and Tomislav Vukina analyse the optimal regulation of a vertically integrated industry in which the level of pollution is generated by a joint production process involving many parties, namely a contractor (as principal) and independent sub-contractors (as agents). Hence, the responsibility for environmental pollution must be shared between the principal and the agents since the provision of production inputs comes from both.

The model developed represents the current situation in the swine and poultry industries in the U.S. where a three-tier decision structure is present: the Environmental Protection Agency (EPA) is the regulator using the contractor or firm as an intermediary agent (as principal) in order to monitor the sub-contractors or producers (as final agents). It is assumed that while output is observable and verifiable and hence contractible, the pollution is observable but non-verifiable so that one cannot contract upon. The information structure is either a one sided moral hazard or double sided moral hazard structure depending on whether only the agents or both the agents and their principals' inputs are non-observable.

An important result is that, regardless of whether there is a single-sided or a double-sided moral hazard problem, the principle of equivalence across regulatory schemes obtains: any sharing of the tax burden between the contractor and the sub-contractors would generate the same solution as the contractor and the sub-contractors will negotiate around this sharing to reach a unique optimal solution as a function of the total tax bill imposed on the industry. Hence, the EPA's only job is to establish the optimal total tax revenue in each state. The relative degree of risk aversion of the contractor and the agent is the ultimate factor that will determine the optimal division of the tax bill the industry as a whole must pay. This is an important extension of the earlier work of Segerson and Tietenberg (1992).

However, when the impacts of such regulation on the endogenous organizational choices of the industry are explicitly taken into account, the equivalence principle disappears and the design of the optimal regulatory scheme becomes much more complex. For instance, when the regulator wishes to foster contracting as a dominant mode of organizing livestock production, the optimal taxation scheme sets the minimal and the maximal portions that the producer and contractor have to disburse. In the event that the EPA would need to simultaneously regulate independent producers and principal-agent contracts in the industry without the possibility of discriminating among those arrangements, a unique optimal division of the cumulative tax load between the principal and the agent emerges.

In the context of new concentrated animal feeding operation (CAFO) regulations, the equivalence principle of regulatory schemes will always fail when rigidity in the implementation of the optimal integrator-grower contracts is present. Furthermore, when environmental externalities and grower's bankruptcy constraints are both present, the legal incidence of regulation becomes relevant for efficiency. For the internalization of animal waste externalities where contract operators are judgment-proof entities, co-permitting may be required.

Contracts between a stakeholder and a judgment-proof firm may not only be affected by adverse selection and moral hazard but also by enforceability problem. This is particularly true in the case of financial contracts when refinancing of projects is at stake. In *Environmental Risks: Should Banks Be Liable*, Karine Gobert and Michel Poitevin consider the possibility that after an environmental accident which would bankrupt the firm under the usual assumptions made in the literature on extended liability, the bank may decide to refinance the firm to capture the future benefits associated with the firm's activities. Indeed, if it was profitable for the bank to finance the firm in the first place, it may still be profitable to do so after an accident has occurred under the principle that "bygones are bygones". Gobert and Poitevin show that full extended liability may induce the bank not to refinance the firm as often as it would otherwise under a partial extended liability rule.

The model considered differs from some former studies, which have concluded that banks should only be partially liable under the assumption that the investments in prevention are not observable (moral hazard), by bringing in two critical elements: dynamics and risk aversion of the firm. In this case, the bank provides two important services in financing the firm: insurance within a given period and income smoothing across periods. Gobert and Poitevin assume that investment in prevention is observable and made at the beginning of the first period and that an environmental accident occurs or not during this period. If an accident occurs, the firm would encounter financial distress given that its revenues in this first period are not large enough to compensate for the harm caused. This is not to say, however, that the bank would not be interested in refinancing the firm following the accident. If the bank were liable at the moment of the accident, refinancing the firm would mean that the later would not file for bankruptcy, thus reducing its liability.

The firm's risk-aversion induces her on the one hand to overinvest in environmental prevention to diminish the probability of bankruptcy but, on the other hand, to underinvest thanks to the incomplete internalization of the cost of accident. Hence two opposite effects arise with an ambiguous outcome. The authors show that a partial bank liability may be necessary to achieve the optimal level of investment in safety care even if such an investment is observable. The rationale behind this result lies first in the assumption that the bank cannot commit to a future refinancing policy (it will refinance the firm ex post only if such refinancing is profitable at the time of decision) and, second, in the tradeoff between the willingness of the firm to avoid bankruptcy, that is, to be refinanced when an accident occurs and the level of risk that she bears under limited liability.

One of the major questions in the public management of risk is to understand the respective scopes for regulation and liability rules. Control policies to reduce the probability and severity of industrial and environmental accidents are discussed by Marcel Boyer and Donatella Porrini in *Sharing Liability between Banks and Firms: The Case of Industrial Safety Risk*. According to Boyer and Porrini, the determinants of such environmental accidents can be regrouped into three different sets: technological, eventual, and organizational. A proper understanding of the specific effects of these factors is essential if proper control policies are to be implemented. Although all three groups of determinants are of great importance, the paper focuses on the organizational factors which include among others the impacts of diverse institutional and informational constraints in the implementation of the proper public policies.

The model developed by Boyer and Porrini has the following features: limited liability (judgment proofness) of firms; limited capacity of governments to intervene in business decisions and transactions; limited power of the court system to properly collect all the facts relevant to a judgment; asymmetric information between the four main agents or players, namely governments, firms, banks and courts, whose decisions and conduct affect the observed probability and severity of environmental or industrial accidents. The authors also discuss what distinguishes the institutional frameworks of the American and European liability systems and the specific roles of each main player in the determination of environmental or industrial accidents.

Boyer and Porrini characterize the socially efficient liability sharing formula and standard of safety care, founded on the complex interactions between the banks and the firms and on the efficiency of the courts in assessing the level of care exerted by the firm. Their results illustrate that the incomplete information and partial control social welfare maximizing liability or accountability share of the banks (alternatively the social profitability of extended lender liability regimes) increases as the profitability of the firm decreases, as the recommended level of care increases, as the cost of care increases, as the effectiveness of care increases, as the efficiency of courts increases, and as the social cost of public funds increases.

References

Beck, U., *The Risky Society: Towards a New Modernity*, (1992).

Bergstrom, T., Blume L. and Varian, H., 'On the Private Provision of Public Goods', *Journal of Public Economics*, 29 (1986): 25-49.

Boyer, M. and Laffont, J.J., 'Environmental Risks and Bank Liability', *European Economic Review*, 41 (1997): 1427-1459.

Kleindorfer, P. and Ordts, E., 'Informational Regulation of Environmental Risks', *Risk Analysis*, 18 (1998): 155-170.

Laffont, J.J., 'Regulation, Moral Hazard and Insurance of Environmental Risks', *Journal of Public Economics*, 58 (1995): 319-336.

Laffont, J.J. and Martimort, D., *The Theory of Incentives I : The Principal-Agent Model*, (Princeton University Press, 2002).

Laffont, J.J. and Tirole, J., *Incentives in Procurement and Regulation*, (Cambridge: MIT Press, 1993).

Myerson, R., 'Two-Person Bargaining Problems with Incomplete Information', *Econometrica*, 52 (1984): 461-488.

Pitchford, R., 'How Liable Should the Lender Be? The Case of Judgement-Proof Firms and Environmental Risks', *American Economic Review*, 85 (1995): 1171-1186.

Segerson, K. and Tietenberg, T., 'The Structure of Penalties in Environmental Enforcement: An Economic Analysis', *Journal of Environmental Economics and Management*, 23 (1992): 179-200.

Shavell, S., 'The Judgement-Proof Problem', *International Review of Law and Economics*, 6 (1986): 45-58.

Part 1

Basic Principles of Risk Regulation, Beliefs and Awareness

Chapter 1

The Case for a Procedural Version of the Precautionary Principle Erring on the Side of Environmental Preservation*

Alessandra Arcuri

Erasmus University Rotterdam

"Primum non nocere."
Salernitan School of Medicine

1.1 Introduction

The development of technological activities, which create serious risks for the environment and human health, is a prominent characteristic of modern societies. Genetically modified organisms, exposure to electric, magnetic and electromagnetic fields, exposure to toxic chemicals, emissions of greenhouse gases, etc., are classical examples of such risky activities. A crucial feature of these activities is that the risks that they involve are partly unknown. In face and in spite of such uncertainties, societies have to choose whether and how to regulate these activities. One of the grand principles of environmental

*This paper has been written during my stay at New York University School of Law as a Global Research Fellow of the Hauser Global Law School Program. The stimulating academic environment of the NYU Law School has provided the excellent conditions for completing this work. In particular, I am grateful to Joseph Weiler, Christian Joerges, Richard Stewart and to all the participants to the Global Fellow Forum for helpful comments. I also wish to thank the participants to the 1st CIRANO-IDEI-LEERNA Conference on Regulation, Liability and the Management of Major Industrial/Environmental Risks, in Toulouse 2003, and in particular my discussant Nicolas Treich, the participants to the JSD colloquium at NYU Law School with whom I discussed the paper at an early stage, as well as Ellen Hey, Giuseppe Dari Mattiacci, Roger Van den Bergh and Olav Velthuis for helpful comments. The usual disclaimer applies. Financial support for this research was provided by the Veregining Trustfonds, Erasmus University Rotterdam, which I gratefully acknowledge. Email: arcuri@frg.eur.nl.

policy aimed at guiding public decision-making under uncertainty is the so-called Precautionary Principle. Notably the application of the Precautionary Principle is going to affect major industrial activities; the result of it may be heavy costs on the part of industrial companies and indirectly on the whole society. It is therefore crucial to have a clear understanding of the Principle and its various interpretations, and to assess them in terms of their possible consequences.

Unfortunately, the debate on environmental policy in general, and on the Precautionary Principle in particular, is often dowsed in two opposite narratives.[1] On the one hand, we find technological pessimism and warnings against "playing God"; on the other, technological optimism and faith in progress. The two narratives support two symmetrically opposite approaches toward the adoption, the interpretation and the implementation of the Precautionary Principle: on the one hand, the pessimistic narrative urges us, in its most extreme manifestation, to stop all industrial activities; on the other hand, the optimistic narrative suggests that any type of control on industrial activity is detrimental. The major flaw of the participants in this duet is that they are often too confident in supporting deterministic conclusions and too dismissive of the complexities and scientific controversies underlying some problems. The paper tries to de-entrench the debate from these two stories by scrutinizing what the available options for the decision-makers and what the underlying rationales are. It is not the goal of the paper to find a magic formula or a final answer to clear up the many doubts on the possible interpretations of the Principle and on its possible applications. The paper is meant to be a modest contribution on such a debated Principle, with the hope of falling in the category of rational constructive dialogue.

The Principle is adopted by European law and an overwhelming number of international treaties and national laws. Its main tenet is that regulatory intervention should not to be postponed only because of uncertainties. However, a precise definition of the Principle is still missing. Ambiguities remain as whether a so-called weak or strong version is more desirable. Can it be used as a mere defense by governments in case of regulatory actions under uncertainty or should it be interpreted as a prescriptive rule on the course of action to be taken? Next, does the Principle tell us when intervention is required? Does it tell who has the burden of proof? Does it indicate whether a particular type of regulation is better than another? Ironically, a Principle that is born to cope with uncertainties seems itself condemned to fluctuate in uncertainty. Partly due to the ambiguities surrounding its meaning, the Principle has attracted many criticisms.

---

[1] John S. Applegate has brilliantly captured such rhetorical duet in his recent analysis on the Regulation of Genetically Modified Organisms. Applegate, J. S., "The Prometheus Principle: Using the Precautionary Principle to Harmonize the Regulation of Genetically Modified Organism" in *Indiana Journal of Global Legal Studies* (2001), downloadable at http://ijgls.indiana.edu/archive/09/01applegate2.shtml (last visited, Feb 2002).

Notwithstanding such critiques, I argue that the Precautionary Principle is necessary to inform the legal system of societies characterized by the continuous development of new hazardous technologies. My paper contributes to the academic debate on the Precautionary Principle in a number of ways. First of all, I develop an original taxonomy of the constitutive elements of the Principle. Such a taxonomy is necessary to elaborate a meaningful and clear definition of the Principle and it is a precondition for sound legal-economic analysis. In addition, I originally recast the debate over the relation between the Precautionary Principle and the practice of cost-benefit analysis. Above all, the paper indicates that the Precautionary Principle, to be operational and desirable at the same time, should be interpreted not only in substantive terms but also along the lines of procedural norms. The problems addressed by the Principle are characterized by the presence of uncertainty and, as I argue in the paper, such type of problems cannot be solved within an analytic deterministic framework that presupposes the existence and the predictability of optimal outcomes. Hence, the paper suggests that we should not look for solutions only in relation to outcomes, but also in relation to procedures. In the absence of optimal outcomes, we should find procedural rules aimed at the reduction of uncertainty. Such procedural norms should stimulate the production of information and allow different qualified viewpoints in the analysis. In addition to procedural norms, I propose that some substantive norms should be considered as interpretative guidelines of the Principle as well; relying on some insights derived from economic theory, I argue that decisions should err on the side of environmental preservation.

Limits of the Paper and Suggestions for Future Research

The debate about the Principle raises a substantial amount of questions. There are indeed entire books already written about it. Given the limited scope of the paper, many important issues are not analyzed and discussed. To begin with, the paper does not flesh out the details of the procedural and substantive interpretative canons suggested, focusing mostly on the reasons why we need them and describing only some minimal and general requirements. Moreover, issues relating to the different needs in diverse legal systems and their implications for the possible interpretations of the Principle are not investigated. Likewise, the paper does not discuss to what extent it is desirable to have a very precise and unambiguous Precautionary Principle. In this respect, it should be observed that, while the Principle does not need to be the same across different national legal systems, its national implementation is likely to have transnational implications. Indeed, one of the most prominent reasons underlying the acrimonious debate over the adoption and the interpretation of the Principle is that its interpretation may have, *inter alia*, serious repercussions on international trade. The so-called *Hormone* case that was disputed before the World Trade Organization (WTO) is just one example of the ways in which the Precautionary Principle may break through the domain of international trade law. Notably, the questions linking the Principle to issues of international trade are not analyzed in this paper.

From a methodological perspective, the Precautionary Principle can be studied in several ways. This paper, falling mostly in the realm of normative analysis, searches for an interpretation of the Principle that shall improve the overall well-being of society. Alternatively, the Principle can be scrutinized through the analytical lenses of so-called private interest theories. According to these theories the regulator is captured by private interest groups. Such a theoretical framework may suggest that, on the one hand, the adoption of the Principle is the result of pressures of special interest groups such as environmental associations pursuing their own agenda. On the other hand, the legislator's reluctance to adopt the Principle may be due to pressures from multinational companies, whose productive activities could be restricted as a consequence of the inclusion of the Principle into the legal system. Such an analytical approach, focusing on regulatory capture, remains outside the scope of this paper.

The relationship between the Precautionary Principle and international trade law, on the one side, and the analysis of the Principle trough private interest theories, on the other side, are two themes that deserve separate analyses. Notwithstanding the fact that both themes have already been discussed in the literature, to my knowledge, there is not yet a comprehensive treatment of either of them. Hence, even if rivers of ink have already been devoted to a better understanding of this contested Principle, there is still room for future research.

Structure of the Paper

The paper is divided in three parts: the first introduces the relevant questions that should be asked to reach a deeper understanding of the Principle; the second discusses the theoretical insights that are crucial to answer the delineated questions and, eventually, the third part provides a proposal for a sound interpretation of the Principle. More specifically, Section 1.2 discusses the legal dimensions of the Principle. After a brief overview of its origins and its adoption at the international level, the Principle, as embodied in European law, is delineated. Next, it is argued that the nature of "legal principles" as opposed to "legal rules" is to be understood for an appraisal of the Principle. These considerations are necessary if we want to use economic insights for legal appraisal; sophisticated economic analyses overlooking legal structures might indeed turn out to be sterile. Finally, as the precise content of the Principle is still unclear, various versions of the Principle are prospected; in this context, the paper develops a taxonomy with two extreme versions of the Principle: α-Precaution (or no-Precaution) and Radical Precaution. It is relatively easy to show that these two extremes are not desirable for society;

hence, attention is devoted to spell out the potential constitutive elements of the Principle, the combination of which might give body to more acceptable versions of it.

Section 1.3 unfolds various arguments to test the relative desirability of different versions of the Principle. Its role and its inter-relation with other general principles, such as the proportionality principle, is shortly addressed. In spite of the shared view among economists that the Precautionary Principle should be interpreted within the framework of standard cost-benefit analysis (CBA), the paper contends that it is difficult to reconcile the Precautionary Principle with CBA as the situation that it means to address is specifically characterized by uncertainty. Some authors argue that cost-benefit analysis can still be used, as the problems introduced by the presence of uncertainty can be obviated by the use of Bayesian analysis. However, this solution seems problematic because employing Bayesian calculation to guide decision-making can constitute a sort of "arbitrary arithmetic" as the choice of *subjective* probabilities may lead to very different outcomes. In this context, I claim that risk-risk tradeoffs and consideration of the ancillary benefits of regulation should still play a role but more adequate methods than cost-benefit analysis should be developed. The paper looks for other theoretical insights that might form the basis for the elaboration of an acceptable version of the Principle; in particular, it reviews the branch of economic literature emphasizing the importance of erring on the side of environmental preservation. Economic arguments to support the "err on the side of preservation" option can be found in the analysis of the conceptual categories of "irreversibility" and "option value", initially developed in 1974 by Arrow and Fisher. Next, reasons are presented to show that, even if CBA is not endorsed, Cost Effectiveness Analysis can still be used. To conclude this section, the paper reflects on the complex relations between science and the law. This relation becomes more complex when science does not have unambiguous answers to the issues that the law deals with. The paper urges the democratization of scientific processes, it warns against the risks of "expertocracy" and advocates for interactive dialogue between experts and lay man.

In Section 1.4, I draw some tentative conclusions in the light of the analysis done in the paper and of the results of previous studies. The elements outlined in Section 1.2 are here re-elaborated on the basis of the considerations done and a specific version of the Precautionary Principle is presented. My main suggestion is a procedural version of the Principle based on sound science and rational analysis, while being open to democratic scrutiny.

1.2 Legal Dimensions of the Principle

1.2.1 The Origin and the European Legal Context

"Community policy on the environment [...] shall be based on the
precautionary principle."
Art. 174 (2) EC Treaty

The introduction of the Precautionary Principle into a legal system can be
traced back to the so-called *Vorsorgeprinzip* - translatable as foresee and fore-
stall principle - developed in the early eighties in German environmental law.[2]
Under the German influence, the Principle has found his way into international law, in particular in the area of sea protection.[3] From the eighties
up to now it has been extensively adopted internationally and locally.[4] The
Principle has entered the legal scenario in a variety of formulations. Still
in many occasions it is only mentioned and referred to, and only few texts
define it. Probably the best-known definition is the one contained in Principle 15 of the 1992 Rio Declaration. "In order to protect the environment,
the precautionary approach should be widely applied by States according to
their capabilities. Where there are threats of serious and irreversible damage,
lack of full scientific certainty shall not be used as a reason for postponing
cost-effective measures to prevent environmental degradation".[5]

The Principle has been introduced into European law by the Treaty on
European Union (Maastricht Treaty) in 1992. Today, art. 174 (ex 130 R)
of the EC Treaty refers to it as one of the principles on which Community
policy on the environment should be based.[6] Moreover, Directive 2001/18/EC

---

[2]Most authors agree on tracing back the Precautionary Principle to the *Vorsogeprinzip*
of the eighties. Frank Cross deems instead that the Principle was coined earlier in the
sixties by German bureaucrats; see Cross, F. B., "Paradoxical Perils of the Precautionary
Principle", in 53-3 *Washington and Lee Review*, (1996), pp. 851-925.

[3]See for instance the Oslo and Paris Convention for the Protection of the Marine Environment of the North-East Atlantic, so-called OSPAR, adopted in 1992 and entered into
force in 1998. See Marr, S., (2003) *The precautionary principle in the law of the sea: modern decision making in international law*, The Hague; New York Martinus Nijhoff; Hey. E.
(2002) "The International Regime for the Protection of the North Sea: from Functional Approaches to a More Integrated Approach", in *International Journal on Marine and Coastal
Law*.

[4]The Principle is adopted or recognized by a series of international treaties and documents such as UN Framework Convention on Climate Change, the 1995 UN Straddling
Stocks Agreement, the African Bamako Convention etc.; some authors and judges of the
International Court of Justice have even argued that the Principle is part of customary inter-
national law relating to the environment. For an overview of the wide range of international,
regional and national rules endorsing the Principle see Cameron, J., "The Precautionary
Principle", in Sampson, G. P. and W. B. Chambers (eds) *Trade, Environment and the
Millennium*, (2002), pp. 287-319.

[5]See Principle 15 of 1992 Rio Declaration on Environment and Development, available
at UNEP website.

[6]Recently the Principle is discussed also by the annex of the Treaty of Nice.

of the European Parliament and the Council for the deliberate release into the environment of genetically modified organism, repealing Council Directive 90/220/EEC, states that "the precautionary principle should be and has been taken into account in drafting the Directive and must be taken into account when implementing it". However, as I argue in this section, the EC fails to provide a detailed explanation as to what the Principle exactly means. Given that the Precautionary Principle is one of the core principles upon which EC environmental policy should be founded, this obscurity is puzzling to say the least. The lack of a precise definition within legal texts has given rise to an unwavering debate on if, when and how to use the Principle. The uncertain status of the Precautionary Principle is not peculiar to European law only; indeed, the status of the Principle has so far remained unclear around the world.[7] In the light of its potential impacts on international relations, it should not come as a surprise that the Precautionary Principle is highly disputed.

The Communication from the Commission on the Precautionary Principle and some Jurisprudence

In order to shed some light on the question, the Commission has issued a Communication, with the goal of setting out the Commission's approach to the application of the Principle.[8] In spite of this, a clear definition is not elaborated in the Communication and the boundaries of the Principle remain deliberately vague. "Like other general notions contained in the legislation, such as subsidiarity or proportionality, it is for the decision-makers and ultimately to the Courts to flesh out the principle. In other words, the scope of the precautionary principle also depends on trends in case law, which to some degree are influenced by prevailing social and political values". Yet, from the Communication we can depict a "minimal core" of the Principle: it is meant to justify intervention also in absence of scientific certainty. In other words, uncertainty *per se* cannot preclude regulatory actions undertaken to protect human health and the environment (non-preclusion rule). So far, however, the most controversial questions are left answered. Does the non-preclusion rule apply to the potential of any harm or only to significant ones? Should regulatory authorities engage in risk assessment before issuing a regulatory measure over a certain activity or should producers and industrial operators do that before engaging in any activity? Does the precautionary principle provide any guidance in deciding which level of risk is acceptable?

---

[7]In this respect see Majone, G., "What Price Safety? The Precautionary Principle and its Policy Implications" in 40-1 *Journal of Common Market Studies*, (2002), pp. 89-109, at pp. 93-95.

[8]Communication from the Commission on the Precautionary Principle, Commission of the European Communities, Brussels, 02.20.2000, Com (2000), 1.

The Communication of the Commission and the related opinion of the Economic and Social Committee on the "Use of the precautionary principle"[9] do not give precise answers to such questions. The Communication devotes some attention to risk assessment and risk management. In particular, the Commission states that the Precautionary Principle pertains to the domain of risk analysis and, more specifically, of risk management. This is *per se* an important indication. However, it does not shed light on when, by whom and eventually on the modalities according to which risk assessment and risk management should be carried out. Similarly when the Communication describes general principles of risk management, such as the use of cost-benefit analysis, it fails to specify the exact relationship between the two. As this paper argues in Section 1.3, cost-benefit analysis is not the ideal answer to problems posed by the presence of uncertainty. Before discussing the modalities of risk analysis, it should be clear when and who has to carry it out, or stated differently, it should be established who has the burden of proof. Unfortunately, the Community notes on the burden of proof are not too innovative, as they either bounce back to existing legislation or remain quite ambiguous.[10] On the one hand, it is stated that when there is a regime of prior approval it is up to the business community to carry out the risk analysis (and this is almost a truism), without further specifying when it might be appropriate to have a regime of prior approval. On the other hand, when it is not otherwise disciplined, citizens or public authorities have to demonstrate the danger; it is specified that, on a case-by-case basis, there can be a reversal of the burden of proof. However, no criteria or guidelines are established to take such decisions. In other words, it emerges that the Precautionary Principle is *also* about shifting the responsibility for producing scientific evidence.[11] At the same time, it is unequivocal that the clause reversing the burden of proof cannot be turned into a general principle.[12] Unfortunately, it is yet unclear when and on the basis of which criteria the shift should occur.

European case law on the issue, tough quite limited in the years preceding the Communication, is on the rise. One of the first cases based on the rationale of the Precautionary Principle, while not directly referring to it, was Case C-405/92 *Etablissments A. Mondiet v. Armement Islais,* in which the Court held that the absence of scientific certainty did not preclude the Council to adopt environmental conservation measures.[13] Later, in one of the

---

[9]Economic and Social Committee on the "Use of the precautionary principle" (2000/C/268/04).

[10]See point 6.4. of the Communication.

[11]This can be inferred from the analysis of the Communication text. Point 6.4 reads: "This is one way of applying the precautionary principle, by shifting responsibility for producing scientific evidence".

[12]On this point see also the opinion of the Economic and Social Committee on the "Use of the precautionary principle" (2000/C/268/04).

[13][1993] ECR I-6133. For a description of the case in relation to the precautionary principle see Hancher, L.,"EC Environmental Policy - a Pre-cautionary Tale?" in Freestone

cases related to the Commission's decision to ban the export of beef from UK, due to the risk of diffusion of the so-called "mad cow disease", the Court held that "[w]here there is uncertainty as to the extent of risks to human health, the institutions may take protective measures without having to wait until the seriousness of those risks become fully apparent".[14] While the rationale underlying the Precautionary Principle seemed to have provided grounds for the Court decision, the Principle itself was not directly invoked.

Eventually, in some recent jurisprudence, the Precautionary Principle has been explicitly considered for the assessment of the cases. The most significant are probably the *Pfizer* and *Alpharma* judgements.[15] In both cases, the decisions of the Council and the Commission to revoke the authorization for the use of certain antibiotics as growth promoters were challenged. The authorization, in turn, had been revoked because of the danger of developing resistance to antibiotics in animals and a subsequent transmission of such a resistance to humans; since such a danger is yet uncertain, the decisions of both the Council and the Commission relied on the Precautionary Principle. The Courts' judgements are richer in their interpretation of the Principle than previous cases: beside the basic idea that scientific uncertainty cannot preclude environmental and health measures, the Principle has been interpreted in relation to the practice of risk assessment. In particular, in the *Pfizer* case, the Court of First Instance held that it is not necessary to carry out a full scientific risk assessment before taking preventive measures.[16] In addition, it was clearly affirmed that while the opinion of scientific committees should be taken into account, the final decision is a political one and is the sole responsibility of the decision-maker.

In spite of the important insights emerged from the recent jurisprudence, the interpretation of the Precautionary Principle remains fragmented and in need of more systematic analysis. On its side, the Communication of the EC Commission states that it "does not claim to be the final word - rather the idea is to provide input to the ongoing debate both at Community and international level". In the remainder of the paper I try to contribute to this debate by elaborating on the possible meanings of the Principle. In doing so, I attempt to incorporate legal arguments as well as new insights from the

---

D. and E. Hey (eds), *The Precautionary Principle and International Law - The Challenge of Implementation*, Kluwer International, (1996), pp. 187-207.

[14]See case C-180/96 United *Kingdom vs. Commission*, § 99, ECJ, Judgement of 5 May 1998. In this case, UK and Northern Ireland challenged the validity of Commission Decision 96/239/EC of 27 March 1996 on emergency measures to protect against bovine spongiform encephalopathy (BSE), which substantially banned the exportation of beef from the United Kingdom to reduce the risk of BSE transmission; the Court dismissed the actions brought by the applicants. Few other cases are in line with the latter jurisprudence; in this respect, see Annex I, Ref. 5,6,7 of the Communication for more details.

[15]See Case T-13/99, *Pfizer Animal Health SA v. Council of European Union*, CFI, Judgement of September 02, and Case T-70/99, *Alpharma Inc., v. Council of the European Union*, CFI, Judgement of 11 September 02.

[16]Case T-13/99, §§ 160 and 381-383.

economic discipline. Before jumping to a review of economic literature that might be relevant for our purposes, some more general issues related to the legal dimensions of the Principle that are the bases for further analysis should be investigated.

1.2.2 The Precautionary Principle as a Restrictive Belt on the Decision Maker

First of all, it should be emphasized that the Precautionary Principle is a "legal principle".[17] On a logical ground "legal principles" should be distinguished from "legal rules". Let me emphasize from the outset that, as with all taxonomy or elaboration of typologies, the analytical distinction made here between legal rules and principles might not always be so clear-cut. Some rules seem in fact to be a combination of legal rules *strictu senso* and principles and *vice versa* principles may have elements of rules. In spite of this limitation, taxonomies are useful to analyze problems in a systematic way. To illustrate the difference between rules and principles, I borrow the analytical distinction elaborated by the legal-philosopher Ronald Dworkin: "[...] [legal rules and legal principles] differ in the character of the direction they give. Rules are applicable in all-or-nothing fashion".[18] Attached to legal rules are precise consequences that follow automatically when certain conditions are met. On the contrary, the legal principle "states a reason that argues in one direction, but does not necessitate a particular decision".[19] Principles work as a sort of guidelines, implying that different outcomes might result from the application of the principle.

Legal principles are a major component of legal systems, both in a quantitative and in a qualitative way. Quantitatively, we might think of principles present in most Western legal systems such as the principle of *bona fide*, of equity, of equal treatment of man and woman, of diligence, of *bonus pater familae*, of subsidiarity, of proportionality etc. Qualitatively, we should note that a law without principles is likely to be underinclusive as it is impossible to foresee all the circumstances that require intervention, the types of intervention, etc. As a consequence, a legal system without principles would be based very much on discretion in its strongest version (exercised by competent authorities, be it judges, regulatory bodies, or State legislators), as a

---

[17]A related issue not discussed in this paper is the question of the degree of precision of norms that is desirable for Societies. This topic is discussed in the law and economics literature by Kaplow, L., "General Characteristics" in Bouckaert, Boudin. & Gerrit De Geest, (eds), XI. *Production of Legal Rules - Encyclopedia of Law and Economics*, (2000), pp. 502-528.

[18]Dworkin, R., *Taking Rights Seriously*, London, (1981), at p. 24. Notably, in the analysis of Dworkin this framework is instrumental to his attack on positivism. It is beyond the scope and goal of this paper to take a position on such debate. Hence, the citation in this paper is de-contextualized.

[19]Dworkin, R. (1981), op. cit., at 26.

legal vacuum would dominate an overwhelming number of situations where legal decisions are to be taken. It would probably be conceivable only in very primitive societies to have a legal system without principles. However, in complex, modern societies such a system seems highly undesirable.

At this point it may be argued that legal principles lead to a strong discretionary system as well. Given the definition that principles merely indicate a direction to be followed but not decisions to be taken, this seems a natural conclusion. However, such a conclusion is flawed. Even if principles are tightly related to the concept of discretion they do not imply a discretionary system. On the contrary, they counter discretionary tendencies. The essence and need of principles lies on the necessity to *bind* discretion that inevitably should be exercised in a complex legal system where not everything can be specified by legal rules. As Dworkin argues: "Discretion, like the hole in a doughnut, does not exist except as an area left open by a surrounding belt of restriction".[20]

These brief considerations about principles in general can be now applied to the analysis of our Principle. The Precautionary Principle can be seen as a *restrictive belt on decision-makers*. A very large set of cases, which consists of uncertain risk of serious and irreversible harms, should be decided in the direction of "precaution". This does not imply that the same decision will be reached in different circumstances. As mentioned, the distinction between principle and rules is not always so clear-cut; however, from this distinction, it should be clear that the Precautionary Principle should not be interpreted as a mere rule of thumb. This point may seem trivial but is not because many criticisms of the Principle stem from treating implicitly or explicitly the Principle as a rule of thumb.[21] In the following I will elaborate on the meaning of the Precautionary Principle indicating the ways in which the Principle can function as a restrictive belt on the decision-maker.

1.2.3 The Range of Precautionary Principles: From α-Precaution to Radical Precaution

At this moment, the Precautionary Principle lacks one clear identity. It is beyond the scope of this paper to review the myriad of Treaties, Convention, Conferences and the like referring to the Principle; excellent overviews are already provided by the literature on international environmental law.[22] This

---

[20]Dworkin, R. (1981), op. cit., at 31.

[21]For instance, Immordino writes: "The precautionary principle, proposed by international treaties as a *rule of thumb* to be used in situation of scientific uncertainty, could indeed be inefficient" (italics is mine). See Immordino, G., "Looking for a guide to protect the environment: the development of the Precautionary Principle", (1999) Working Paper FEEM. In spite of the considerations done and in spite of the ambiguities of many Treaties, it should be clear why it is incorrect to conceive a Principle as a rule of thumb.

[22]See Freestone, D. and E. Hey (eds), *The Precautionary Principle and International Law - The Challenge of Implementation*, (1996) Kluwer International; more recently, see

section sketches out several "identities" of the Principle that depend on its possible interpretations. The different versions of the Principle can be shaped around the way we answer various questions. The aim of this section, then, is to flesh out the relevant questions that should be asked. Different answers to such questions lead to different versions of the Principle. Most of the answers will be provided only in the final part of the paper where, on the basis of the theories discussed below, a proposal for a particular interpretation of the Principle is presented. Before framing such questions let me briefly outline two extreme versions of the Principle that can be immediately dismissed as undesirable for society.

On the one side, in line with the Law & Economics approach,[23] we can consider a negative rule: the complete absence of a Precautionary Principle, what I call α-Precaution (or no-Precaution). This rule implies that no activity can be regulated in presence of uncertainty; it entails an uncritical approach of the type 'anything goes', elsewhere characterized as "cornucopian".[24] On the other side, we can envisage a Radical-Precautionary Principle: according to this version any activity which presents uncertain hazards, irrespective of the seriousness of the hazards, of the social benefits of the activity and of the social costs of the regulation, should be banned. This approach relies on an apocalyptic vision of progress.[25] For various reasons, these two extreme versions should be dismissed. First of all, and in line with previous considerations, these extreme versions should be rejected because, according to them, the Principle would be conceived as a rule of thumb. In addition, it is quite intuitive to conclude that such extreme versions of the Principle are not desirable for society.

The α-Precaution rule could be used to halt any type of regulation no matter how devastating the consequences of an activity could be, and without even considering the possible benefits from regulating. In this context, procrastination can be conceived as a real policy option where inadequate information is the best excuse for delaying regulation. It is plausible to argue that such type of rule induces a rational proponent of activity to hide information about the risks of its activity because the lack of knowledge about risk would imply no regulatory measures. Notably, if we agree that α-Precaution is not desirable for society we should welcome some versions of the Precautionary Principle. Symmetrically, Radical Precaution is to be criticized because, in practice, it

O'Riordan, T. and J. Cameron, (eds), *Reinterpreting the Precautionary Principle*, (2001) London: Cameron Mat.

[23] In many Law and Economics analyses the rules compared range from the absence of the rule to different types of rules. The most significant example is Economic Analysis of Liability rules. Here strict liability, negligence and other combinations of the two are compared also with "no liability".

[24] Stirling, A. (2001), "The Precautionary Principle in Science and Technology" in O'Riordan, T. and Cameron, J. (eds) *Reinterpreting the Precautionary Principle*, (2001), pp. 61-94.

[25] Stirling, A., op. cit., (2001), at 66-67.

would result in banning uncritically a great number of activities. Any project, no matter how certain and large its benefits would be, would be banned if there is the suspect of hazardous consequences, no matter if and how small would be its worst case scenario. Paradoxically, the Radical Precautionary rule, if applied literally, is paralyzing and it would lead in no direction at all because "in the relevant cases, every step, including inaction, creates a risk to health, the environment, or both".[26] In conclusion, both α-Precaution and Radical Precaution should be dismissed.

In between the two extremes, we can envisage a great variety of versions of the Precautionary Principle. Instead of enumerating all the possible versions that lie in between these two extremes, the paper briefly discusses the main issues that are at stake when opting for one version or another. In other words, it fleshes out the main aspects that can be found in the Principle. These aspects can be combined in several ways and these combinations shall constitute the possible versions of the Precautionary Principle. As shown below, the possible elements of the Principle are shaped around various questions. Different versions of the Principle correspond to different answers to these questions.

Four elements constitute the texture for the versions of Principles. The first element deals with the question: what circumstances trigger the application of the Principle? The second element of the Principle concerns its timing: when can/should a regulation be adopted? The third element concerns the question of the burden of proof; in particular, who has to prove that an activity is dangerous, the regulator or the operator? Intertwined to these two issues is the fourth element, which concerns the question of how the regulatory measure (including no measure at all) should be. Should a regulation comply with Cost-Effectiveness Analysis (CEA) or with Cost-Benefit Analysis (CBA)? Should it have a bias in favor of some type of errors (e.g. errors on the side of preservation or errors on the side of technological development)? To what extent should experts contribute to the decision? Finally, should a regulation be prohibitory or have some particular content?

Circumstances Triggering the Application of the Principle: Toward a Definition of the Concepts of "Serious", "Irreversible" and "Uncertain" Harm

The first element is relatively uncontested. Unless we embrace a "Radical Precaution" version, a possibility dismissed earlier in this section, there is

---

[26]Sunstein, C., "Beyond the Precautionary Principle", *Chicago John M. Olin Law & Economics Working Paper*, (2002), No. 149. Actually the author argues against the Precautionary Principle in general. However, it is my contention that what he calls Precautionary Principle, is what in this paper goes under the name of Radical Precaution. Moreover, the author treats the Principle mainly as if it applies to cases of risk. Only a short section of his analysis focuses on uncertainty. However, uncertainty seems to be the central element of the Principle.

quite some agreement that the application of the Precautionary Principle is triggered by the presence of "serious" and "irreversible" threats to human health and/or to the environment. Moreover, such hazards should be accompanied by poor understanding, more specifically by "uncertainty". For instance, the Principle should not be invoked to regulate an activity entailing well-understood risks such as car driving. In fact, there are quite accurate statistics about the number of fatalities due to car accidents; in most countries we have a mixed system of safety regulation, liability rules and compulsory insurance aimed at controlling car accidents. Such legal regimes exist independently of a Precautionary Principle. However, there are many other cases in which harmful consequences of human activities, such as the introduction of GMOs in agriculture or the installation of antennas for mobile phones, are not very well understood. When there are reasonable suspects that such technologies might have serious and irreversible consequences for Society, the Precautionary Principle can be invoked for the enactment of laws regulating such activities.

The problems begin when we want to define more precisely the concepts of "serious", "irreversible" and "uncertainty". In this paper I define *"serious"* in relation both to the quantitative and to the qualitative dimensions of risk. These two dimensions are combined; this means that a hazard should not only be catastrophic in terms of its magnitude but should also possess some qualitative attributes, such as being involuntarily born by people. Why should we take qualitative dimensions of risk into consideration as well? Because characterizing risk only on the basis of numbers can lead to a mischaracterization. Several studies have shown that people often characterize risks on the basis of some of the following qualitative dimensions: risk is voluntarily or involuntarily taken; risk is chronic or catastrophic; it is common or dread; it is known or unknown to those exposed; it is known or unknown to science; it is old or new; it is controllable or uncontrollable by those exposed.[27] A risk that is voluntary undertaken is rated as less dangerous than a risk involuntarily taken; a chronic risk is feared less than a catastrophic one and so on. These factors might explain why risk created by nuclear power, which is involuntary taken, catastrophic and dread, is considered to be very serious by laymen, in spite of the fact that the "risky numbers" are very low.[28] As this literature has shown, qualitative dimensions of risk are an important aspect too. Therefore, the dimension of "seriousness" is given by very large potential consequences

---

[27] In psychology, this type of research, aimed at systematically understanding people perceptions of risk, is also called "psychometric paradigm". For a discussion of the "psychometric paradigm" see Slovic P., Fishoff, B. and Lichtenstein S., "Facts and Fears: Understanding Perceived Risk" in Schwing, R. C. and Albers, W. A., (eds) *Societal Risk Assessment: How Safe is Safe Enough?*, New York-London, (1980), pp. 181-214.

[28] Evidences of the fact that people consider nuclear power more risky than other activities and technologies are reported in Slovic, P., Lichtenstein, S. and Fishoff, B., *"Images of Disaster: Perception and Acceptance of Risks from Nuclear Power"* in Goodman, G.T.and W. D. Rowe (eds), *Energy Risk Management*, (1979), pp. 223-245.

of hazardous activity coupled with the fact that the risk has some of the qualitative characteristics that make the risk more undesirable, such as the fact that it is involuntarily taken or uncontrollable by those exposed.

In addition, the threat should have "*irreversible*" consequences. This term implies that it is not possible to restore the status quo ante, once the consequences manifest. Unfortunately, the term might give rise to some confusion because if it is taken literally almost any action has irreversible consequences. Killing a chicken to have a nice meal is an action that has irreversible consequences. Irreversible is then a term that should be further defined in relation to an object. If we think of animals, for instance, we often look at the species and not at single animals. Therefore, killing one chicken (or thousands) is not irreversible if the object of irreversibility is the species of chicken. It should be obvious, then, that the object of irreversibility has to be specified on a case by case basis.

Finally, to define "*uncertainty*" one should distinguish the case when the distribution of probability that a particular event will occur is known and when it is not. Knight distinguishes between measurable and unmeasurable uncertainty, using the term "*risk*" to designate the former and the term "uncertainty" for the latter. "The practical difference between the two categories, risk and uncertainty, is that in the former the distribution of the outcome in a group of instances is known (either through calculation *a priori* or from statistics of past experience), while in the case of uncertainty this is not true, the reason being in general that it is impossible to form a group of instances, because the situation dealt with is in a high degree unique".[29] *Uncertainty*, therefore, will be used to refer to a situation characterized by a certain degree of ignorance. Uncertainty in the knightian sense is typically invoked with respect to the ignorance over the distribution of probabilities of an outcome. However, uncertainty can also be over the outcome itself. For instance, it is not yet clear whether hormones used to treat the female menopause are beneficial or dangerous for heart related diseases. Hence, a further distinction needs to be made: uncertainty over the probability that an event will take place only, and ignorance that implies absence of knowledge about the effects of an action. We can then elaborate a fourfold distinction. First of all, we speak of certainty when we know the effects of a certain action and the probability that they will occur equals 1. Secondly, when this probability is less than 1 then we speak of risk. Thirdly, if we know the effects but not the distribution of probability, then we are under conditions of uncertainty. Finally, if we lack accurate information over the probability as well as over the possible effects as such, we speak of ignorance. This fourfold distinction is summarized in Table 1.1.[30] In the remainder of the paper, for reasons of

---

[29]Knight, F. H., *Risk, Uncertainty, and Profit*, Boston, New York, Houghton Mifflin Company, (1921) at p. 233.

[30]The academic literature about risk distinguishes also between legal-moral, institutional, societal, situational, proprietary and scientific uncertainty. These distinctions are relevant

brevity, I use mostly the term uncertainty. The arguments are applicable also for cases of ignorance that indeed might turn out to be the most significant cases.

Having outlined and defined in some details the circumstances triggering the application of the Precautionary Principle, a last consideration is due. In spite of the precision of the definitions offered, there will still be cases for which it will be difficult to decide whether the Principle should be applied. In other words, under the depicted framework for the delimitation of the field of application of the Principle, gray areas remain. Notwithstanding the existence of such gray areas, the framework elaborated seems enough precise to circumscribe the application of the Principle to a limited set of cases.

When or When Not? That is the Question!

"The element of time [...] is the center of the chief difficulty of almost every economic problem."
Alfred Marshall[31]

As it has been pointed out elsewhere, "[t]he new element is the timing of, rather than the need for, remedial action. Preventive action once damage has been determined is a long-standing requirement of international environmental law. In fact, the essence of the precautionary concept, the precautionary principle, is that once a risk has been identified the lack of scientific proof of cause and effect shall not be used as a reason for not taking action to protect the environment".[32]

The element of timing is sometimes overlooked in the literature about the Precautionary Principle, as the attention is focused mostly on whether we need a certain type of remedial action. This might be due to the fact that the questions "what type and when we need a remedial action" are intricately related and seem difficult to disentangle. However, it is worthwhile to make this analytic distinction as timing might turn out to be the essence of the Principle. Then, what is implied by the term "timing" when discussing the Principle? To shed some light on this issue, let's consider a simplified case showing the relation between the initiation of an activity and the adoption

---

for the discussion of the regulatory frameworks to manage industrial risk. For reasons of simplicity, in this paper, I address only scientific uncertainty. For a discussion of the implications on the management of disasters of these different types of uncertainty see De Marchi, B., S. Funtowicz, and J. Ravetz, *Seveso: A paradoxical classic disaster*, in Mitchell, J. K. (eds) *The long road to recovery: Community responses to industrial disaster*, (1996), United Nations University Press, The United Nations University, available at: http://www.unu.edu/unupress/unupbooks/uu21le/uu21le00.htm#Contents.

[31] I am in debt to Guido L. R. Munzi for this quote. Marshall, A., *Principles of Economics*, (1961), 9th edition, Macmillan for the Royal Economic Society.

[32] Freestone, D. and E. Hey, "Origins and Development of the Precautionary Principle" in Freestone, D. and E. Hey (eds) *The Precautionary Principle and International Law - The Challenge of Implementation*, (1996), Kluwer International, at p. 13.

of regulation in a purely temporal perspective. The question is not whether the regulator should regulate, but when she should regulate. The example considers a case where at time T_1 an activity might be initiated; nevertheless, there are uncertainties about its potential harmful effects. The activity can be initiated at time T_1, without any regulatory measures being adopted. Alternatively, regulation might be adopted at time T_1, banning or restricting the activity. At time T_2 the decision can be revised. In Table 1.2 some possibilities are presented. Of course, several combinations can be envisaged and the process can unfold in different ways than those prospected. Moreover, the process can be repeated several times (up to infinity).

What is important to emphasize at this stage is just that the timing of decisions might affect the content of decisions themselves. The questions pertaining to the element of timing that will be answered in the last part of the paper are the following: 1. How does the sequencing of the decision-making process matter (if it matters at all) for the outcome reached? 2. Can the Precautionary Principle give any guidance for the dynamic of the decision making process?

Onus probandi incumbit ei qui facit aut ei qui regulat? [...] That is Not the Question!

Besides the question of timing, a widely discussed element of the Principle is the question of the burden of proof, which indeed is the most controversial one. To believe that the Principle implies a so-called shift of the burden of proof, namely that the proponents of the activity have to proof the safety of their activities or products, means to adhere to a "strong version" of the Principle. However, characterizing this question as a *shift* of the burden of proof implies that the physiological situation is one in which the non-safety of an activity is proven by the regulator. Only when this is proven, regulation can be enacted. Is this assumption plausible? It depends on the way we define the physiological case. On the one hand, if we refer to the general idea underlying liberal systems and embodied in many western legal regimes that private enterprise is free, then it seems fair to denote the question as a "shift" in the burden of proof. On the other hand, if we look at figures, in liberal legal systems we find an overwhelming number of regulated activities under a system of prior approval, where the burden of proof typically falls on the proponent of the activity. In particular, in the context of environmental and safety law, where the Principle is more likely to be applied, activities under prior approval are the majority. To mention a few major examples, think of the regulation of pharmaceutical products, the regulation of hazardous installations, the regulation of pesticides etc.[33]

---

[33]In USA we might mention the Federal Food, Drug and Cosmetic Act and the Federal Insecticide, Fungicide, and Rodenticide Act.

I believe that most controversies about the question of the burden of proof stems from misconceived questions. Opponents of the Precautionary Principle are particularly keen in showing how vexing it is to shift the burden of proof on the proponents of an activity. How can we possibly expect that the proponent demonstrates that an irreversible damage is not going to occur or the probability that is going to occur is low enough, something which by definition is uncertain? *Probatio diabolica*, and consequently a certain halt to any kind of development! However, such reasoning seems to prove nothing as, symmetrically, one might wonder how the regulator can demonstrate the opposite, i.e. that the damage is going to occur with certainty or with a sufficiently high probability? *Probatio diabolica*, and consequently a certain halt to any kind of regulation! The criticism, as formulated, introduces a paradox. This paradox might be overcome by rethinking the critical questions to be asked. The first question to be asked is: what should be proven? Then, it will be easier to tell who has to provide the evidence. Here, it seems important to give a feasible task to both parties.

In addition to such question, other relevant issues should be taken into account for an appraisal of the rule of the burden of proof. Several criteria might play a role. From an economic perspective it seems relevant to ask who incurs the least costs in gathering information. Moreover, it should be asked what types of incentives are created by the rule on the burden of proof on the production of information. If proponents have to prove that their activities meet certain safety standards, as it is the case in regimes of prior approval, they are forced to produce information about their activity. On the contrary, parties might have few incentives to produce information over the risks of their activities when the regulator has to prove that the activity is unsafe: the less is known, the less is regulated and the less likely is liability to fall upon them.[34] This trend toward the suppression of the creation of new information about hazards is clearly undesirable.

In conclusion, it does not seem wise to stigmatize a so-called "shift" of the burden of proof as part of a too radical version of the Principle. Instead, it seems wiser to understand the nuances of the question of the burden of proof in a wider context where the relation between technical knowledge and the law is in continuous evolution and where incentive streams related to the production of information might affect the knowledge stock of our society significantly. The last section of the paper suggests a solution to the question

---

[34]An interesting example in this respect is the case of the use of Vinyl Chloride in industrial production. In this case the behavior of the chemical industry was very dubious in relation to the disclosure of information about the toxic properties of Vinyl Chloride. Such behavior might be understood as a way to postpone safety regulation and to avoid liability. See Bettin, G. and M. Danese, *Petrolkiller*, (2002), Feltrinelli, with an Appendix containing several confidential letters of chemical companies such as Union Carbide, Shell and Montedison concerning the toxic properties of Vinyl Chloride and the willingness to maintain this information secret. Some of these documents can be also found at the following web sites: http://chemicalindustryarchives.org/ and http://vinylchloride.com/.

of the burden of proof that takes into account both the paradox introduced by a double *probatio diabolica* and the problems related to the production of information.

How Should the Regulation be?

The fourth question asks how the Precautionary Principle should shape regulatory measures conceived to deal with uncertain hazards. As a preliminary issue, it should be clarified whether the Principle dictates a particular decision (e.g. a ban on every activity posing a serious threat to the environment). This question can be immediately answered in a negative way. If the Precautionary Principle would dictate *ex ante* a specific decision, it would be a rule and not a principle.[35] Therefore, from the outset, we can conclude that the Precautionary Principle does not prescribe any particular measure. The Precautionary Principle is more about the direction of the decision and how the decision is adopted than about the exact content of the decision.

Many legal economic scholars seem to consider the Precautionary Principle only viable if it remains within the paradigm of cost-benefit analysis (CBA).[36] In addition, some legal texts do refer to CBA and some to Cost Effectiveness Analysis (CEA). It is, however, not clear from these legal documents to what extent these techniques of economic appraisal are to be used. The answer to this question requires a theoretical discussion of CBA as a way to appraise regulation and of the viability of CBA in the context of uncertainty. Partly related to these issues is the question of how science can enter the decision-making process. For instance, in USA a two step analysis where risk is scientifically assessed in the first place (so called risk-assessment) and managed by the policy maker in a second moment is endorsed. However, the presence of scientific uncertainty challenges such a deterministic stance. Strictly inherent to the presence of uncertainties is the question whether it is desirable to err systematically in favor of environmental preservation or in favor of development. The following section discusses different theories on which it is possible to ground the answers to such questions.

In the light of the above reasoning, the two extreme versions of the Principle, i.e. α-Precaution and Radical Precaution, can be dismissed as undesirable for society. To see what versions are more desirable for society, it is necessary to scrutinize in a more analytical fashion what the *pro* and the *contra* of the outlined elements are. Therefore, in the next section of the paper, some ma-

---

[35]See Section 1.2.2, above.

[36]See, for instance, Stweart, R. B., "Environmental Regulatory Decisionmaking Under Uncertainty" in *An Introduction to the Law and Economics of Environmental Policy: Issues in Institutional Design - 20 Research in Law and Economics*, (2002), pp. 71-126; Sunstein, C., (2002), op. cit. and Gollier, C., "Should we beware of the Precautionary Principle?" in *Economic Policy*, (2001), pp. 302-327; Hammitt, J. K., "Global Climate Change: Benefit-Cost Analysis vs. the Precautionary Principle", in 6-3 *Human and Ecological Risk Assessment*, (2000), pp. 387-398.

jor theoretical contributions relevant to our analysis are discussed. Finally, in the last section of the paper, these theoretical findings are used to propose a particular version of the Precautionary Principle that seems beneficial for Society and operational at the same time.

1.3 "Precautionary Economics" in Essence and Beyond

1.3.1 Do we Need Precaution on Top of Prevention and Proportion?

Up to now we have seen how the Principle has entered the legal scenario and underlined that so far the Principle lacks clarity; it has been emphasized that in order to suggest a more clear interpretation of the Principle it is important to hold in mind the very nature of principles; at last, the most relevant questions that lead to various interpretation of the Principle have been delineated. Now I turn to discuss some theories that can be helpful to elaborate a proposal for an interpretation of the Principle. I start by briefly sketching out the Law and Economics theoretical approach.

In Law and Economics legal systems are mainly conceived as means to reach efficient outcomes. In more general terms, the law is seen as a corrective measure for some kind of problem and ideally this correction should be optimal. In this cadre we should ask the following interrelated questions. What is the problem that is to be solved by the Precautionary Principle? And, why do we need the Principle? To put it simply, the Principle pertains to the domain of risk regulation, where the policy maker is confronted with the issue of how to control hazardous activities in our society. Risks posed by industrial activities are often regarded as externalities. Hence, from an economic perspective, some form of control over such risks is necessary. Notably, reducing risks to zero would imply in most of the cases to ban the activity *tout court*. However, most of these activities are beneficial for society. Law and Economics scholars argue that, in order to have a rational risk policy, decision-makers should appraise the costs and benefits of regulatory measures.[37] Even tough there might be some controversies on how to carry out a full-fledged cost-benefit analysis, the idea that the major tradeoffs which the

---

[37]See for instance, Pildes, R. H. e C. R. Sunstein, "Reinventing the Regulatory State", in 62 *The University of Chicago Law Review*, (1995), pp. 1-129; Ogus, A., "Regulatory Appraisal: A Neglected Opportunity for Law and Economics" in 6 (1) *European Journal of Law and Economics*, (1998), pp. 53-68; Zerbe Jr., R. O., "Is Cost-Benefit Analysis Legal? Three Rules", in 17-3 *Journal of Policy Analysis and Management*, (1998), pp. 419-456; Arrow, K. J., M. L. Cropper, G. C. Eads, R. W. Hahn, L. B. Lave, R. G. Noll, P. Portney, M. Russel, R. Schmalensee, V. K. Smith, R. N. Stavis, "Is there a role for Cost-Benefit Analysis in Environmental, Health and Safety Regulation?" in 272 *Science*, (1996), pp. 221-222.

regulatory process entails should be taken into account, seems uncontroversial among most legal economists. The cost-benefit analysis approach is "softly" embodied into European law by the so-called Proportionality Principle (art. 5 EC Treaty); Clinton's Executive Order No. 12866 directly implements it into American law-making system.[38] Hence, in the light of the fact that we do have a major guideline for such circumstances, why do we need the Precautionary Principle? From an economic perspective, isn't cost-benefit analysis sufficient to guide decision-making?

The basic condition to carry out a cost-benefit analysis is that information about the costs and benefits of a certain action is available. Absent this information cost-benefit analysis is not viable. It is worth noting that, in the context of risk regulation, the most difficult information to obtain is about the benefits of a regulation. In the case of environmental regulation, these include but are not limited to the decrease in risk due to the proposed measure, which in turn is calculated in terms of lives saved or environment preserved. For instance, among the benefits that the US Environmental Protection Agency considered in order to assess the benefits of the Great Lakes Quality Guidance, there was improved water quality and the reduction of direct mortality risk of many aquatic, avian and mammalian species of concern.[39] Leaving aside the disputes over the commodification of life or of nature, these valuations are difficult albeit not impossible to conduct.[40]

However, many hazards are of uncertain nature. The problem is that, at first sight, benefits (e.g. in terms of diminished hazards) cannot be assessed under conditions of uncertainty; therefore, in presence of uncertainty cost-benefit analysis does not seem a feasible option. Then, under conditions of uncertainty, we need an alternative guiding principle for decision-making in the field of environmental and safety law. It is in this context that the Precautionary Principle might enter the scene to play an important role.

Eventually, Precaution is different from prevention and proportion, as it deals with uncertainty, a case not considered by the other two approaches. To cut a long story short, the economic counterpart of Prevention is externalities economics, still a dominant paradigm among environmental economists; the economic counterpart of Proportion, as seen, is cost benefit analysis in a broad

---

[38]For the Proportionality Principle see Art. 5 of the EC Treaty. Moreover there is a Communication from the EC Commission in the direction of a systematic use of Regulatory Impact Analysis see the Communication from the Commission on Impact Assessment, Brussels 5.6.2002, COM (2002) 276 final. For United States see, *Executive Order No. 12866, Regulatory Planning and Review*, 58 F.R. 51735, September 30, 1993.

[39]The Great Lakes Quality Guidance established water quality standard, antidegradation policies and implementation procedures for the Great Lakes Basin. For a discussion of the Regulatory Impact Analysis carried out in that case see Castillo, E. T., M. L. Morris, et al., "Great Lakes Water Quality Guidance", in *Economic analyses at EPA: Assessing regulatory impact*. R.-D. e. Morgenstern. Washington, D.C., Resources for the Future, (1997), pp. 419-54.

[40]For general critics of CBA see for instance Kelman, S., "Cost-Benefit Analysis: An Ethical Critique" in *Regulation*, (1981), at p. 33.

sense. Is Precaution conflicting with these principles? No, because, as seen, it is intended to solve problems not addressed by the other two principles and therefore it becomes eligible to be a necessary complement. The question that will be investigated in more details in the following is the relationship between the implementation of the Precautionary Principle and the use of cost-benefit analysis.

1.3.2 Should the Precautionary Principle be Embedded in the Paradigm of Cost-Benefit Analysis?

Let us now investigate in more depth whether it is feasible to carry out CBA in presence of uncertainty or of ignorance. As mentioned above, at first sight, it is not possible for the simple reason that uncertainty undermines the preliminary condition for carrying out CBA, namely to have information about the costs and the benefits. However, scholars in the field of economics and of law and economics, argue that CBA can be used in presence of uncertainty and that the Precautionary Principle should still fulfill a cost-benefit test. The method to overcome the problems introduced by uncertainty is the so-called Bayesian updating.[41] Here "it is assumed that the decision-making agents know the probability distribution for the complete set of possible outcomes, and that they act to maximize an 'expected value' of the outcome. [...] In a relaxation of the informational assumptions, it may be proposed that, rather than knowing all the probability distributions, the decision maker makes a *subjective* estimate of probabilities attaching to each possible outcome. These subjective probabilities may differ from the true (objective) probabilities, and this discrepancy is seen as introducing the possibility of learning through time, following Bayesian decision rules".[42] Let me now try to single out different ideas constituting the core of Bayesian theory.[43] Two main ideas are implied in a Bayesian framework: the possibility of updating our knowledge and the formulation of subjective estimation of probabilities. The idea of updating is crucial for a process of decision-making under uncertainty. Decision-making process should be dynamic in the sense that previous decisions should be systematically revised on the basis of new information. Dynamic is the precondition for learning. This dimension of Bayesian theory seems crucial and, as will be argued later, should be taken into serious considerations in a discussion of the Precautionary Principle. The idea of assigning subjective probabilities is,

---

[41] In this respect see Stewart, R. (2002), op. cit. at pp. 91-92.

[42] O'Connor, M., S. Faucheux, G. Froger, S. Funtowicz, and G. Munda, "Emergent Complexity and Procedural Rationality: Post-Normal Science for Sustainability" in *Getting Down to Earth: practical application of Ecological Economics*, Island Press, (1996), pp. 223-48, at pp. 232-33.

[43] Notably, the following analysis is instrumental to appraise the role of CBA applied to regulation in situations of uncertainty. It is absolutely beyond the scope of this contribution to appraise Bayesian theory as such or applied in other contexts; besides, it would be beyond the author's competence to carry out such analysis.

instead, very problematic. The problem is that Bayesian theory does not tell *how* to build the initial subjective probabilities. This leaves room for highly discretionary choices that have little to do with rationality. In fact, choosing a high or a low probability is not indifferent in relation to the outcome of our decisions. Without specifying a rule on how to assign subjective probabilities, a Bayesian decision-making process under uncertainty will not guarantee that the same results will be reached by different decision-makers.

On the basis of these considerations we can reach two different conclusions: a stronger one, rejecting the use of cost-benefit analysis and another, more mild, advocating for the use of cost-benefit analysis with some adjustments. The endorsement of one of these two conclusions depends, on the one hand, on the way the arguments presented so far are interpreted; on the other hand, it depends on other more general views about the necessity, the viability and the acceptability of cost-benefit analysis. Let me briefly explore these two lines of reasoning.

In line with the first conclusion, CBA is not an appropriate decision criterion in the presence of uncertainties because assigning subjective probabilities when using CBA is a way of eluding a problem not of resolving it. As put by O'Connor et al. "[...] in practice the employment of the Bayesian approach amounts to adoption of an 'empty box'. Depending on the subjective probabilities and states of the world incorporated into the calculations, the procedure may serve to legitimate decisions that are substantially arbitrary. [...] Seen in this light, employing Bayesian calculation to guide decision-making amounts to an arbitrary choice of decision-making form, ironically masquerading as substantive rationality".[44] It should be emphasized that what is mostly criticized by these authors is the deterministic spirit of cost-benefit analysis[45] By adhering to such spirit, the complexity and the uncertainties that characterize new environmental problems are overlooked. The reductionism implicit in cost-benefit analysis seems then counterproductive because it creates an illusion of knowledge. Finally, these authors emphasize that there are various alternative to cost-benefit analysis, as multi-criteria decision analysis and the sustainability tree, which here I merely note and don't discuss.[46]

In contrast, other scholars deem that the use of cost-benefit analysis is reconcilable with the presence of uncertainty. A way to solve the problems introduced by the assignment of subjective probabilities is to allow for mechanisms that give due account of the presence of uncertainty. This is the position of Prof. Stewart who argues that cost-benefit analysis is still the best decision making criterion. In this respect, he suggests that subjective

---

[44]O'Connor, M. et al., (1996), op. cit., at pp. 233-34.

[45]See O'Connor, M. et al., (1996), op. cit. at p. 235; from a very different intellectual perspective, other authors have criticized the determinist spirit of cost-benefit analysis, see Resnick, S. and R. Wolff, *Knowledge and Class: A Marxian critique of Political Economy*, (1987), Chicago and London: University of Chicago Press.

[46]O'Connor, M. et al., (1996), op. cit., at pp. 238-244.

probabilities should be based on *best judgements* of the decision-maker. As put by him: "In order to promote accountability and transparency, regulators should make their best estimation of the probability distribution for the uncertain risk in question through a process that invites public and expert input and makes explicit and public the relevant uncertainties and the bases for the regulators' determinations. [...] Because of limitations in data and scientific understanding, very substantial uncertainties are likely to remain in many cases. They should then exercise their best judgment in resolving the remaining uncertainty, estimating a probability distribution that they should then treat as if it were a known distribution for purposes of the later stages in the decisional process."[47]

Both conclusions seem supported by valid arguments. Content wise both theories acknowledge the importance of clearly fleshing out the uncertainties in the decision-making process and the importance of public and expert input to the debate. I believe that the substantial difference is more of a philosophical kind. In particular, the first conclusion is based on a "post-normal" conception of science, whereas the second is based on a traditional approach to science.[48] I am more sympathetic to the first conclusion because I believe it best captures the essence of our main problem. The paradigm of cost-benefit analysis is still centered around outcomes. Reasoning in terms of net benefits, expressed by a precise number, means reasoning *as if* there were no uncertainty. In contrast, by emphasizing that in presence of uncertainty it might be unfeasible to predict outcomes, the attention is shifted from best outcomes to best ways of reaching decisions. This does not mean that outcomes as such are not discussed or that traditional science is abandoned. Nor that costs and benefits should be neglected. It means, however, that all the steps where uncertainty is dominant are made explicit and central in the process of decision making. The process should guarantee that different stakeholders are allowed to contribute to the creation of knowledge in the decision making process, enhancing transparency and reducing the risk of regulatory capture. Most notably, the intertwined relationship between facts and values, typical of decisions concerning very serious social risks, is not hidden behind the veil of the objectivity of dry numbers.

---

[47]See Stewart, op. cit. at pp. 91-92.

[48]For an elaboration of the concept of post-normal science see Funtowicz, S. O. and Ravetz, J. R., "Science for the Post-Normal Age", in *Futures*, (1993), pp. 739-755. The concept of post-normal science is elaborated against the background of Kuhn's idea of science. As put by the authors, for Kuhn "'normal science' referred to the unexciting, indeed anti-intellectual routine puzzle solving by which science advances steadily between its conceptual revolutions. In this "normal" state of science, uncertainties are managed automatically, values are unspoken, and foundational problems unheard of". Post-modernism is critical to the normal science paradigm; however its alternative is nihilism and despair. The authors introduce the idea of post-normal science, where "enriched awareness of the functions and methods of science is being developed". In post-normal science, "uncertainty is not banished but is managed, and values are not presupposed but are made explicit". Notably in the conception of post-normal science, uncertainty gains central importance.

Having concluded that cost-benefit analysis cannot be considered as the ideal solution for decision making under uncertainty, we might look for other theories that can give guidance in interpreting the Principle. In the next section I elaborate on the question of whether a lack of cost-benefit analysis entails a disregard of the benefits of development.

1.3.3   Risk Tradeoffs Analysis and Consideration of Ancillary Benefits

One of the thorniest issues in the discussion of the Precautionary Principle turns around the fact that technologies that might have catastrophic consequences carry with them the promises of tremendous benefits. Probably the most salient critique to the Precautionary Principle is that the forgone benefits of technological development are systematically disregarded. The only solution seems to interpret the Precautionary Principle within the framework of cost-benefits analysis. I have already outlined the main reason why cost-benefit analysis is not an ideal answer to the problem under investigation.

It might be counter-argued that cost-benefit analysis is the only alternative that does not neglect consequences of regulatory measures. Such argument, however, is not convincing. First of all, cost-benefit analysis cannot be considered comprehensive in terms of the accounted consequences. Given the fact that any activity has an infinite numbers of effects, in the process of cost-benefit analysis the analyst has to select what are the effects to be considered. In reality, cost-benefit analysis of regulation has started as a means to deregulate. In this respect, it is interesting to note that cost-benefit analysis of regulation, in the beginning (i.e. during the Nixon Administration), was focused on regulatory costs for firms only, and the benefits of regulatory actions accruing to citizens were systematically disregarded.[49] The antiregulatory bias that has accompanied the use of cost-benefit analysis and its proxy, the so-called risk tradeoff analysis, in the American administrative State, has been acknowledged by prominent legal-economic scholars.[50] In addition, cost-benefit analysis considers the consequences of regulatory measures only in a very specific way, i.e. by translating them into money figures. In this respect,

---

[49]For an historical overview of the development of the use of CBA in the American administrative law see Morral III, J. F., *An assessment of the US regulatory impact analysis program*, in OECD *Regulatory Impact Analysis, Best Practices in OECD Countries*, Paris, (1997) at pp. 71-87. The antiregulatory bias characterized the experience in United Kingdom as well. Here, cost-benefit analysis started in 1985 in the limited form of Compliance Cost Assessment (CCA) under the so-called Deregulation Initiative initiated by Margaret Thatcher; in this case, Governmental Departments were asked to present alongside their law proposals a document estimating the costs to business of complying with the proposed measures.

[50]Rascoff, S. J. and R. L. Revesz, "The Biases of Risk Tradeoff Analysis: Towards Parity in Environmental and Health and Safety Regulation", in 69-4 *The University of Chicago Law Review*, pp. 1763-1832.

it seems plausible to acknowledge that the consequences of regulatory measures can be discussed also in terms different than money, especially when uncertainty would exponentially complicate the analysis.

The considerations relating to the weaknesses of cost-benefit analysis should not imply the abandonment of rational comparison of the consequences of a regulation. Comparing the consequences of different regulatory measures is still a procedure that should be welcome. Alternative frameworks for such comparisons should be developed. Given the presence of uncertainty, ignorance and high stakes involved, it is important that in democratic contexts the civil society is involved in the "selection" and "definition" of costs and benefits. Of course, scientific knowledge should play an important role in the decision-making process as well.[51] It is beyond the scope of the paper to elaborate such a framework. Here it suffices to say that the considerations of the consequences of regulatory measures should be part of the implementation of the Precautionary Principle but cannot be confined to the reductivist approach of cost-benefit analysis, at least as it is practiced now, without due account to uncertainties.

In this context, the possible uses of the so-called risk tradeoffs analysis should be mentioned. Such analysis that is gaining support in American administrative law, both between academics and between judges, considers the ancillary negative effects of regulation.[52] The idea underlying risk tradeoffs analysis is that regulations aiming at reducing certain risks may create other risks and instead of reducing the overall risk for Society, do increase it. Such reasoning is particularly appealing in the context of the Precautionary Principle, as precautionary regulations might have undesirable effects that create risks even more serious than those targeted. In this respect, risk tradeoffs analysis can be welcome to counter the danger of tunnel visions in which the focus is only on the risks targeted by regulatory measures while the side effects of regulation are disregarded. Unfortunately, both academic scholars and judges who support the use of such analysis show tunnel vision themselves, because they consider only the ancillary negative effects of regulation, while systematically neglecting the ancillary benefits of regulation.[53] In this respect risk tradeoffs analysis has developed as a methodology with a strong antiregulatory bias. However, there is no reason to consider only ancillary

---

[51]See the Section 1.4.3 of this paper, where a threefold model, including a phase for a scientific appraisal of risk, is elaborated.

[52]Several legal and legal-economic scholars have endorsed this approach. For notable examples see: Breyer, S. (1993), *Breaking the Vicious Circle: Toward Effective Risk Regulation*, Harvard; Graham, J.D. and J. B. Wiener (eds), (1995), *Risk versus Risk: Tradeoffs in Protecting Health and the Environment*, Harvard; Sunstein, C. R., "Health-Health Tradeoffs" in 63 University of Chicago Law Review, (1996), pp. 1533- ; Viscusi, W. K., "Mortality Effects of Regulatory Costs and Policy Evaluation Criteria", in 25 *Rand Journal of Economics*, (1994), pp. 94-109.

[53]This issue has been extensively and convincingly discussed by Rascoff and Revesz (2002), op. cit.

risks without taking into account ancillary benefits as well. Therefore if risk tradeoff analysis is carried out, symmetrically ancillary benefits of regulation should be given due account.

To conclude these brief notes on risk tradeoffs analysis I want to bring the attention to unnecessary juxtapositions. Many commentators often reason in terms of environment versus development.[54] However, adopting measures protecting the environment might not necessarily entail less development, but only "different" development. Adopting as a bottom-line a principle that is biased in favor of the environment can boost development that is environmentally friendly. In this respect the Precautionary Principle might be seen as stimulating creativity and innovation. In the following section I present other reasons why erring on the side of the environment is rational.

1.3.4   Irreversibility, Option Value and Flexible Decisions

In 1974, economists Kenneth Arrow and Anthony Fisher have analyzed the effects of uncertainty on the criteria for choice between two alternative uses of a natural environment, namely preservation or development.[55] Their model relies heavily on the idea that many decisions concerning the introduction of new dangerous activities or technologies share the characteristics of being irreversible. In these cases, the authors conclude, the decision is likely to be optimal when it is flexible; the larger the discrepancies in opinions and the potential for serious irreversible damage, the larger should the flexibility be. The idea is very simple. The inaction leaves open the possibility of acting later, while more information is collected. On the contrary, acting first, in case of irreversible consequences, does not allow going back at a later stage. Following previous models[56] the authors assume irreversibility, technically characterized as a development that would be infinitely costly to reverse. Given the restriction on reversibility and the plausible assumption that realization in one period affects expectation in the next, they discover the

---

[54]In the same spirit a subcategory of risk tradeoffs analysis, called "health-health trade-offs", has recently emerged. The assumption underlying this analysis is that there is a positive relation between health and wealth: the wealthier, the healthier. Therefore, health regulations that impose costs on society, indirectly decrease society overall level of health. The problem with this analysis is that, as discussed for risk tradeoffs analysis, it does not take into account the secondary benefits of regulation. For instance, technological innovation might spur from regulation as companies will develop better strategies to comply with regulatory requirements and this, in turn, will increasing societal wealth/health. For an excellent analysis of this issue see Rascoff and Revesz, (2002), cit., pp. 1778-1780 and pp. 1809-1811.

[55]Arrow, K. J. and Fisher, A. C., "Environmental Preservation, uncertainty, and irreversibility" in 88 (2) *The Quarterly Journal of Economics*, (1974), pp. 312-319.

[56]Fisher, A. C., J. V. Krutilla, and Cicchetti, "The Economics of Environmental Preservation: A Theoretical and Empirical Analysis" in 62 *American Economic Review*, (1972), pp. 605-19; Arrow, K. J., "Optimal Capital Policy and Irreversible Investment", (1968), in J. N. Wolfe, ed., *Value, Capital and Growth*; Arrow, K. J. and M. Kurz, "Optimal Growth with Irreversible Investment in a Ramsey Model" in *Econometrica*, 38, (1970) , pp. 331-44.

existence of a "quasi option value". This is the value of the options that would be lost with the development; such effect, while goes in the same direction of risk aversion, should not be confused with it.[57] In short, they find that the expected value of benefits under uncertainty is less than the value of benefits under certainty. "An interpretation of this result" the authors argue "might be that, if we are uncertain about the payoff to investment in development, we should err on the side of underinvestment, rather than overinvestment, since development is irreversible". At the same time, similar results were found by Henry who coined the expression "irreversibility effect".[58]

While in the seventies the Precautionary Principle was mostly unknown, today economists are aware of its existence. Recent economic studies have tried to relate directly to it. In particular, the work of economic scholars Christian Gollier, Bruno Jullien and Nicolas Treich is an attempt to give an interpretation of the Precautionary Principle within a standard Bayesian framework.[59] Their model shows that it is desirable for Society to take stronger preventive measures today in the presence of more scientific uncertainty as to the distribution of future risk (i.e. a larger variability of beliefs). An important point in their contribution is that "environmental problems deal with the management of a limited stock of a good. Varying this stock through early consumption generates an environmental externality for future generations". They seem to be the first to deal not only with irreversibility but also with accumulation phenomena (e.g. past emissions of CO_2 increase current exposure to greenhouse risks due to the accumulation of gas in the atmosphere).[60] In conclusion, the model predicts that more uncertainty should induce Society to leave more options open for the future.[61]

To conclude, it should be noted that the above-mentioned studies do not tell exactly what to do in presence of uncertainty; they give a general guideline. They don't tell neither what type of decision-procedure will lead to efficient decisions nor if certain types of decision will be best in every situation. In fact,

---

[57]Their model assumes risk neutrality.

[58]Henry, C., "Investment Decision under Uncertainty: the irreversibility effect", in 64 *American Economic Review*, (1974), pp. 1006-1012.

[59]See Gollier, C., B. Jullien and N. Treich, "Scientific Progress and irreversibility: an economic interpretation of the Precautionary Principle", in 75 *Journal of Public Economics*, (2000), pp. 229-253.

[60]They find that accumulation yields two distinct effects called "precautionary and rigidity effect". In this case more prevention is desirable when prudence is larger than twice the absolute risk aversion or when the inverse of marginal utility is concave.

[61]This survey is very limited. The literature mentioned, however, seems the most relevant for its generality. For a recent systematic survey of the economic literature focusing specifically on the Precautionary Principle see Gollier, C. and N. Treich, "Decision-making under scientific uncertainty: The economics of the Precautionary Principle", (2000), forthcoming in the *Journal of Risk and Uncertainty*. In addition to the contribution discussed in the paper, other relevant works are the following: Kostald, C. D., "Fundamental Irreversibilities in Stock externalities", in 60 *Journal of Public Economics*, (1996), pp. 221-233 and Ulph, A. and D. Ulph, "Global Warming, Irreversibility and Learning" in *The Economic Journal*, (1997), pp. 636-650.

these models are very sensitive to the assumptions made, assumptions that are often very difficult to test (e.g. the form of the utility function or the degree of prudence). In spite of this, from the studies reviewed two main generalizations can be drawn. First, decision-making under uncertainty should be dynamic in the sense that the passage of time is the critical factor that allows for learning and hence for improving our previous choices on the basis of new information. Second, that under a large set of conditions, decisions should err on the side of environmental preservation, in other words "pre-caution" should be applied. This insight relies on the idea of option value and flexibility in decision-making. The economics of Pre-caution can then be captured in these two main results.

1.3.5   Cost-Effectiveness Tests

So far we have argued that cost-benefit analysis is problematic and that "erring on the side of environmental" as a bottom line hermeneutic canon is meaningful from an economic point of view. A potential criticism to these first preliminary conclusions is that there is no consideration of costs. However, it would be fallacious to conclude that the Precautionary Principle invites to neglect regulatory costs. Cost-effectiveness analysis (CEA) is still an assessment procedure that can be welcome. As seen, the Rio Declaration expressly advocates CEA as a dimension of the Precautionary Principle. CEA is an analytical technique that looks for the least-cost means to achieve fixed regulatory goals. To put it differently, given certain goals the costs incurred to reach them have to be minimized. Thus, CEA is about comparing costs but does not relate to the calculation of benefits. CEA comes into play when a goal has already been established; in other words, it is not about what goals should be pursued but about what are the best means to achieve them.

The reason why CEA is less problematic than CBA is that, in the context where the Principle is advocated, the uncertainties are most likely to be about the benefits of a regulatory measure than about the costs.[62] Let me illustrate this statement with an example. Consider, for instance, the case in which toxic chemicals are discharged into the sea. A possible side effect of this discharge is the extinction of the fish population of the area affected by the discharge. The benefits of a regulation limiting the discharge of toxic chemicals into the sea would be mostly assessed in terms of forgone losses as the fish population would not die. However, as there are not uncontested evidence that the fish population affected by the toxic discharge will probably die, benefits are uncertain. On the other side the costs would be mostly compliance costs of the affected industry and administrative costs to implement the regulation that cannot be considered uncertain (e.g. it will be known

---

[62]For the sake of simplicity in this section I focus only on compliance costs, disregarding costs in the form of forgone benefits. This type of costs have been dealt with in Section 1.3.3.

what are the costs of a certain technical device to purify the discharge, it will be known the costs of some enforcing measures for the public authority and so forth). Given that the uncertainty falls only on the benefit side of our equation and that we might have information about the costs, it seems indeed rational to endorse CEA.[63] Moreover, from an ethical and practical perspective CEA seems easier to be widely accepted.[64] This is not to say that CEA would be unproblematic. It is beyond the scope of the paper and beyond the competence of the author to discuss the critique that can be done to this technique on an ethical basis and I will not embark on a discussion about the ways this analysis is most properly carried out. However, for the purpose of the paper, some problematic issues should be raised. First, should CEA be considered a constitutive element of the Precautionary Principle? If yes, when should CEA be carried out?

The answers to these questions might have important consequences at a practical level. To see why, let's consider Art. 5.1 and 5.2 of the Sanitary and Phytosanitary Agreement (SPS) 1994 where it is provided that SPS measures should be based on risk assessment.[65] The failure of carrying out risk assessment before adopting a SPS measure is likely to imply that the measure is in violation of the SPS rules.[66] The WTO Appellate Body both in the

---

[63]Of course there might be risk-risk tradeoffs that might complicate the analysis. These are briefly discussed in Section 1.3.3.

[64]The reason why CEA would be easier to accept from an ethical perspective is that only costs have to be compared. As mentioned in the text the costs of a regulation are mainly compliance costs for firms and enforcement costs for the public administration. These services are usually goods that are already exchanged in the market. Therefore, criticisms of "commodification" will not arise. Still, redistributive issues are not taken into account in CEA; however, if CEA is considered only to be informative and not prescriptive (as suggested in the text) redistributive decisions can be implemented in spite of CEA.

[65]Art. 5.1. and 5.2. of the SPS Agreement titled *Assessment of Risk and Determination of the Appropriate Level of Sanitary or Phytosanitary Protection* reads as following:

1. Members shall ensure that their Sanitary or Phytosanitary measures are based on an assessment, as appropriate to the circumstances, of the risks to human, animal or plant life or health, taking into account risk assessment techniques developed by the relevant international organizations.

2. In the assessment of risks, Members shall take into account available scientific evidence; relevant processes and production methods; relevant inspection, sampling and testing methods; prevalence of specific diseases or pests; existence of pest- or disease-free areas; relevant ecological and environmental conditions; and quarantine or other treatment.

[66]Here I use the rules of the SPS Agreement as an example, without discussing the relation between the SPS Agreement and the Precautionary Principle. This relation has been discussed by Prof. Ellen Hey: see Hey, E., "Considerations regarding the Hormones case, the precautionary principle and international dispute settlement procedures", in 13 *Leiden Journal of International Law*, (2000), pp. 239-248. See also Cameron, J., "The Precautionary Principle in International Law", in O'Riordan, T. and Cameron, J. (eds) *Reinterpreting the Precautionary Principle*,(2001), pp. 113-142; Majone, G., (2002), op. cit., at pp. 95-98. The delicate role to be played by risk assessment in international relations has been discussed by Walker, V. R., "Keeping the WTO from Becoming the "World Trans-

so-called Hormones case and in the Australia-Salmon case gave this interpretation of the SPS rules. These cases can be illustrative of what would be the consequences if CEA were considered a constitutive feature of the Precautionary Principle. *Mutatis mutandis*, we could say that the failure of carrying out a CEA when adopting a "precautionary" measure would be considered a violation of the Precautionary Principle.

In the light of such consequences, is it desirable to consider CEA constitutive of the Precautionary Principle? A tentative answer might be yes, with some important adjustments. From an economic perspective it seems obvious that CEA is a minimal requirement to have rational measures. Why if you can achieve the same result at a lower cost would you choose the more costly solution? From other perspectives, there seem no major motives for not having CEA. Most notably, the fact that a CEA is carried out should not imply that the decision should be adopted accordingly. In other words, CEA should provide information for the decision-maker but not the decision itself.[67] A decision-maker might deviate from the result suggested by the CEA if there are valid reasons to do so (e.g. costs are borne only by a particular group of society and the next measure available, slightly more costly, is more just in term of redistribution). In other words, CEA should *not be prescriptive*; it should be merely *informative*. The fact of carrying out a CEA would improve not only the rationality of the decisions but also the accountability of the decision-making process. Moreover, such an interpretation would be consistent with the idea that principles bind discretion.

In addition, an important adjustment is required when considering the factor of timing. There might be cases in which the regulator deems appropriate to adopt an urgent decision. Carrying out a CEA costs time and, as seen time is crucial. Procrastinating the decision because a CEA has not been carried out is itself a decision (no regulatory action until CEA is undertaken); however, the no regulatory action option does not have to satisfy any tests as it is just not recognized as a decision. However, as it emerged from the economic literature reviewed above, in most cases, it is efficient to bias our decision in favor of environmental preservation. No regulatory action would go in the opposite direction because development would be allowed during the time a CEA is carried out. In order to avoid it, it seems plausible to accept decisions made at T_1 without CEA. As it will be clear in the third section of this paper the version of Precautionary Principle suggested mandates, *inter alia*, that the decision-making procedure should be iterative, i.e. decisions should be regularly reviewed. Within such a framework, it can be proposed that the CEA should be carried out at least at T_2 (i.e. where the decision is

---

science Organization": Scientific Uncertainty, Science Policy, and Factfinding in the Growth Hormones Dispute", in 31 *Cornell International Law Journal*, (1998), pp. 251-320.

[67]For a similar argument in the context of CBA see Zerbe Jr., R. O., "Is Cost-Benefit Analysis Legal? Three Rules", in 17-3 *Journal of Policy Analysis and Management*, (1998), pp. 419-456.

reviewed for the first time). In conclusion, CEA is to be considered constitutive of the Precautionary Principle, in the sense that CEA has to be carried out, even if the outcome should not be considered binding on the decision-maker. However, it should not necessarily be carried out when the measure is adopted for the first time; it should be carried out only before the decision is reconsidered at time T_2.

1.3.6 Science, Economics and the Law: Rethinking the Paradigm

The Precautionary Principle has emerged in the interstices between the formation of new scientific knowledge and the formation of new legal rules. It is only in this conceptual space that the real essence of the Principle can be captured. Actually, the Principle is the offspring of an apparently paradoxical relation between science and risk law: scientific information is normally the basis of consensus for the adoption of risk law; however, in our case, scientific information turns into the basis of disagreement.[68] Acknowledging that in certain circumstances science does not provide unambiguous answers does not mean that a "scientific approach" should be dismissed. The question, then, is the following: how can decisions remain scientific and accountable when the scientific results are in doubt?

This question, in turn, begs another question. When in the scientific community multiple answers are given to the same problem, how can one science-based decision be better than another? Even if we believe that there is a "true" answer ex post, at the moment of decision we cannot conclude that there is one "best" decision. The best decision would, indeed, depend on the "best" science; however, if our assumption is that there is controversy within an established scientific environment we cannot assert, by definition, that there is a "best" science *ex ante*. Therefore, it is not in the substantive dimension of the question that we can find an answer. Aware of the circularity of the problem, we can overcome it by looking at what would be the "best way" to take decisions on the basis of controversial science.

One solution would be to leave the problem in the hands of experts and, therefore, endorse an "expertocracy" model. In this model the decision would directly follow from the judgement of experts. However, if this model is to be welcome in other circumstances (if at all), there are clear disadvantages to endorse it in our case. To begin with, we have said that experts have different

---

[68] For a more sophisticated discussion of such issue in a broader philosophical context see, von Schomberg, R., "The Erosion of Our Value Spheres: The Ways in which Society Copes with Scientific, Moral and Ethical Uncertainty", in von Schomberg, R. and K. Baynes, (eds) *Discourse and Democracy: Essays on Habermas's Between Facts and Norms*, (2002), State University of New York Press. The traditional Law and Economics approach is also locked in this paradox, as the major paradigm endorsed for risk regulation is cost-benefit analysis that in turn relies on risk assessment. When the results of risk assessment are highly controversial, the outcome of cost-benefit analysis will depend on the results chosen.

views on the same problem. Therefore, by choosing one group of experts, the decision-maker takes already a direction. The decision of which experts to choose is not scientific but political, as the decision-maker has no scientific expertise to choose experts. However, the decision based on the experts' judgements will finally be considered as legitimized by science. As a result political decisions are masked behind the façade of neutral science. Related to this, it should be reminded that the neutrality that is commonly associated with experts is not always there. Many times experts are strictly linked to industry or other groups, because for instance their salary comes from there; at the same time these groups have direct interests in the findings of science. Where scientific uncertainty is present, experts that are not independent will likely present an answer favoring the group sponsoring them directly or indirectly. This consideration brings into focus another element that is very important when discussing the role that should be played by experts, which is their independence.

This reasoning should not lead to the conclusion that experts have no role to play in the decision-making process. The role to be played by experts can be better discussed in the light of the two-fold process of risk evaluation commonly divided in risk assessment and risk management. "Risk assessment is a methodology for making predictions about the risk attached to the introduction, maintenance or abandonment of certain activities [...] is a way of ordering, structuring and interpreting existing information with the aim of creating a qualitatively new type of information, namely estimations on the likelihood of occurrence of adverse effects".[69] Both in Europe and in USA risk assessment for toxic substances consists of four steps: 1. hazard identification, 2. dose-response assessment, 3. exposure assessment and 4. risk characterization.[70] Risk Management is the public process of deciding what to do when risk has been determined to exist. This distinction has originated in the USA where, from the early eighties, a quantitative, formalized approach to science-based regulation has been preferred.[71] The dominant position maintains that

---

[69]Heyvaert, V., "Reconceptualizing Risk Assessment", in *Review of European Community & International Environmental Law*, (1999), pp. 135-143, at 1. Another well-known definition of risk assessment is given by Ruckelshaus, who in the early 80s has been leading EPA and has strongly defended the position of a clear separation between risk assessment and risk management. "Risk assessment is the use of a base of scientific research to define the probability of some harm coming to an individual or a population as a result of exposure to a substance or a situation" in Ruckelshaus, W. D., "Risk, Science and Democracy", in *Issues in Science and Technology*, (1985), pp. 19-38.

[70]The guiding principles have been set in Europe by Commission Directive 93/67/EEC of 20 July 1993 laying down the principles for assessment of risks to man and the environment of substances notified in accordance with Council Directive 67/548/EEC, and in USA by the Committee on the Institutional Means for Assessment of Risks to Public Health, National Research Council, Risk Assessment in the Federal Government: Managing the Process (1983).

[71]One significant decision that marked a shift toward the use of formalized risk assessment is the Supreme Court's benzene decision of 1980. In that case, the United State Supreme

the two processes should be clearly separate as the first relates to science and the second to politics.[72]

This position has merits as well as drawbacks. The uncontested merits are that some objectivity, in regulatory processes pertaining to the domain of science, is preserved. Unless we want to embrace a cultural relativistic position that denies any degree of objectivity to risk assessment and conceives of risk as nothing more than a "social construct", we have to grant some credit to "purely" scientific evaluation.[73] Indeed, acknowledging that risk judgements are to some extent value-laden should not entail a complete dismissal of rational analysis of risk. In this respect, cultural relativists' reasoning can be labeled as reductivist. As Shrader-Frechette put it: "[r]isk may be defined primarily in terms of probabilities, and probabilities are always (to some degree) theoretical. But this does not make [...] evaluation a mere construct. Construct do not kill people, but faulty reactor and improper stored toxics do."[74]

Considering risk assessment as completely free from value judgments is similarly reductivist because when carrying out risks assessment many assumptions, conservative or less conservative, determine the final outcome of the analysis.[75] In this respect, the kind of risk assessment institutionalized in the American regulatory process has received many criticisms. As put

---

Court turned down a regulatory standard set up by the Occupational Safety and Health Administration (OSHA), because the agency did not provide enough scientific evidence to prove the existence of significant risk. See *Industrial Union Department, AFL-CIO v. American Petroleum Inst.*, 448 U.S. 607 (1980).

[72]See Ruckelshaus (1985), op. cit.

[73]The most important and well-known contribution of cultural relativists is the book of Douglas, M. and A. Wildavsky, *Risk and Culture. An Essay on the Selection of Technical and Environmental Dangers*, (1982), University of California Press. The main claim of the book is that all judgements about risks are value-laden; therefore, improvement in knowledge about risk has no value and risk assessments are not much different than judgements in aesthetics.

[74]Shrader-Frechette, K., "Reductionist Approaches to Risk" in *Acceptable Evidence - Science and value in risk Management*, (1991). Mayo, D. G and Hollander, R.D., (Edited by), (1991) NY, at p. 219. For a painstaking analysis of the logical faults incurred by Douglas and Wildavsky read the whole article.

[75]There are many problematic passages in risk assessment. An example is cancer risk assessment. Among the most contentious judgements in carcinogenic risk assessment is how to extend the dose-response curve from the high doses to which animals are exposed in the laboratory to the lower doses to which humans are exposed in the environment. There are several statistical models for fitting the animal data and extrapolating the dose-response curve to low doses. Even if these models often have some plausible basis in biology, they might yield low-dose risk estimates for the same chemical that vary enormously by factors of hundreds or even thousands. As a default position based primarily on policy considerations, EPA requires use of linearized multistage model (LMS), because this is considered to be a conservative method of estimating low-dose risks. However, some authors argue that so far this is the best method we have and it is not necessarily conservative, see, for instance, Finkel, A. M., "A second opinion on an environmental misdiagnosis: the risky prescriptions of *Breaking the Vicious Circle*" in 3 *New York University Environmental Law Journal*, (1995), pp. 295-381.

synthetically by Tickner and Raffensperger "[u]nfortunately, risk assessment often ask wrong questions (acceptable rather than preventable risks); is based on numerous assumptions that are often policy or value judgement disguised as science; are narrowly defined, limiting considerations of multiple disciplines and lines of evidence, frequently leading to overly precise answers to the wrong questions; fails to consider cumulative and interactive effects; and is cost and time consuming, for analysing problems rather than solving them".[76] In the light of these criticisms and of the previous considerations about expertocracy, it seems that the dichotomy risk assessment/risk management is not the best solution to our problem because value judgements remain disguised as science, uncertainty does not clearly emerge, and complexities are not given proper account.

It seems crucial to find solutions that can account for the uncertainties in the scientific analysis and for the preferences of citizens in the face of these uncertainties. In other words, processes that democratize science are to be welcome, especially when we are in presence of great uncertainties. The paper presents a possible solution to this problem by suggesting a threefold decision-making process; in this framework, risk assessment is separated by risk management and a third intermediary phase, where uncertainties can be debated and lay knowledge and public participation are introduced, is added.

1.4 A Proposal for a Mixed Version: The Procedural and Substantive Nature of the Principle

In this section, I try to see whether it is possible to identify one or more versions of the Principle that seem most desirable for society. The theories and ideas discussed so far will be the basis for the elaboration of a new proposal. Before making any concrete proposals, it is important to remind that, as discussed above, the subject of analysis is a principle; hence we cannot expect to find an all-or-nothing fashion optimal rule, but only a set of reasons arguing in one direction. Yet, to have a Principle that is operational and capable to bind the discretion of the regulator, the content should be clear. Moreover, the fact that the Principle is discussed both at national, federal and international level might mean that it is not desirable to have only "one" version of the Principle (i.e. one world-wide accepted Principle). This issue is only noted here and not discussed. Still, it can be emphasized that most of the disputes around the formulation of the Principle emerged because the Principle, even when locally adopted, has transnational impacts. For an example of a "local"

---

[76]See Tickner, J. and C. Raffensperger, "The American View on the Precautionary Principle" in O'Riordan, T. and Cameron, J. (eds) *Reinterpreting the Precautionary Principle, London: Cameron Mat*, (2001), pp. 183-214, at p. 199.

Precautionary Principle that has transboundary relevance, just think of the European law of GMOs, which is based on the Precautionary Principle. In spite of the fact that the Principle is endorsed at European level, the legal effects are of transcontinental nature because the trade of GMOs is a world-wide business. An analysis of possible EU violations of general principles of commercial international law bounces back to a discussion of the international acceptability of the Precautionary Principle, adopted at EC level.[77]

Given that the Principle is widely adopted internationally and that when adopted locally it has transboundary effects, it seems important to have a transnational dialogue about its content. This analysis outlines a "minimal core" of the Principle: a set of norms capable of binding the discretion of the decision-maker. The version suggested contains minimal requirements that seem acceptable to Western legal systems, such as the European and American ones.[78] Of course, more elaborated versions of the Principle can be specified. However, here it is suggested that the features of the Precautionary Principle described below should be a necessary point of departure for the elaboration of more sophisticated versions of the Principle.

In the following, I sketch out the elements that constitute the suggested version of the Precautionary Principle. The first element specifying *the circumstances triggering the application of the Principle* has been already discussed. In Section 1.2, I concluded that the Principle should be applied only when the possibility of a *serious* and *irreversible* harm, tough *uncertain*, is attached to a particular activity. In order to avoid ambiguities, I have defined the concepts of serious, irreversible and uncertain harm in some details in the same section. Various aspects of the other relevant elements have been discussed as well; however, I left several questions unanswered. In the remainder, I propose some answers to the questions set out in Section 1.2 so to build a proposal for an interpretation of the Precautionary Principle. I anticipate that the type of Precautionary Principle advised in this paper is two-dimensional, in the sense that it is characterized both by a substantive and by a procedural dimension.

1.4.1 The Question of Time

The set of economic theories discussed above seems to point to the same conclusion: learning trough time should be possible as a way of reducing uncertainty and minimizing errors. How can we translate such theoretical in-

---

[77] In this respect see Scott, J. and E. Vos , "The Juridification of Uncertainty: Observations on the Ambivalene of the Precautionary Principle within the EU and the WTO" in Dehousse, R. and C. Joerges (eds.), *Good Governance in an Integrated Market*, OUP (2001).

[78] The assumption to state that we can have norms acceptable to Western legal systems is that they share common values. For the reasoning of this paper, it is necessary that the Countries rely on democratic values and accept to some extent consequentialist evaluations of law.

sights in practical suggestions to give more body to the Principle? In order to allow learning trough time, I propose that decision-making procedures under uncertainty, to be in compliance with the Principle, should be *iterative*.

To come back to the example of GMOs, it can be envisaged a system where decisions are reviewed within predetermined periods of time. For instance, the decision to ban or to allow controlled field trials or to allow uncontrolled culture of Genetically Modified Seeds could be reviewed every five years. In this respect, moratoria are not illicit as such, but moratoria without revisions are not acceptable. In this manner, it is avoided the problem of moratoria that are misused as a means of procrastination.[79] The iterative dimension of the decision making process should be in place, not only when a regulatory decision (banning or limiting a certain activity) has been adopted, but also when a decision not to regulate has been taken. Besides being iterative, the procedures should provide for appropriate institutional structures allowing for the evaluation of new information. This is because the iteration has meaning only if it stimulates the acquisition of new information; in other words, the iteration should be functional for updating. In this sense the procedures should be *informative*. The ways by which an institutional structure allows for the acquisition and assessment of new information might be several. A possibility would be to arrange meetings between decision-makers, committee of experts and interested parties before the "updated" decision is adopted. Committee of experts and interested parties will function as informing parties. The decision-maker should then be forced to motivate his decision in reference to the information provided by the informing parties. Coming back to our initial questions about timing,[80] we have now the answers. First, the sequencing matters in that it might allow learning through time. To the second question I gave a positive even if very general answer by proposing that the Principle should be interpreted in a procedural way in the sense that it prescribes a decision making process that is both iterative and informative. This answer will become richer in the following section, where something more is said about the steps that should be taken to reach a decision.

The goal of this paper, to give more body to the Precautionary Principle, is partly fulfilled by interpreting it as a Principle requiring the implementation of *iterative and informative procedures*. Versions of the Principle lacking this procedural dimension are therefore to be rejected. This side of the Principle seems indeed not too problematic as the relation between the theory and the norm can be considered quite straightforward. Strangely enough this

---

[79]The fact that the decision process is iterated does not tell anything about the content of the decision. This will depend on the information gathered. Let's consider, for instance, the case in which a certain product is banned at time T_1. At time T_2, the decision-maker can either withdraw the ban or reiterate it. This will depend only on the information accumulated.

[80]In Section 1.2.3, we asked respectively: 1. How does the sequencing of the decision-making process matter (if it matters at all) for the outcome reached? 2. Can the Precautionary Principle give any guidance for the dynamic of the decision making process?

dimension of the Precautionary Principle is lacking from legal texts that focus mostly on the substantive dimension.

However, if we want the Principle to be operational, an interpretation based only on such procedural dimension is still too laconic. The direction of the decision to be taken remains highly unspecified. Therefore, to complement this proposal we should look at other dimensions of the Principle as well.

1.4.2 The Question of the Burden of Proof

As mentioned in the first section, the question of the *onus probandi* should be reshaped. The question of who has to prove something can be recast in the light of what has to be proven. Here, it is suggested that the regulator should demonstrate the existence of the triggering factors: namely, the possibility of very serious and irreversible harms. In this sense the regulator is burdened with a "minimal" *onus probandi*. She has only to prove that the hypothesis of the possibility of very serious and irreversible harm has some acceptance in the scientific world. Considering the fact that a painstaking *ad hoc* assessment of risk might take quite long time, the regulator can simply rely on shared scientific knowledge.[81] For example the regulator that wants to impose a moratoria on the use of some GM seeds can adopt the measure without carrying out an *ad hoc* risk assessment. Still, she has to refer to some relevant work in the field of science that shows the possibility of serious and irreversible harms. Such works can be in the form of articles published in established journal (e.g. Science, Cancer Research, etc.) or reports of respectable committees (e.g. Royal Society of Canada etc.).

The theoretical support for such a "minimal" burden of proof for the regulator comes from the economic literature reviewed that favors the "err on the side of environmental preservation" option.[82] In fact, putting on the regulator a heavier burden of proof would imply that during the period that the *ad hoc* scientific evaluation is carried out the regulator cannot act. Moreover, not all the States/regulatory bodies have the resources to establish each time an *ad hoc* risk assessment and therefore some States with less resources will have their decisions always skewed on the side of regulatory inaction. In other words, we would favor the "err on the side of development" option. Such decision lacks the flexibility element. Given the urgency of the decision, it seems justifiable to allow regulators to take a decision on the basis of general scientific knowledge and this could satisfy the burden of proof on the regulator.

In sum, the regulator has to show that, even if uncertain, possibilities of serious and irreversible harm are attached to the activity proposed. The proofs can be provided by relying on general scientific knowledge or on *ad*

---

[81] In this respect see Section 1.4.3, where it is argued in favor of SAR that has a broader meaning that Risk Assessment as embodied in policy making in USA.

[82] See Section 1.3.4.

hoc evaluation of risks. In this way part of the burden of proof is placed on the regulator while what has to be proven is circumscribed; setting limits on the object to be proved has the uncontested advantage that it transforms the otherwise *probatio diabolica* into a feasible task.

At the same time, depending on the decision of the regulator, part of the burden of proof might be shifted on the initiator. In all systems of prior approval the proponent of an activity should meet certain requirements in order to obtain an authorization. For instance, in the case of pesticides, European legislation provides that in order to obtain an authorization the proponent should provide certain information that show the compliance with the conditions specified in the law.[83] The fact that the information needed in order to obtain an authorization is specified in the regulation means that the burden of proof for the initiator becomes a feasible task, solving the issue of the *probatio diabolica* also in this case. Within this cadre, the burden of proof is shared and the issues to be proved are to some extend limited. As mentioned earlier in the paper, the literature on the Precautionary Principle treats the issue of the burden of proof as a clear-cut dichotomy, where the only relevant question is about "who has such a burden". This solution has advantages. Most significantly, it overcomes the impasse created by the double *probatio diabolica*.

However, my solution is not unproblematic either. The fact that the initiator of an activity has to carry out a series of studies to meet his own burden of proof means also that the costs of his activity increase. Therefore, the question remains whether, in respect to "the costs of proof", it is desirable to have such allocation of burden of proof between the regulator and the initiator. Several arguments point in the direction of the desirability of such solution.

First, in this context it is likely that the initiator has the knowledge at a lower cost because he is the one developing the technology. Second, the costs of gathering specific information about the new technology are spread between the proponents. Consider, for instance, that in Europe some twenty thousands agricultural pesticides are in the market. It seems an unfeasible task to put on the regulator the burden of carrying studies and gather the information about the effects of each product. Indeed, a task that can be feasible for the operator, who has to proof the safety of his own product only, might be unfeasible for the regulator, who has to proof the safety of all the products that are submitted for circulation in the market. Moreover, as mentioned in Section 1.2.3, a system where the proponent is free from any burden of proof is likely to provide perverse incentives on the production of information, in the sense that the producers will be encouraged to hide information about the potential risks of his products.

---

[83] See Directive 91/414 concerning the placing of plant protection products on the market, OJ L230/1.

1.4.3 The Questions Pertaining to the Content of the Decision

I have already argued in Section 1.2 that the Precautionary Principle does not dictate any specific decision *ex ante*. The questions concerning the use of cost-benefit and cost-effectiveness tests to appraise the decision were left unanswered. As discussed in Section 1.3, cost-benefit analysis is not ideal because it cannot really give due account of the existence of uncertainties and it can lead to arbitrary arithmetic. This does not necessarily mean to disregard regulatory costs as argued Section 1.3.3 and cost-effectiveness analysis seems still viable. In delicate decisions where values at stakes are high and might involve future generations, it seems important to create procedures that guide the decision maker in adopting decision that are respectful of scientific findings and citizens preferences at the same time. These procedures, as said, should be able to create information. In order to reach scientific and democratic decisions my suggestion is to have a threefold framework constituted respectively by: 1. Scientific account of risks;[84] 2. Process of Public Participation; and 3. Risk Management.

A Threefold Framework to Recast the Relation Between Science, Economics and Law

Scientific Account of Risks (SAR)

The first phase recalls the process of risk assessment but relies on a broader paradigm. Scientific account of risks (henceforth SAR) can be carried out in *many ways* and not only in the form of traditional risk assessment as carried out in the American legal system. As seen in the previous section, the regulator has only to investigate that the possibility of very serious and irreversible harm from the to-be regulated activity is scientifically plausible. The object of SAR is, therefore, limited; for example, substance-specific studies will generally not be necessary, as what the regulator should show is a more general risk.

Some examples of SAR are here presented. In its most basic version, SAR can be a simple report compiled by scientists that review the results of studies already published in academic journal or in the form of official reports. Alternatively, it can be a process of deliberation of a group of scientists. In this case various experts belonging to the recognized scientific community are invited to present their studies and their scientific evidences on the problem under scrutiny. The experts appointed by the Public authority should be *independent*. Affected parties (industry, local communities, etc.) can present their own analyses carried out by scientists selected by them. A final document

---

[84] In this context it should be held in mind what has been said in relation to the burden of proof. Hence the Scientific Assessment of Risk is limited to what should be proven by the regulator.

summarizing the main findings of the independent group of scientists and of the group of scientists hired by interested parties should be submitted to the regulatory agency. The major controversies should be clearly described in the final document. Finally, SAR can take also the form of *ad hoc* risk assessment as typically carried out in USA. For all of these assessments, to the extent possible, cumulative and synergistic effects should be taken into account. Moreover, it is important that the scientists selected are independent and that the results of these studies are made available to scrutiny of other stakeholders.

Process of Public Participation

In the second step, public decision-makers should establish institutional *fora* where a conversation between experts and non-expert can take place. The goal of such *fora* is to reach consensus on highly controversial topics. As put by Habermas: "There are no questions so specialized that they cannot be translated when it is politically relevant to do so, and even adapted in such a way as to make it possible for the alternative experts discuss to be rationally debated in a broader public forum as well. In a democracy, expertise can have no political privilege".[85] Examples, unfortunately not always successful, of institutions already used to enhance public participation in decision-making are citizens jury, consensus conferences and citizens panels.[86] In spite of the good intentions behind the creation of such institutions trying to directly involve the public, it should be noted that they are problematic as well.[87] In suggesting the involvement of citizens in the decision making process, hence, it is important to acknowledge that democratization of science is not unproblematic. As it has been said elsewhere: "Once science enters a realm of direct democracy it becomes inbred with the failings as well the triumphs of democracy".[88] In this context, it should be observed that citizens are often highly unqualified to discuss specialized topics. Moreover, people are very sensitive to the way questions are framed, which may increase the risk of manipulation. Such institutionalized discussions are also time consuming and many don't want to invest their time in understanding complex issues having the feeling that their opinion will not count anyway. In short, the direct participation of citizens can be very costly for Society.

---

[85]Habermas, J., "A Conversation about Questions of Political Theory", in von Schomberg, R. and K. Baynes, (eds) *Discourse and Democracy: Essays on Habermas's Between Facts and Norms*, (2002), at p. 246.

[86]For a discussion of various forms of public participation see Stewart, J., *Innovations in Public Participation*, (1996), London.

[87]For a discussion of the problematic dimension of involvement of citizens in scientific debate see Weale, A., "Science advice, democratic responsiveness and public policy", in 28-6 *Science and Public Policy*, (2001), pp. 413-421.

[88]O'Riordan, T., "The Precautionary Principle and Civic Science", in O'Riordan, T. and Cameron, J. (eds) *Reinterpreting the Precautionary Principle*, (2001), ch. 4, at p. 104.

A way to overcome all the problems related with the direct involvement of the citizens could be to create *fora* with qualified representatives of society. For instance, when adoption of controversial technologies such as GMOs is discussed, these *fora* might involve only representatives of industry groups, environmental NGOs, trade unions, religious groups, scientists and groups of people that have local knowledge. To enhance transparency, the outcomes of these *fora* should be accessible to the large population. Given the limited scope of the paper, these brief notes do not suggest a particular solution. They just point at the problem, suggesting that more should be done in order to involve the civil society when the decision-making process concerns technology whose consequences are highly disputed in science and are likely to be born by the population at large.

Risk Management

Finally, the Risk Management phase is carried out by the decision-maker; in setting the goals of the measure to be adopted the decision-maker should take into account the results of previous steps. In the light of the theories discussed in the second Part of the Article, the decision should be flexible and, hence, skewed in favor of the environment. This means that it is legitimate to adopt safety regulation even when the adverse effects of the regulated activities remain uncertain. Finally, a CEA can be carried out to further specify what measures reach the set goals in the most efficient manner.

In relation to the use of cost-benefit analysis for the appraisal of regulatory measure as, I propose a binary solution. On the one hand, I suggest that cost-benefit analysis should not be used at all. As it should be clear from previous sections, I privilege this solution. I send back to Section 1.3.2 for a detailed review of the arguments in support of this position. Above all, the rejection of cost-benefit analysis is dictated by the fact that, under conditions of serious uncertainty, political decisions would be concelead under the false precisions of numbers. Such processes could dangerously degenerate into a demagogy of numbers that is all but desirable in democratic societies.

On the other hand, if it is endorsed the thesis that cost-benefit is still desirable, some major adjustments are necessary to improve the reliability of this practice. In particular, uncertainty and the steps taken to address it should be made explicit and intelligible to the public.[89] In the same spirits, the analysis should be transparent and the experts that carry out cost-benefit analysis should be independent (e.g. not connected in any way with special interest groups that will be affected by the decision). Moreover, the Precautionary Principle could be interpreted as a hermeneutic canon to guide cost-benefit analysis itself. This would imply, for instance, that the 'measures' chosen in presence of uncertainty should err on the side of environmental preservation, discount rates for long-term benefits of regulation should be very low if not

---

[89] See Stewart, R. (2002) op. cit., at pp. 91-92.

zero, etc.[90] Similarly, in line with some findings of cognitive psychology, when valuing lives, the involuntary nature of risks and the catastrophic component of it should be accounted as well, leading to a likely upward adjustment of the value of statistical lives in most of the cases under investigation.[91]

Looking at these procedural norms, from a temporal perspective, we should note that the decision taken at time T_1 has not to comply with all the steps proposed. Only at time T_2, failings to carry out such procedures might be considered in violation of the Principle. This is, for the same reasons, articulated in the context of CEA.[92] As noted, procedures consume time and during this time regulation is absent. It would, indeed, be paradoxical if procedural dimensions of the Principle would be used as antiregulatory devices.

The threefold decision-making process, as presented, is still in a very rudimentary and abstract form, and if the idea is given any credit it needs more sophistication. If my proposal is rejected, it seems that the problem presented remains and other solutions should be found in order to have a more balanced relation between science and law making. In any case, either if the proposal is considered to be a good point of departure or if it is rejected, interdisciplinary analysis bringing together the knowledge of political scientists, economists, scientists and philosophers, is needed to give more practical guidance in reshaping the relation between science and the making of laws when major hazards of uncertain nature are involved.

In conclusion, the Principle cannot be interpreted only in substantive terms, but should be reinterpreted to include a procedural dimension as well. On the procedural side, the decision should follow procedural rules that guarantee the adoption of scientific, democratic and effective decisions. On the substantive side, the final decision should be adjusted in order to take the due account of the "erring on the side of environmental preservation" bias and of cost-effectiveness considerations.

1.5   Concluding Remarks

"We consider it to be a mythical concept, perhaps like a unicorn", John D. Graham, the administrator at the American Office of Management and Budget, recently said.[93] Notwithstanding its large adoption by legal texts, the Precautionary Principle still has many critics. The most prominent critiques

---

[90] At the moment, however, cost-benefit analysis is skewed in an opposite direction, undervaluing human lives and environmental benefits. For a discussion of the techniques that undervalue regulatory benefits see Revesz, R. L. (1999), "Environmental Regulation, Cost-Benefit Analysis and the Discounting of Human Lives", in 99-4 *Columbia Law Review*, pp. 941-1017.

[91] This is because the Precautionary Principle applies only under the circumstances specified in Section 1.2.3.

[92] See Section 1.3.5.

[93] Reported by Loewenberg, S. in "Precaution Is for Europeans" in *NY Times*, (May 18, 2003), at p. 14 (wk).

are that the Principle is not scientific and that to be acceptable it can only be reinterpreted in the framework of cost-benefit analysis. In this paper I have shown that the Principle is both scientifically sound and that it is meaningful outside the paradigm of cost benefit analysis.

Because the regulation of major industrial hazards is likely to be significantly affected by the future implementation of the Precautionary Principle, it is important to agree on the basic content of the Principle. The Precautionary Principle, as suggested in this paper, is characterized by four elements. First, the application of the Principle is circumscribed to cases in which the potential hazards are characterized by serious, irreversible and uncertain consequences. Second, in order to allow learning through time, the Principle should entail dynamic decision-making processes that are both iterative and informative. Third, the burden of proof should be shared between the regulator and the proponent. On the one hand, the regulator should show the existence of the triggering circumstances for the application of the Principle. On the other hand, the proponent should carry out his own research prescribed by the regulators to provide information about the consequences of his activity. Such rule has two advantages: it solves the problems of a double *probatio diabolica* and it gives incentives on the production of information relating to the risk properties of the alleged hazardous activity. Finally, in relation to the content of the decision, it is argued that the Principle does not prescribe any particular decision *a priori*. Because it is not possible to interpret the Principle in a deterministic framework, the paper shifts the attention from optimal solutions to procedures necessary to reach acceptable decisions for society. I have, hence, proposed a decision-making framework that gives due account of the need of basing decisions on science without denying the presence of large uncertainties. Therefore, a process of public participation follows the phase in which independent experts carry out a scientific appraisal of risk. Eventually, on the basis of facts and values emerged in the previous phases the decision-maker will adopt a decision.

By shifting the attention to the procedural dimension of the Principle, I believe to have captured the essence of the Principle. In a Society dramatically characterized by complexities and environmental interdependencies coupled with uncertainty and ignorance, it is crucial to shape a new body of environmental and safety law taking these features into account. This is possible only by introducing dynamic decision making processes that rely on scientific knowledge and are open to the scrutiny of civil society. The Precautionary Principle, in spite of its detractors, seems to be the cornerstone of this new body of law.

Table 1.1: A four-fold distinction

Certainty	Knowledge about certain outcomes
Risk	Knowledge about probabilities and outcome
Uncertainty	Knowledge about outcome; ignorance about probability
Ignorance	Ignorance about outcome and probability

Table 1.2: Sequencing in decision making

T_1	T_2	T_3
Activity is initiated without restriction as no Regulation is adopted	Regulation is adopted and Activity banned or restricted	Regulation is Revised and Activity is banned or restricted
Regulation is adopted and Activity is banned or restricted	Regulation is withdrawn and activity is undertaken without restriction	New regulation is Adopted and Activity is banned or restricted

Chapter 2

Informational Regulation of Industrial Safety - An Examination of the U.S. Local Emergency Planning Committee*

Nathalie De Marcellis-Warin
École polytechnique de Montréal

Ingrid Peignier
CIRANO

Bernard Sinclair-Desgagné
HEC Montréal and École polytechnique Paris

2.1 Introduction

A recent study conducted by the U.S. Chemical safety Board (CSB) estimated that, on average, 55 000 chemical incidents happened every year between 1987 and 1996 in the United States, or more than 150 per day. On average, chemical accidents kill about 250 people in the U.S. alone every year. These figures obviously tend to enhance the value of industrial safety.

---

*We wish to thank, without implicating, Éric Clément and Jean-Paul Lacoursière, Craig Matthiessen (EPA, Washington (DC)), and the following Vermont LEPC members : Tim Bouton and Matt Fraley (Addison County - Middlebury and Lee Sturtevant and Michele Boomhower (Lamoille County - Morrisville). We also acknowledge valuable discussions with Mary K. Moses, Larry A. Mabe and Douglas W. Richmond (Harford County Division of Emergency Operations, Forest Hill (MD)), Mikal Shabazz (US-EPA REGION 3 Philadelphia (PA)), Robert A. Barrish (State of Delaware, New Castle (DE)), James Belke and William J. Finan (US-EPA Headquarters, Washington (DC)), John A. Jurcsisn (Clorox Products Manufacturing Company, Cleveland (OH)), Chuck Marzen (Clorox Products Manufacturing Company, Aberdeen (MD)), and finally Wally Mueller and Geoffrey L. Donahue (State Emergency Operations Center, Reisterstown (MD)).

Since the Bhopal disaster and Basel incident of the 1980s, a major trend in the regulation of industrial risks to human health and the environment has been to empower all stakeholders and risk bearers. Eighteen years ago, for instance, the United States government introduced the legal concept of "community right-to-know", which stipulates that all residents be entitled to access specific information concerning the hazardous chemicals in use or in storage in their respective neighborhood.[1] In 1990, furthermore, section 112 (r) was added to the Clean Air Act (CAA), asking the U.S. government's Environmental Protection Agency (EPA) to set requirements for facilities to reduce the likelihood and severity of major accidental chemical releases. The EPA soon launched its Risk Management Program, in which about 60 000 facilities had to convey their respective Risk Management Plan to the public (via the EPA's website).[2]

The means identified by the 1986 Emergency Preparedness and Community Right-to-know Act (EPCRA) included the mandatory creation of organizations at both the state (the State Emergency Response Commissions - SERCs) and local levels (the Local Emergency Planning Committees - LEPCs), whose tasks would be to develop emergency plans in case of catastrophic releases of toxic chemicals and to supply information about covered facilities. There are now approximately 4000 LEPCs in the United States. It is widely believed that these are key items in fostering cooperation between federal and state agencies, the chemical industry and the general public, thereby making the EPCRA, the RMP, and the general informational regulation of industrial risks effective. In the aftermath of the AZF plant explosion in Toulouse, which on September 21, 2001, left 30 people dead and more than 2500 injured, the French government is now considering the creation of local committees of information and prevention similar to the LEPCs around each Seveso facility.[3] In Québec, Canada, a new law on civil protection adopted in December 2001 and its subsequent bylaws will also promote the implementation of such committees.

The purpose of this paper is to provide a concrete look at the functioning of the LEPCs. Our presentation is based on many documents available on the Internet and in the literature on informational regulation, on personal interviews carried out at selected LEPCs in the states of Vermont and Maryland, and on preliminary results from a recent survey we just completed of

---

[1] In Europe, the Seveso II Directive also granted the public greater access to information concerning industrial hazards and plant safety. And for some years the *Canadian Council of Major Industrial Accidents* has recommended extensive and sustained information transfers through joint committees involving people from industry as well as from local communities.

[2] Following the terrorist attacks of September 11, however, those Risk Management Plans have been removed from the EPA's website.

[3] Material damages from the Toulouse explosion amount to more than US$2 billion. By all accounts, this industrial accident is now viewed as the worst one that ever happened in France.

more than 2900 LEPCs.[4] The following section first summarizes the underlying rationale of right-to-know regulation. Section 2.3 next presents our main findings. Section 2.4 contains an appraisal of the actual LEPCs. And Section 2.5 concludes the paper.[5]

2.2 Why Regulating by Informing?

Many arguments support a right-to-know approach to regulating environmental risks. They respectively relate to ethics, epistemology, economics and public policy.

- *From an ethics perspective*, it is necessary to distinguish between a risk that someone is informed about and willing to bear, like that a (possibly insured) car driver takes when entering a highway, from a risk that is generally unknown and unsuspected (for some time period, at least), like the exposure to a potential epidemic of mad-cow disease that red-meat consumers in Europe have only recently become aware of. In a democracy, the former is morally superior to the latter. Each risk bearer must then be enabled to assess and decide for him or herself the hazards he or she wants to face, which means in particular that all the relevant information be made available.

- *From an epistemology viewpoint*, past experience reveals that science-based assessments of environmental and industrial risks often convey a significant margin of error, for they must commonly cope with missing data and involve accordingly subjective judgments with respect to permissible analogies, extrapolations and methodology. Moreover, science-based scenarios of catastrophic events would often constitute rough approximations of what may truly happen (especially when there can be human casualties), because any catastrophe generates (almost by definition) a politicized, media stirring, highly emotional, and therefore unpredictable situation. It thus seems reasonable to require that risk assessment ultimately incorporate as many relevant and well-founded opinions as possible, which presupposes open questioning by informed stakeholders.

- *For economists*, informational regulation is usually justified by invoking the so-called "Coase Theorem" (from the Nobel-prize winner Ronald Coase), which says that socially optimal risk sharing can obtain if all stakeholders can negotiate at no transaction costs. This assertion underlies the use of tradeable permits for controlling pollution. When it

---

[4]We sent a questionnaire to 2935 LEPCs, 858 through e-mail and 2077 per regular mail. A total of 288 LEPCs responded, which gives a response rate of about 10%.

[5]The reader who wishes to benchmark the U.S. experience against that of other countries, namely Canada and France, may also consult Appendix 2.1.

comes to technological risks, it supports open disclosure of information in order to foster the parties' mutual trust and limit wasteful conflicts. As the current pace of innovation increases, furthermore, and new products and processes having an ambiguous impact on human health and the environment are constantly being introduced, it stresses that risk management is most likely to yield efficient outcomes if the risk bearers themselves are well-informed.[6]

- Finally, many recent studies, conducted in some developed and emerging countries (Indonesia, for example), illustrate that the information of local communities puts effective pressure on companies to significantly reduce the risks to human health and the environment. Most owners of hazardous facilities do indeed care about their reputation and want to have a good relationship with their neighbors. The latter may then act as "surrogate regulators", thereby lowering the enforcement costs of safety regulation.

To fulfill its objectives, however, a right-to-know approach presupposes that the disclosed figures remain credible and of good quality. Relevant information must then be organized, formatted, and (especially after 9/11) made available without compromising on security. This is where nonprofit, independent, and representative bodies like the LEPCs are expected to play a key role.

2.3   What do LEPCs do?

What role do LEPCs actually play in informational regulation? This section first sketches the legal framework of LEPCs. We next turn to their actual modus operandi and draw some international comparisons.

Our basic sources are several existing surveys conducted at various universities and research centers.[7] Although these were made before September 11th, 2001, and are essentially descriptive, they provide valuable information about respective staffing, activity portfolios, available resources, and training programs of LEPCs, and about their common relationship with the EPA, state agencies, industrial firms, local communities and citizens. This information is now supplemented by systematic interviews done at some selected LEPCs in

---

[6]The implementation of health and environmental risk disclosures raises of course many questions, pertaining for instance to the scope and content of reports, the quality and credibility of information, the design of communication strategy, and whether disclosures should be voluntary or mandatory. The recent paper by Sinclair-Desgagné and Gozlan (2003) represents a first attempt (in environmental economics) to address these issues. This model is summarized in Appendix 2.1.

[7]These surveys have been conducted by The George Washington University in Washington, by the National Institute for Chemical Studies in Charleston and by the Graduate School of Public Affairs of the University of Colorado at Denver.

Vermont and Maryland, and by a home-made survey seeking to understand how much attention was recently allocated to risk communication and hazard reduction.

2.3.1 The Legal Framework

On October 17, 1986, a Federal legislation enacted the Superfund Amendments and Reauthorization Act (SARA). A major part of this law was Title III, the Emergency Preparedness and Community Right-to-know Act (EPCR A), which was a direct reaction to the Bhopal disaster. The EPCRA seeks, first, to provide a basis for each local community to develop suitable ex ante emergency measures, and second, to entitle people to identify and quantify the chemical hazards present in their neighborhood.

Section 301 of the EPCRA mandated the creation of organizations at both state and local levels that would foster cooperation between local communities and the industry. State governors were henceforth required to set up and appoint a state commission - the SERC - that had in turn to divide the state into "Local Emergency Planning Districts" and appoint a Local Emergency Planning Committee (LEPC) in each one of them. Each LEPC would include representatives of: (1) elected state and local officials, (2) law enforcement, (3) emergency management (civil defense), (4) firefighting, (5) emergency medical services (first-aid), (6) health care, (7) local environmental groups, (8) hospitals, (9) transportation personnel, (10) broad-casted and printed media, (11) community groups, and of course (12) the owners and operators of targeted facilities.

The LEPCs constitute thereby the primary linkage between citizens, the industry and the various government layers. They must

(1) prepare a comprehensive emergency response plan, submit it for approval to their respective SERCs, and update it each year. Such a plan would have to include the name and location of hazardous materials, procedures for immediate response to a chemical accident, and ways to notify the public about what to do and whom to get instructions from;

(2) organize and store the information about hazardous chemicals which is supplied by local facilities;

(3) make this information available to the public upon request; and

(4) increase community awareness about, and safety from, hazardous materials, by means of public releases, formal training, general simulations and drills.

In 1990, section 112 (r) of the Clean Air Act Amendments (CAAA) asked the EPA to set requirements inducing hazardous chemical facilities to decrease the likelihood and severity of major accidents. Under the subsequent Risk

Management Program (RMP), covered facilities must now provide hazard assessments, which include worst-case scenarios, alternative release scenarios, a five-year accident history, a prevention program, an emergency response program, and evidence of an effective management system to oversee the implementation of RMP elements. Compliance with those requirements was first expected by June 21, 1999.

Although it did not explicitly mention the LEPCs, section 112(r) has lead to a significant extension of its role, from a planner of emergency response to a key player in the prevention of industrial accidents (EPA, 1998). But because the significant impact of an accident is local, effective accident prevention activities must of course have a local focus. Delegating the responsibility of developing local emergency plans to the LEPC ensures that communities will develop personalized, need-specific, and effective emergency plans. Indeed, in designing the Risk Management Program, the US Congress and the EPA anticipated that public scrutiny would help regulate the behavior of hazardous chemical facilities to a greater extent than regulatory requirements alone.

To make the EPCRA and RMP effective, the LEPCs are supposed to constitute a formal linkage between citizens, the industry and the government, receiving information from local industry, using this information to prepare an emergency response plan for the community, submitting this information to their respective SERC, and responding to public inquiries about chemical hazard and releases. Figure 2.1 gives a summary of this.

To be sure, the LEPCs are given three key roles in accident prevention:

(1) They collect the data submitted by covered facilities. Industry must be involved in prevention. Concretely, every regulated facility is, among other things, responsible for reporting HAZMAT inventories (quantities and locations of hazardous chemicals) annually to the LEPC, and for supplying a material safety data sheet (MSDS) or a list of hazardous chemicals and an annual report of toxic chemicals releases to the EPA and the State. Moreover, those facilities covered by the emergency planning provisions must designate a facility emergency coordinator to assist the LEPC in emergency planning. The LEPCs must also be given updated emergency information as soon as practically possible. They may initiate civil action against the owner or the operator of a facility for failing to supply information under Section 303(d) or to submit Tier II information under Section 321(e)(1) (EPCRA, Public Law 99-499).

(2) The LEPCs are also chief actors in emergency response: Section 303 of the EPCRA required the LEPCs to prepare comprehensive emergency response plans for their community by October 1988, to submit them to their respective SERC and to update the plans annually. The EPCRA also demanded the LEPCs to increase community hazardous materials safety through responder and emergency medical personnel training and the coordination of emergency measures with local communities. The EPCRA thereby estab-

lished the LEPC as a basic "forum" for neighborhood discussions and for local action on matters pertaining to hazardous material management.

(3) The LEPCs are a source of information to the public and local governments. They have the obligation to make the information concerning risks in their district's covered facilities available to the public upon request, and to consequently set and publicize procedures to handle those requests. An LEPC must respond to a request for Tier II information no later than 45 days after reception of the request (EPCRA, Public Law 99-499). It is deemed more efficient to encourage people's access to an effective single information window instead of letting them inquire each facility.

Following this description of the LEPCs' legal framework, let us now turn to what exactly LEPCs do in practice. Our aim is to understand how much attention the LEPCs actually devote to both outreach/risk communication and the promotion of hazard reduction, or to what extent the LEPCs have met their expected role up to now.

2.3.2 Modus Operandi

Let us first look at the geographical distribution and the composition of LEPCs.

• *Geographical distribution of LEPCs* (see Table 2.1). Each state currently averages about 81 LEPCs, organized by counties, large towns or relatively large portions of territory. There are significant discrepancies across states, however: twenty states have less than 50 LEPCs, while four states - New Jersey, New Hampshire, Massachusetts, and Texas - have more than 200. According to our survey, the average number of facilities reporting to an LEPC (that is to say, facilities subjected to SARA Title III and facilities subjected to the RMP) is 109, but there can also be significant differences as the number of facilities per LEPC may go from 1 to 2552. The correlation coefficient between the number of LEPC and the number of RMP facilities is only 0.0121.

Clearly, the actual distribution of the LEPCs reflects that of hazardous facilities and is meant to insure that emergency planning is done at the most useful level, but the approach varies across states. The state of New Jersey, for example, uses city and township lines as boundaries to define emergency planning districts. In Maryland (see Figure 2.2), districts are defined per counties plus two cities - Baltimore (a highly industrialized city) and Ocean City (well exposed to floods and hurricanes), given the greater risks that these two cities must bear.

Composition. The 1994 Nationwide LEPC Survey by Adams, Burns and Handwerk locates the median size of a "functioning" LEPC at 20 members. Our own survey now locates the median size at 26 members.[8]

According to Adams (1998), "While Congress perhaps thought requiring a broad base of members for the LEPCs would automatically create community-wide interest in their tasks, the reality is that many LEPCs are managed by county sheriff, fire departments, and industry representatives, who, in many states, were managing the emergency response function before Title III became law". This observation corroborates a previous one by Hadden (1989), who reported that "SERCs and LEPCs are dominated by emergency response personnel rather than by citizens or environmental personnel." This imbalance in LEPC composition is also confirmed by our survey. Indeed, the groups that are the most represented are industry representatives (an average 7 members per LEPC), fire departments (4 members), and elected state and local officials (3 members) per LEPC; on the other hand, the citizens and the media are the least significant, with less than 2 members per LEPC on average.

Let us now consider how the LEPCs actually perform their mandated duties - i.e. enhancing accident mitigation, emergency planning, public training and the coordination of respondents, together with informing the stakeholders about industrial risks. Since this paper mainly seeks to examine how informational regulation works in practice, more attention will accordingly be put on the latter.

• *First task: coordination of emergency planning.* The 1994 Nationwide Survey estimated that 81% of LEPCs had submitted a complete emergency response plan - i.e. one dealing specifically with planning, training, and exercising as continuous and intertwined activities - to their respective SERC. At that time, another 11% of LEPCs had "almost" set their own plan.

Exercises, simulation and drills appear to be the best (yet most expensive) means to foster public awareness and involvement, as well as to articulate emergency plans and procedures, and to assess the preparation of all interested parties. As we noticed in our survey, emergency exercises are widely used by the LEPCs since only 6.3% of the respondent don not perform them. The average frequency of emergencies simulations is 1.4 per year. There are two main types of exercise, referred respectively as tabletop and full-scale.[9] The purpose of a tabletop exercise is to have participants practice problem

---

[8]The term "functioning" in that nationwide survey refers to those 79% of all LEPCs that are "quasi-active", "compliant", or "pro-active"; excluded are the remaining 21% that are inactive.

[9]Several guidebooks, training and evaluation documents currently exist to support exercise planning and making. Frequently used ones are the Federal Emergency Management Agency (FEMA)'s "Hazardous Materials Exercise Evaluation Manual", and NRT- 2's "Developing a Hazardous Materials Exercise Program" which is available on the internet (See http://ntl.bts.gov/DOCS/254.html for the entire guide).

solving and resolve questions of coordination and assignment of responsibilities in a non-threatening format, under minimum stress. In 1999, for example, the Harford County LEPC built a reduced yet exhaustive model of the district that covered one hundred square-meters. All buildings (houses, schools, hospitals, plants, warehouses, etc.), roads and landscape features (hills, parks, etc.) were represented. Different scenarios were then simulated, in front of LEPC members as well as all residents willing to attend. This experiment was widely described as a success. It certainly represents a valuable intermediary step towards a functional or full-scale exercise. The latter can be used to test an emergency plan under a still higher degree of realism; however, it mobilizes significant physical and human resources (e.g., State and local Emergency Operation Centers, incident command posts, mass care centers, medical facilities equipment staging areas, and lay citizens) and involves an increased level of stress. Yet, in 1996 the state of Florida's District 3 LEPC, for example, conducted such an exercise. It simulated a leakage of chlorine from a one-ton container located between two schools.[10] A few minutes after the incident, Gainesville Fire Rescue staff discussed with the Hazmat team to coordinate an intervention. The former subsequently undertook to knock down the noxious cloud, and the Hazmat team rushed on the spot to mend the leaking container before proceeding to decontaminate the area. Students, teachers and other exposed people had meanwhile been explained the ongoing situation and brought to protective shelters. According to our survey, 50.2% of the LEPCs use the two main types of exercise - tabletop and full-scale; 75.3% arrange a table-top exercise while 68.6% organize a full-scale one each year.

• *Second task: risk communication.* A common way to reach out to people is through *public meetings*. These have the advantage of being inexpensive and of allowing direct interaction between the interested parties. In order to enhance public participation, such meetings are usually advertised through local media (i.e. newspapers, brochures, and local television). According to our survey, the average frequency of LEPC meetings per year is 6,2, and a proportion of 84% of these is advertised. Some LEPCs hold public meetings jointly with the targeted facilities, in order to bring stakeholders closer to their respective local plants and to convey RMP information more effectively. At the Whatcom County LEPC in the State of Washington, for example, three advertised public meetings were recently held, where the dean of the Huxley College of Environmental Studies provided a broad overview of the RMP process and intent; people were next allowed to look around and meet with attending local facilities personnel.[11] As might be expected, however, meeting frequency generally correlates with LEPC activism and compliance

---

[10]For more on this, see http://ncflepc.org/exercises/xrsz96gv.html.

[11]It is worth noticing that this process was meant exclusively for local residents: the Whatcom County LEPC systematically refused to allow non-residents in.

with the mandates of SARA III, as well as with an LEPC's having or planning to have computerized data inventories. In their 1997 study, Adams, Morgan and Viana identify nevertheless a positive trend in the frequency of meetings, which suggests increasing LEPC activism.

Another popular channel is of course the *Internet*.[12] This medium constitutes a key instrument to quickly reach a large number of people without having to figure out time and location constraints. Hence, many LEPCs now have a website. In accordance with our survey, 28% of the respondents use a website to reach to and inform the public before September 11th, and a little more after September 11th (29.7%). Such websites are accessible to a variety of users with differing needs (see Viridescent Inc., 2002). For instance, people may visit an LEPC website to get information about what to do during an emergency situation, how to get real-time information when there is an accident, how to handle household hazardous materials, whom to contact at their respective LEPC, and what are the upcoming meetings or exercises schedules.[13] For example, on the website of the Waltham LEPC,[14] one can view the emergency response plan of the town and information on the facilities that manufacture, use, store, sell, or otherwise handle hazardous materials. Webpages provide facility contact information, a listing of chemicals and their CAS numbers, and a listing of populations within one half mile that could be impacted by a release. Following those listings is a map showing the location of the affected populations within a half-a-mile radius. Some facilities have additional information pages because of the number of different chemicals they handle. Figure 2.3 shows an example from this website.

Businesses, on the other hand, would likely visit an LEPC website to know what emergency response resources are available to the business community, where to get "Tier II" inventory forms and how to complete them, what are the current reporting requirements and contacts in case of accidents and toxic releases, and how to get involved in LEPC activities. The media, finally, would likely visit an LEPC website to find out whom to contact and what/how to communicate with the public in case of an emergency.

Information brochures, finally, are a standard way for an LEPC to also reach the public. For example, at the end of each year Maryland's Harford County LEPC and Vermont's Addison County LEPC send a free new calendar

---

[12]According to the Department of Commerce, more than half of America's households currently have an Internet connection. The study A Nation Online from the U.S. Department of Commerce, which uses census data to track internet usage, reports that 143 million Americans, or 54% of the population, were using the internet as of last September, up 33% from three years ago. Two million people were going online for the first time every month. And among younger people Internet usage is even higher.

[13]Such information can be found on the websites of several LEPCs, such as Maryland's Harford County LEPC (http://www.co.ha.md.us/lepc/) or Texas's City of Deer Park LEPC (http://www.deerparklepc.org/). The latter also provides special information for children, parents and teachers.

[14]See at http://www.walthamlepc.org/.

to all district members, which contains summary information about local industrial hazards and the risks associated with different chemical substances; also included are the meeting dates of those associations involved in emergency planning, basic instructions for emergency situations, and a map showing where industrial accidents could happen. Such a calendar is not costless, however. Our survey found that about 40% of the answering LEPCs use brochures as a way to reach the public. After September 11th, we noticed a slight reduction in the use of brochures, TV/radios and newspapers but an increase in the use of the Internet and public meetings.

An additional activity that is worth mentioning is the mailing list[15] created by and for LEPCs Chairpersons in order to keep them abreast of any law or guidelines changes and to enhance technical assistance and the transfer of best practices. One LEPC Chairman, for example, once used it to make the following request: "[...] We are now in the early stages of planning a tabletop exercise and I'd like to hear any stories, tips or would appreciate any guidance or training resources regarding this subject. I am in the process of putting an orientation together for our LEPC members so they can better understand what will be happening". This e-mail quickly elicited many helpful hints and references to key documents.[16] This mailing list doesn't exist anymore but it was replaced by other ones like NASTTPO-mailing list or the environment news service (ENS). The National Association of SARA Title III Program Officials, or NASTTPO, is made up of members and staff of State Emergency Response Commissions (SERCs), Tribal Emergency Response Commissions (TERCs), Local Emergency Planning Committees (LEPCs), private industry, and various federal agencies. Membership is dedicated to working together to prepare for possible emergencies and disasters involving hazardous materials. The IAEM-list is sponsored by the International Association of Emergency Managers and is designed to provide a public forum for emergency managers to exchange ideas, thoughts, problems and solutions that relate to the emergency management profession. There is also another example of mailing list, the LEPC *Information Exchange*,[17] to allow LEPCs to learn about what other LEPCs are doing nationwide, for facilities, responders and the public. The LEPC Information Exchange also volunteers much of its time to promote information exchange to, from and among LEPCs nationwide. The goals of the LEPC Information Exchange are:

- To allow LEPCs to learn about what other LEPCs are doing nationwide, for facilities, responders, and the public;

- To provide LEPCs with outreach ideas and contact information;

---

[15] lepc@list.uvm.edu.

[16] Such as the one entitled, *Developing a Hazardous Materials Exercise Program: A Handbook for State and Local Officials*, that had just been set by the EPA Region VI and the U.S. Trade and Commerce Department.

[17] See their website : http://www.lepcinfoexchange.com/.

- To provide information on conferences and training; and

- To provide LEPCs with recent regulatory information.

In general, however, it seems that the LEPCs receive few information requests from the general public. The 1994 Nationwide Survey estimated that many (41%) had received no public inquiries and a mere 25% had received more than six inquiries. Our survey reveals a similar situation: in the year preceding September 11th, 52% of the respondents had received no public requests, and a mere 50% did receive some request after this. Among those which usually receive requests, nevertheless, we noticed a slight increase in numbers after September 11th. So before September 11th, an average LEPC received 2.84 requests per year while it would receive 3.81 requests a year after the catastrophe happened.

Public inquiries may concern the meaning of some technical words or expressions, the overall risk figure emerging from complex data, the various scenarios described in the Risk Management Plan, the interpretation of the community hazard map and its vulnerability zones, the possible health effects associated with exposure to hazardous substances, etc. Requests at the Harford County LEPC, for instance, pertained more often to emergency response and facilities risk assessment (36 requests per year) than to MSDSs - material safety data sheets (only 10 requests per year). Many LEPCs ask for written requests only.[18] People may also be required by law to come to the LEPC's office, show an identification document (ID), and pay a fee if the information they seek is for commercial purposes.

- *Extended tasks.* In 1995, a one-day focus group conducted by the National Institute for Chemical Studies with a sample of LEPCs coordinators, industry personnel, citizens and community lobbies revealed that the expectations raised by the EPCRA extended far beyond the minimum requirements of the law. The list of "should-do" responsibilities included *pro-active* identification, analysis, and especially communication of industrial hazards. According to the National Institute for Chemical Studies' study (2001), several LEPCs are therefore undertaking some tasks to encourage hazard reduction that go beyond legal requirements, such as

(1) helping to identify facilities covered by RMPs,

(2) assisting facilities, mainly smaller ones, to design and submit their respective RMPs,

(3) supporting public disclosures,

(4) providing a forum in which local plants industries present their respective RMPs to each other and exchange information on safety programs,

---

[18] In the state of Arizona, for example. For more details, see http://www.azleg.state.az.us under title 26, chapter 2, article 3.

(5) surveying companies to find out what changes were necessary for compliance with the RMP,

(6) participating in community fairs to promote hazard reduction and encourage public awareness.[19]

According to Adams, Morgan and Viana (1997), furthermore, 34% of the LEPCs use EPCRA data for zoning and land-use planning. And our more recent survey finds that 23% of the LEPCs are involved in land-use planning.

An illustration of point (2) is the Harford County LEPC's assistance to small facilities in completing their Risk Management Plan and in submitting it to the EPA. The Harford County LEPC also currently supports facilities in risk assessment and auditing. Item (5) could mean in fact that an LEPC takes part in the compliance audits of facilities. They could thereby provide advices and help to the facilities about risk management. Indeed, 35.2% of the LEPCs in our survey are involved in inspecting and/or auditing of large businesses and 29.4% in inspecting and/or auditing of small ones.[20]

2.4 An Appraisal

Across the United States, the operation and effectiveness of LEPCs vary widely. A single LEPC can cover an entire state or be centered on only one city. Some LEPCs are well funded through state, local or industry contributions while others have none. The current status of America's LEPCs cannot be generalized as either utter failure or phenomenal success. In many rural areas, LEPCs exist in name only; at the other end of the spectrum, some LEPCs have integrated emergency response plans and display some very high technical sophistication within a narrow range of skills.

The Adams, Burns and Handwerk (1994) nationwide survey gives a list of ten key provisions to evaluate an LEPC's compliance with the law. An LEPC should have:

(1) an LEPC Chair,

(2) an Emergency Coordinator,

(3) an Information Coordinator,

(4) members representing at least 12 of 13 specified stakeholders,

---

[19]For examples of LEPC pro-activism, see Appendix C, *National Institute for Chemical Studies*, 2001.

[20]Our survey indicates differentiated LEPC involvement across large and small businesses. LEPCs involvement is greater with large businesses, 56.5% and 62% of them being respectively involved with large firms in conveying chemical hazards information to the public and in reducing chemical hazards, while the figures for small and medium enterprises are respectively 48.6% and 54.4%.

(5) regular formal LEPC meetings,

(6) regular publicly advertised meetings,

(7) an emergency response plan submitted to the SERC,

(8) a plan incorporating at least 9 of 10 key SARA III elements,

(9) a plan that has been reviewed in the past year,

(10) and published newspaper notice that the plan and local hazardous substances data are publicly available.

The Adams, Burns and Handwerk (1994) survey also proposes five criteria to assess an LEPC's activism:

(1) whether it has simulated an emergency during the past 12 months,

(2) whether its emergency plan was updated during the past 12 months,

(3) whether its plan takes natural hazards into account,

(4) whether it meets at least on a quarterly basis,

(5) and whether it uses its EHS data to make hazard reduction or prevention recommendations to local governments or the industry.

Based on this, the survey found that one fourth of the LEPCs strictly complied with their legal mandates and even took numerous proactive steps that went beyond the minimum required by the law. Although they were not as proactive, most other LEPCs were either highly compliant (16%) or mostly compliant (39%) with the law. However, a significant proportion (21%) of LEPCs - disproportionately more in less populated rural areas - are inactive or, if once active, are now idle. The National Institute for Chemical Studies (1995) generally corroborates this picture: at least one quarter of the LEPCs do not comply with minimal requirements to simply make right-to-know data available to the public upon request, but an equally significant percentage are being quite proactive in interacting with the public.

This heterogeneity and the significant proportions of both impressive successes and blatant failures can be attributed to a number of causes (Adams, 1998). First, there might be some natural resistance to transparency: for example, industry representatives who, along with emergency planners, dominate most LEPCs, do not generally welcome public scrutiny of their work in managing hazardous chemicals. Second, companies already working with other risk management programs, such as the Chemical Manufacturers Association's CAER, sometimes prefer these programs (for example, the CAER puts the emphasis on safety measurements rather than on the amounts and types of chemicals in use). Third, most LEPCs consider emergency response and planning, not prevention, to be their most important task and have little inclination, time and budget for what they often see as peripheral duties.[21] Fourth, the LEPCs follow the leadership of their respective SERCs, so a given LEPC may not be able to organize drills or simulations, for example, because of the grants awarded for this purpose by its SERC are too small.

Clearly, a large majority of LEPCs do not take full advantage of existing means of communication to reach the public. In all surveys, public communication tends to be their weakest area. The following is a list of sensible explanations.

- *Lack of time.* LEPC members are generally volunteer people holding an outside full-time job. Time consuming public relation operations are thus often unfeasible. As revealed by our survey, 21.2% of the LEPC responded that lack of time was their major problem in developing the emergency response plan.

- *Lack of expertise.* Two types of expertise are usually scarce among LEPC members: technical expertise and communication expertise. According to the National Institute for Chemical Studies (2001), several LEPCs expressed the belief that they are limited in their ability to encourage facilities to reduce hazards because they lack the necessary engineering knowledge or expertise to identify how chemicals or processes in a plant could be changed. According to another study from the state of Colorado (Adams, 1998), few LEPC members have an extensive background in public relations, social dynamics, and citizens' participation. Consistent with our survey, about 20% of the LEPCs replied that there is not enough expertise among their LEPC members.

- *Lack of chemical threats.* According to Adams, Burns and Handwerk (1994), perceived lack of actual chemical threats was singled out by 34% of LEPCs as the reason for their inactivity. It is worth noting, for instance, that LEPCs in areas that have already had toxic releases in the past five years are 10 to 15% more likely to have updated their plans, based on vulnerable zone hazard analysis, than LEPCs located in areas with no recent accident history (Texas A&M University's Hazard Center, 1992).

- *Lack of financial support.* 38% of those interviewed in the Adams, Burns and Handwerk (1994) survey identified inadequate financial support as their

---

[21]The Rich, Conn and Owens (1993) report concludes that "most LEPCs in our study have focused on the technical aspects of their job and have not made a concerted effort to bring hazardous materials issues to public attention. This is quite understandable given the constraints under which they labor. They generally run entirely on volunteer effort and have little or no independent budget or staff. Their mission has been defined primarily in terms of developing a technically adequate emergency response plan. As a result, they have few members with extensive background in public relations, citizen participation or communications. Most make risk communications a low priority and do not know how to go about obtaining public involvement even when it occurs to them to attempt to do so".

main concern. Several LEPCs wrote that they needed the "resources to go with the mandates". Only 42% of the functioning LEPCs actually have an operating budget. According to our survey, half of the answering LEPCs have an average annual budget lower than $2700, and among them 28% have no budget at all. On average, the annual budget averages $15 103. This shows again the large discrepancies among the LEPCs. According to Adams, Burns and Handwerk (1994), some LEPCs draw direct funding from local governments (34%) and a few from local industry (14%). Our own survey reveals that an average of 18.4% of the funding comes from the county or the town, 18% from other grants, 15.4% from the state, 14.2% from the EPCRA fees, 7.1% from the SERC and 5.6% from industry. The National Institute for Chemical Studies (2001) recently reported that funding is basically not available for hazard reduction activities, even in states that have assumed 112 (r) delegation.[22]

- *Public apathy.* Nearly all the LEPCs contacted through interviews or surveys point out that public apathy towards local chemical hazards made it difficult to generate support or demand for hazard reduction. In Adams, Burns and Handwerk (1994), 67% of the surveyed LEPCs said they were inactive because of the indifference of their local community. Many blamed their small population base for this. According to Adams (1998), for example, in the state of Colorado there is apparently a rather small level of public interest in emergency planning (50% said there is little interest, 40% reported some interest and only 10% indicated a great deal of interest), and in the potential dangers associated with the presence of hazardous chemicals in the area (40% said there is little interest, 20% reported some interest). As revealed by our survey, 8.5% of the answering LEPC responded that public apathy and a lack of involvement from the public was their principal problem in developing the emergency plan and moreover 9% of the answering LEPC finds the fact that members of LEPC are all volunteers is also a great problem to develop the emergency plan. Some LEPC chairpersons attribute this to "[...] the many other facets of life people are busy with and the idea that the public trusts them to do their job, as well as their experience that people typically become involved in local issues when they have a monetary stake in the outcome. People must see a correlation between their safety or quality of life before they take an interest in the dangers posed by chemicals in the community". According to the National Institute of Chemical Studies (2001), "reasons for this apathy varied widely, but it was generally believed that either the public did not perceive the risk from chemical hazards to be great, or they believed the government was taking care of managing these risks and therefore the public does not need to worry about it. As evidence

---

[22]This brings up a more fundamental problem: there is no federal funding of the LEPCs or the SERCs, which are therefore quite short in resources to carry out their EPCRA-mandated responsibilities (National Institute for Chemical Studies, 2001).

of this attitude, nearly all of the LEPCs reported little or no public attendance at RMP rollout events and few or no requests for RMP data or any other hazard information". During our interviews in Maryland and Vermont, we were also told that few people attended the public meetings organized by their LEPC. The 9/11 terrorist attack seems to have somewhat changed this situation, however: no citizen showed up at a public meeting organized by the Addison County LEPC at the beginning of 2001, but a similar event held on November 1st attracted sixty citizens! According to the National Institute of Chemical Studies (1995), "the most significant factor predicting involvement by community interests (environmental, neighborhood, universities and labor) is a higher number of facilities in the area. To a much weaker degree, having recent transportation accidents or contamination incidents are positively related to more community interest membership on LEPCs". To be sure, a quarter of the LEPCs in our survey mentioned that a local accident in their community had changed public perception: for the Somerville LEPC (MA), it was a phosphorus trichloride spill in 1980 which entailed the evacuation of 25 000 people; for the Mohave County LEPC (AZ), it was a butane tank car explosion; for the Live Oak County LEPC (TX), it was on July 9, 2001 a fire and explosion at a local refinery.

Community involvement therefore seems to be the most difficult aspect to organize and control. Because getting the community involved in emergency planning and prevention does offer significant benefit - greater awareness of the local emergency plan, development of an emergency plan that accurately addresses the community's specific needs and concerns, and increased legitimacy for funding local government entities, LEPCs must strive to devise more creative channels to convey information on chemical risks to the public.

2.5 Concluding Remarks

"Information regulation is any regulation which provides to affected stakeholders information on the operations of regulated entities, usually with the expectations that such stakeholders will then exert pressure on these entities to comply with regulation in a manner which serves the interests of the stakeholders" (Kleindorfer and Ordts, 1998). This paper considered one particular implementation of informational regulation in the United States - the *Local Emergency Planning Committees*.

Our overall appraisal of the LEPCs, as far as informational regulation is concerned, is a mitigated one. On the one hand, a fair proportion of LEPCs seems to be working effectively and to have met expectations in terms of promoting public awareness and involvement; furthermore, the fact that other countries are also seeking to implement similar devices at home may indicate that this approach to practicing informational regulation is right and doable. On the other hand, there is considerable heterogeneity in performance across

LEPCs, as an equally significant proportion of them still faces minimal public interest. Clearly, the actual process is a learning one which is subject to trial and error. Further creativity and effort need to be deployed vis-à-vis the transfer of best practices across LEPCs and the design of effective two-way communication channels with and between stakeholders.

Appendix 2.1: International Experiences

Recent international experiences on the informational regulation of industrial risks may also contribute original benchmarks and perspectives on the workings of LEPCs. This subsection focuses on two such experiences currently occurring in Canada (Québec) and in France.

• *The Québec CMMIs*
In 1993, the Canadian government created the *Canadian Council of Major Industrial Accidents* (CCAIM) to deal with the prevention and mitigation of industrial risks. At that time, municipalities were traditionally responsible for their citizens' safety in case of a natural or industrial disaster, while firms concentrated mainly on compliance with safety and health regulations in the workplace. In this context the CCAIM launched some special committees - the CMMIs (*Comité mixte municipal-industriel*) - staffed jointly by municipalities and industries to coordinate prevention and emergency planning, and to develop consistent risk communication. The aims are to promote the safety of workers and citizens altogether, and to therefore decrease property and environmental damages simultaneously, reduce response time, reduce the costs and duration of accident recovery, and foster the confidence of municipal or industrial personnel and of lay citizens. These committees are *not* mandatory, however. They may typically include representatives from municipalities, hazardous facilities, local citizens, the Ministry of Transportation, and the Ministry of Health.

The urban community of Montréal has four CMMIs. Among them, the CMMI of eastern Montréal was the first organization of its kind in Canada. It is composed of personnel from ten of the largest industrial companies operating a plant in the area, together with representatives from municipal, provincial and federal government agencies and from the community. Like a U.S. LEPC, it informs the local community of any identified potential risks and checks that effective procedures are in place to deal with emergencies. A recent simulation of toxic release and intervention based on the current means lead the committee members to formulate several recommendations, of which some - such as the installation a siren able to alert the whole neighborhood - have been implemented.

Created less than a year ago, the CMMI of Saint-Laurent is another Canadian version of an LEPC. It has asked the facilities it covers to document potential accidents. Based on the supplied data, it plans to make careful simulations and set fine-tuned emergency plans for the whole community.

Most of the Québec CMMIs are not that pro-active, however. Yet, the new *Civil Protection Act* (Bill 173) promulgated on December 20, 2001, by the provincial government imposes general obligations of prudence and foresight on all hazardous facilities and requires them to report on their respective hazards and on the specific safety measures they implement. Section III, Art. 55, of this Act's Chapter IV stipulates, for instance, that: "Local and regional authorities must take part in information efforts so that citizens may become involved in the pursuit of the objectives of this Act, in particular by disseminating advice on safety measures [...], and by taking part in committees or information sessions organized in conjunction with businesses". It is thus quite reasonable to expect that subsequent bylaws will grant further responsibilities to the CMMIs.

• *France's DRIREs and CLICs*
The Council Directive 82/501/EEC - better known as the *Seveso Directive* - was adopted in 1982. After the major industrial accidents of the 1980's, this directive was amended twice. In 1996, finally, the so-called *Seveso II Directive* replaced the initial one. Its aim is two-fold. First, it seeks to prevent as much as economically and technically possible all major industrial accidents involving dangerous substances. Second, it wants to limit the consequences on humans and the environment of any such accidents. Like its American counterpart, the Seveso II Directive entitles the public to information and consulting services concerning identified local industrial hazards. Governments of states pertaining to the European Union, for instance, must provide anyone that might be affected by the leakage of some hazardous substance with information about safety measures and the behavior deemed appropriate in the event of an accident. Furthermore, people must be fully consulted when new higher-risk facilities (as defined in the Seveso II Directive) are planned or when modifications are proposed to processes or emergency plans of existing ones.

In France, the *"Législation sur les installations classées"* authorizes the application of the Seveso II directives. The number of higher-risk facilities amounts to 400, out of a total of 1 250 covered facilities. Information about the latter is usually conveyed to the public via some information brochures distributed at the various city halls. In addition, some public administration units called DRIRE, or *Direction Régionale de l'Industrie pour la Recherche et l'Environnement*, are responsible for the prevention of major technological risks, the reduction of pollution and nuisances, and the control and elimination of process waste. A DRIRE is allowed to ask for third-party expertise to supplement a facility's hazard analysis.[23] More importantly, the DRIREs are also ascribed a role of catalyst in some associations, like the ones supervising air quality (39 associations) or industrial pollution (11 SPPPIs), and in the local commissions of information and surveillance (numbering more than 300).

The latter are becoming more numerous year after year, and they allow citizens to participate in the decision process. In Strasbourg, the local commission, which comprises 130 volunteers, has organized some educative and informative activities for children (Pupils manipulate and experiment with some chemical reactions; they also design some posters about what to do in case of emergency), patients in hospitals, and the general population (public meetings, quarterly information bulletins, and outdoor information panels). It also advises local industries on public risk communication (videos, pamphlets, brochures, etc).

The SPPPIs (*Secrétariat Permanent pour la Prévention des Pollutions Industrielles*), on the other hand, are created by the "Préfet", the top regional authority, who usually chairs them, and have no legal status.[24] In Strasbourg, more than 350 people of various backgrounds - civil servants, environmental protection activists, industry employees, and scientists - participate in the local SPPPI. Besides technological risks, the SPPPI deals with water, air and noise pollution.

After the AZF explosion in Toulouse, a large debate concerning better ways to implement and refine the Seveso II directive is currently occurring. Dozens of public workshops, involving owners and operators of hazardous facilities, elected officials, and citizens have been held throughout the country. Besides zoning and prevention within facilities, the issue of public information remains a recurrent one: all parties recognize that greater awareness on the part of citizens would render prevention and mitigation efforts more effective. In the footsteps, a recent law was adopted by the French senate on May 15th, 2003; article 2 of this bill precisely proposes the creation (supervised by the Préfets) of local committees of information and dialogue in charge of preventing and mitigating major technological and industrial risks - the so-called CLIC or *"comité local d'information et de concertation"* - around any industrial basin including one or more installations subjected to SEVESO II.

Sitting on these committees will be some hygiene and security experts, industry representatives, local and national public officials, plant employees, and local residents. The CLICs will have a duty to organize and disclose

---

[23]Expertise is generally supplied by some insurance companies or the INERIS (*Institut National de l'Environnement Industriel et des Risques*) - a multi-disciplinary public-sector think tank.

[24]Since March 1982, the civil organization of France comprises three administrative layers at the local level: the communes, the departments and the regions. Each region is managed by a *"Préfet de Région"*, who is under the direct authority of the country's Prime Minister and implements central government policies regarding regional development and land-use planning. The SPPPIs of Strasbourg and Toulouse are the only ones not lead by a Préfet but by an academic.

information to the public; in particular, they will be entitled to use counter-expertise if necessary. At this time, there are still many unanswered questions about the budget and precise scope of the CLICs. However, the French government is already committed to actively support their implementation and "mise en route".

Appendix 2.2: A Model of Voluntary Environmental Risk Disclosure

This appendix sketches the model and conclusions of a recent article on voluntary environmental disclosures by Sinclair-Desgagné and Gozlan (2003).

Let some stakeholder (I) contemplate a project that has a probability B of being safe (g) and a probability $1 - B$ of being dangerous (b) for human health or the environment. The stakeholder might endorse that project or not. If she does and the project turns out to be safe (dangerous), then she gets a positive payoff H (a negative payoff B). If she does not, then the project is dropped and she gets her status quo payoff, which is assumed to be 0. The project is proposed by a firm (F) that knows a priori whether it is safe or not. The firm's ex post payoff is strictly positive or is equal to 0, depending on whether the project gets endorsed or not; it might therefore seek to convince the stakeholder by disclosing more or less accurate information concerning the nature of the project. Let the endogenous precision (or quality) of the information provided be represented by a number $\exists \in [0,1]$: $\exists$ is the likelihood that the stakeholder's evaluation of the project based on the delivered information will be positive given that the project is truly safe and negative given that the project is dangerous.[25] A number $\exists$ strictly between 0.5 and 1 thus captures the situation where the entrepreneur provides faithful, albeit imprecise, information. In this case it follows from Bayes's rule that studying the firm's announcement makes the stakeholder believe the project is safe or dangerous with respective revised probabilities:

$$\text{Prob (safe, given positive)} = \pi\beta/[\pi\beta + (1 - \pi)(1 - \beta)]$$

and

$$\text{Prob (dangerous, given negative)} = (1 - \pi)\beta/[(1 - \pi)\beta + \pi(1 - \beta)].$$

At this point the model yields a first conclusion: if information is costless and cannot be misleading, then the firm provides fully accurate information

---

[25] In this game, the probability a positive evaluation results from a safe project is always equal to the probability an evaluation turns out negative when the project is dangerous. Hence, $\exists$ corresponds to the likelihood of reaching the right conclusion upon further evaluation of the project, so $1 - \exists$ is the probability of being mistaken, i.e. Prob(positive, given unsafe) = Prob(negative, given safe).

($\beta = 1$) and so reveals the true nature of the project. This is just another statement of the well-known "disclosure principle" of persuasion games. This principle is supported by the following argument: the stakeholder would discount any imprecise information, because the firm has no incentives to conceal data if the project is a safe one.[26] The policy implication is that the quality and content of voluntary disclosures are maximal when the public can veto a given project and information provision is free.

Disclosure related costs, however, which might come from the preparation and dissemination of reports or from the proprietary nature of information, raise a different picture. Partial information in this case cannot be exclusively associated with a dangerous project. One major consequence is that, if the stakeholder holds a positive expected payoff ($\pi H + (1 - \pi)L > 0$) and is thus a priori favorable to the project, then the firm would not deem it useful to produce additional information. This calls for government intervention, which can take the form of mandatory disclosures (like the above EPCRA and RMP) or an advertisement campaign aimed at instilling doubts in the stakeholder's mind. The presence of a reluctant stakeholder with expected payoff $\pi H + (1 - \pi)L \leq 0$ who can block the project, on the other hand, puts pressure on the firm to disclose information.

Let us examine the latter situation in the (most realistic) case where the firm cannot lie, its ex post revenue $R(\cdot)$ is highest if the project turns out to be a safe one (i.e. $R(g) > R(b)$), and the stakeholder can assess the accuracy level of disclosures. Denote $P(\beta)$ and $S(\beta)$ the firm's and the stakeholder's costs of respectively producing and studying information of precision β.[27] Two situations are now possible, depending on the specific configuration of revenues and costs. In the first one, the firm with a safe project produces information of precision β such that $P(\beta)$ is just superior to $R(b)$, and the firm with a dangerous project remains silent. Upon public disclosure by the firm, the stakeholder thus infers immediately that the project is safe and endorses it.[28] In the second case, however, disclosures are the same whatever the safety of the project; their precision is just high enough to induce the stakeholder to undertake some costly evaluation before making a decision. The project is approved if and only if the evaluation concludes positively.[29] Government

---

[26] In the language of game theory, this is a "forward induction" argument. It supports the equilibrium precision level $\beta = 1$ by reasoning that the stakeholder's revised probability that the project is innocuous would be 0 if the firm switched to (out-of-equilibrium) partial disclosure and set $\beta < 1$ instead.

[27] The two functions are strictly convex and increasing on the interval [0.5, 1], with $P(0.5) = S(0.5) = 0$ and $P(1) > R(g)$.

[28] This equilibrium is also based on forward induction arguments. The stakeholder believes that (out-of-equilibrium) additional information of precision β such that $P(\beta) < R(b)$ (resp. $P(\beta) > R(b)$) can only be supplied under a dangerous (resp. safe) project; her revised probability that the project is safe is then set to 0 (resp. 1) in this case and the project is not endorsed (resp. is endorsed).

[29] Denote β^* the accuracy of disclosures at this equilibrium. It is the unique β that maximizes (we assume of course that this maximand exists) the firm's expected payoff

intervention might again be justified in this case, through subsidies to the study of disclosures or enforced standards of higher quality and accuracy.

References

Adams, M.P., 'LEPCs in Colorado: How Does Public Participation Fit Their Mission?', (1998), Working paper, Graduate School of Public Affairs - University of Colorado at Denver, May.

Adams, W.C., Morgan M.B. and Viana, M.M., 'Nationwide Survey of LEPC data Management Practices : Executive Summary', (1997), *Working paper*, The George Washington University - Public Administration Department, May.

Adams, W.C., Burns S.D. and Handwerk, P.G., *Nationwide LEPC Survey*, (1994), Sponsored by USEPA, Chemical Emergency Preparedness and Prevention Office, Department of Public Administration, The George Washington University, October.

Bromley, D.W. and Segerson, K., (eds), *The Social Response to Environmental Risk*, (Boston, MA: Kluwer Academic Publishers, 1992).

Conseil Régional des Accidents Industriels Majeurs du Montréal-Métropolitain (CRAIM), (1996), *Guide de gestion des risques d'accidents industriels majeurs à l'intention des municipalités et de l'industrie.*

Duclos, D., *La Peur et le Savoir. La société face à la science, la technique et leurs dangers*, (Paris: Éditions La Découverte, 1989).

EPA, 'The Role of Local Emergency Planning Committees (LEPCs) and Other Local Agencies in the Risk Management Program (RMP) of Clean Air Act (CAA) Section 112 (r)', Subgroup #7 Report, Report to USEPA's RMP Implementation Workgroup (1998).

EPA-CEPPO, 'RMPs are on the way How LEPCs and other Local Agencies can include information from Risk Management Plans in their ongoing work', (November, 1999).

Essig, P., 'Débat National sur les risques industriels, Octobre - Décembre 2001', Rapport à Monsieur le Premier Ministre, (January, 2002).

Glickman, T.S. and Gough, M., (eds), *Readings in Risk*, (1990), Resources for the Future, Washington DC.

Hadden, S.G., *A Citizen's Right to Know, Risk Communication and Public Policy*, (Boulder, CO: Westview Press, 1989).

---

under stakeholder evaluation, i.e. that maximizes $\beta H - P(\beta)$ under the constraints that the stakeholder prefers to further evaluate the project rather than to either veto it [formally, $\pi\beta G + (1-\pi)(1-\beta)B - S(\beta) \geq 0$] or endorse it right away [$-(1-\pi)\beta B - \pi(1-\beta)G - S(\beta) \geq 0$]. The forward induction argument that supports this equilibrium runs as follows. Disclosures of precision $\beta \neq \beta^*$ such that $P(\beta) < R(b)$ (resp. $P(\beta) > R(b)$) can only occur under a dangerous (resp. safe) project; the stakeholder's revised probability that the project is safe is then set to 0 (resp. 1) in this case and the project is not endorsed (resp. is endorsed).

Kleindorfer, P.R. and Ordts, E.W., 'Informational Regulation of Environmental Risks', *Risk Analysis*, 18(2) (1998): 155-170.

Lindell, M.K. and Withney, D.J., 'Effects of Organizational Environment, Internal Structure, and Team Climate on the effectiveness of Local Emergency Planning committees', *Risk Analysis*, 15(4) (1995).

Magat, W.A. and Viscusi, W.K., *Informational Approaches to Regulation*, (Cambridge, MA: The MIT Press, 1992).

National Institute for Chemical Studies, 'Focus on the Future of LEPCs', Report of a One-Day Focus Group Held at the National Institute for Chemical Studies June 9 and 10, 1995, Charleston (WV).

National Institute for Chemical Studies, 'Local emergency Planning committees and Risk Management Plans : encouraging hazard reduction', (2001), Report.

National Response Team. NRT-2, September 1990, 'Developing a Hazardous Materials Exercise Program - A Handbook for State and Local Officials'.

Pressman, J.L. and Wildavsky, A., *Implementation*, 3rd ed. (Berkeley, CA: University of California Press, 1984).

Rich, R.C., Conn W.D. and Owens, W.L., 'Indirect Regulation' of Environmental Hazards Through the Provision of Information to the Public: the Case of SARA, Title III', *Policy Studies Journal*, 21(1) (1993): 16-34.

Rosenthal, I., McNulty P.J. and Helsing, L.D., 'The role of the community in the implementation of the EPA's Rule on Risk Management Programs for Chemical Accidental Release Prevention', *Risk Analysis*, 18(2) (1998).

Sinclair-Desgagné, B., 'Environmental risk management and the business firm', in Tom Tietenberg and Henk Folmer (editors), *The International Yearbook of Environmental and Resource Economics 2001/2002*, (Northampton, Massachussetts: Edward Elgar, 2001).

Sinclair-Desgagné, B. and Gozlan, E., 'A Theory of Environmental Risk Disclosure', *Journal of Environmental Economics and Management*, 45 (2003): 377-393.

Sinclair-Desgagné B. and Vachon, C., 'Dealing with major technological risks', in Henk Folmer and H. Landis Gabel (eds), *Principles of Environmental and Resource Economics: A Guide for Students and Decision Makers*, (Aldershot, U.K.: Edward Elgar, 2000) (second edition).

Texas A&M University's Hazard Reduction & Recovery Center, 1992, Survey.

Tietenberg, T. and Wheeler, D., 'Empowering the community: Information strategies for pollution control', in Henk Folmer, H. Landis Gabel, Shelby Gerking and Adam Rose (eds), *Frontiers of Environmental Economics*, (Cheltenham, U.K.: Edward Elgar, 2000).

Viridescent, Inc., 'Improving Web Site Usability for Local Emergency Planning Communities of EPCRA, SERCs, and LEPCs', (January, 2002).

Wettig, J. and Porter, S., 'The SEVESO II Directive', (February, 1999).

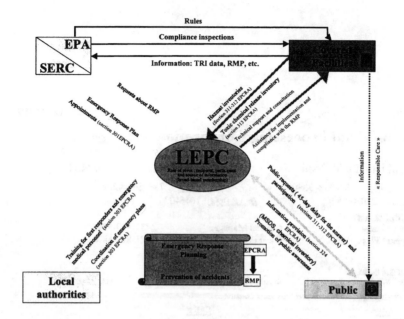

Figure 2.1: The LEPC and its landscape

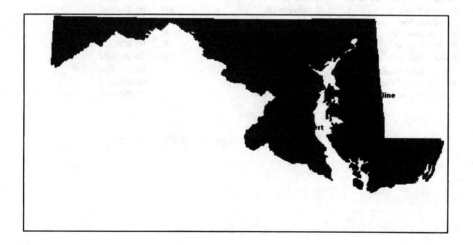

Figure 2.2: Distribution of LEPCs in the state of Maryland

2.1.01   Updated   02/15/2002

Acton Metal Processing Corporation

SIC:

SIC: 3672

41 Athletic Field Road   Waltham   02451

Facility Phone 781-893-5890   **Facility Fax**   781-647-4226
Lat/Long: 320/4695   **Dun & Brad**   1046408

Facility Contact
Martin Flagg   General Manager   781-893-5890
Ruppert Flagg   Owner & President   781-893-5890

Mailing Address
PO Box 671   47 Athletic Field Rd   Waltham   MA   02451

Owner/Operator: Acton Metal Processing Corporation

Chemicals:	CAS #
Hydrofluoric Acid	7664-39-3
Sulfuric Acid	7664-93-9
Sodium Hydroxide [Liquid]	1310-73-2
Nitric Acid	7697-37-2

Special Populations	Address:	Population:	Phone:
Chapel Hill Chauncy Hall School	785 Beaver St	474	781-894-2644
Plympton School	20 Farnsworth St.	372	781-314-5760
Piety Corner Nursing Home	325 Bacon St	44	781-894-5264
Waltham Child Care	764 Main St	55	781-899-2070
Waltham Day Care	50 Church St	51	781-899-0531
Waltham Boys & Girls Club	20 Exchange Street	280	781-893-6620
Waltham Police Department	155 Lexington St		781-314-3600
Waltham Fire Department	175 Lexington Street		781-893-4105
Waltham Emergency Communications	161 Lexington Street		781-893-4105

Figure 2.3: An exhibit from an LEPC website

Table 2.1: Geographical distribution of the LEPCs[30]

States	Number of LEPCs	Number of RMP facilities	States	Number of LEPCs	Number of RMP facilities
Alabama	68	247	Montana	56	128
Alaska	19	28	Nebraska	84	627
Arizona	16	132	Nevada	17	42
Arkansas	77	193	New Hampshire	225	16
California	6	1067	New Jersey	564	128
Colorado	52	251	New Mexico	32	71
Connecticut	157	57	New York	58	233
Delaware	4	41	North Carolina	97	341
Florida	11	575	North Dakota	53	353
Georgia	16	392	Ohio	87	512
Hawaii	5	30	Oklahoma	80	392
Idaho	43	82	Oregon	1	140
Illinois	103	1077	Pennsylvania	67	410
Indiana	92	541	Rhode Island	9	28
Iowa	72	1007	South Carolina	46	216
Kansas	105	815	South Dakota	62	124
Kentucky	117	233	Tennessee	95	245
Louisiana	64	360	Texas	272	1424
Maine	16	37	Utah	30	87
Maryland	25	128	Vermont	10	10
Massachusetts	351	95	Virginia	114	193
Michigan	90	269	Washington	47	265
Minnesota	7	567	West Virginia	54	83
Mississippi	82	185	Wisconsin	72	321
Missouri	90	420	Wyoming	23	77

[30]The number of LEPCs per State was found on the following website of the EPA: http://yosemite.epa.gov/oswer/ceppoweb.nsf/content/lepclist.htm. Moreover, the number of RMP facilities per State was found on the following website : http://d1.rtknet.org/rmp/.

Chapter 3

Regulating an Agent with Different Beliefs[*]

François Salanié
Université de Toulouse and INRA

Nicolas Treich
Université de Toulouse and INRA

3.1 Introduction

How to regulate risks? A typical problem analyzed in this book is the one of a regulator who cannot perfectly monitor some risky activities. For instance, the regulator may lack information on the activity itself and/or because of the judgment-proof issue. The regulatory problem then consists in designing the incentives to give to the various actors (entrepreneurs, experts, politicians...) so that the development of the risky activity is consistent with consumers' welfare. In this paper, we focus on another source of inconsistency. This source has been rarely analyzed in economics as it is related to consumers themselves, more precisely to the fact that consumers may have biased perceptions of risks.[1]

There is a lot of evidence that seem to indicate that people misperceive the risks they face.[2] This may be rooted in people's psychology, or due to

---

[*]We thank Ted Bergstrom, Yann Bramoullé, Bruno Jullien, Jack Marshall, David Martimort and Bernard Salanié for useful discussions, as well as an anonymous referee and participants to seminars in the Universities of California-Santa Barbara, Iowa State, Wageningen, Toulouse, Nottingham and the Harvard School of Public Health. The usual caveat applies.

[1]Our approach does not propose a new model to understand how individuals' subjective beliefs affect their own choices, and does not use non-expected utility models based on systematic distorsions of probabilities. Our paper is about the impact of individuals' beliefs on the regulatory policy within an expected utility framework based on different beliefs.

[2]A famous example is the bias in mortality risks perception. Individuals systematically overestimate the rare causes of death such as cataclysmic storms or plane crashes and underestimate more common causes of death like cancers or automobile accidents (Lichtenstein et al., 1978).

bounded rationality.[3] A direct implication is that individuals' beliefs on hazard risks differ from risks data frequencies and experts' scientific evidence (Slovic, 1987). This may induce a situation where consumers' behavior may likely be judged as faulty by an informed regulator. And, as a result, risks misperceptions may affect, or even trigger, some regulatory intervention. For instance, a regulator may set a maximum fat content for some food products because he may be concerned by the fact that some consumers underestimate the long-term negative effects of eating fatty food. Similarly, a regulator may subsidize airline safety prevention efforts because he may judge that people are too afraid of taking the plane. This paper provides a theoretical analysis of these motives for intervention. It basically investigates one question: under which conditions do public risks misperceptions call for more, or less, regulatory intervention?

From a policy viewpoint, the inability of citizens to accurately evaluate the consequences of their own decisions has often been a justification for extensive regulatory programs (Viscusi, 1998). This justification is actually based on a merit good argument (Musgrave, 1959). Education policies, cigarette taxation or compulsory insurance programs may provide examples of government intervention on the basis of consumers' misperceptions. Merit good arguments for public intervention are however quite controversial since they rely on a paternalistic view of consumers' preferences. They presume that the regulator has a better knowledge of consumer's preferences than the consumers themselves.[4]

The main prediction of merit good theory for public intervention is clearcut. The consumption of merit goods should be subsidized, while demerit goods should be taxed (Besley, 1988). The key regulatory instrument here is taxation (see, e.g., Harris, 1980, Glazer, 1985 and recently within a hyperbolic discounting model, see O'Donoghue and Rabin, 2003, Gruber and Koszegi, 2004). The positive effect of the tax policy is immediate: people

---

[3]In psychology, there is a very well documented literature on risk misperceptions (see, e.g., Tversky and Kahneman 1974). Individuals have difficulties to evaluate small probabilities. Individuals also use heuristics or rules of thumbs that are useful but misleading. For instance, they are subject to 'availability heuristic'. People assess the risks of heart attack by recalling such occurences among one's acquaintances. They 'anchor' their estimates to easily retrievable events in memory such as sensational stories in the medias. People can also manipulate their own beliefs in order to confirm their desired beliefs (see, e.g., Akerlof and Dickens, 1982). This process of beliefs misperceptions can be exacerbated at a collective level by a chain reaction that gives the perception increasing plausibility through its rising availability in public discourse (Kuran and Sunstein, 1999).

[4]Note however that, in the context of risk policy, the controversy may be less vivid, as the difference in preferences comes from a difference in beliefs, not from preferences under certainty per se. However, the model still needs to assume that agents have different prior beliefs. Hence there is a need to motivate why the Harsanyi Doctrine does not apply. In other words, why do agents agree to disagree? We will not say much about this assumption in this paper (but rather focus on its policy implications). Possible theoretical justifications may go from bounded or rival rationalities to shortcuts that capture some communications or trust issues.

consume less demerit goods and consume more merit goods. The effect of a taxation policy is actually straightforward in a risky context as well. The government must tax the consumption of those goods whose risky consequences are underestimated by consumers, and subsidize the consumption of goods whose risky consequences are overestimated. Such a monotonic pattern between the optimal public taxation as a function of the difference in beliefs has been derived in Sandmo (1983).

Nevertheless public intervention is not limited to the use of taxes or subsidies. Many regulatory efforts take the form of direct risk-reduction programs such as safety standards or prevention expenditures; examples abound in areas such as health or food safety.

This paper mostly focuses on these instruments, which we gather under the term of norms. We set up a framework to answer the question of whether public misperceptions support the setting of higher (more stringent) or lower norms by the regulator. This paper builds on some empirical evidence on risks misperceptions to assume that the regulator knows better what is the risk faced by the public, and, as a result, has some legitimacy to correct individuals' suboptimal risk-exposure.[5] The framework assumes that the consumer's decisions are based upon his own beliefs. This introduces an indirect effect that makes norm analysis more complex than taxation analysis. Indeed, the consumer's response to a safer situation may offset the direct beneficial effect.[6]

Our main message is the following. We show that risk misperceptions call for setting higher norms, under plausible conditions on the agents' preferences. This result does not depend on whether the representative agent is pessimistic or optimistic compared to the regulator. In fact, the optimal norm level is shown to be monotonic with the *absolute distance* between beliefs. This property is in sharp contrast with the monotonic pattern for the optimal taxation policy as a function of the difference in beliefs.

The paper proceeds as follows. The next section builds on a simple example in order to illustrate the policy questions and the result just mentioned above. This example is adapted from another paper, Salanié and Treich (2003). Section 3.3 develops the analysis in a more general framework. Section 3.4 concludes.

---

[5] An alternative approach would be to assume that the regulator has superior information about the distribution of the risk but, for some reasons, he cannot signal this information to the public; Spence (1977) offers an early reference on this problem. For a principal-agent approach, see Wirl (1999). For an analysis of the signalling issue in consumption regulation, see the recent paper of Barigozzi and Villeneuve (2002).

[6] Consider the consequences of making automobile seatbelt compulsory. The direct effect is that seatbelts reduce the gravity of injury in case of accident; the associated indirect effect is to make drivers driving faster, maybe offsetting the benefits of the safety regulatory policy (Peltzman, 1975). Similarly, Viscusi (1998) found a correlation between child-resistant packages and an increase in accidental poisonings. He suggested that consumers might have become less safety conscious due to the existence of safety caps.

3.2 Regulation in Happyville

The following problem was introduced by Portney (1992, p. 131). It is called 'Trouble in Happyville':

> "You are Director of Environmental Protection in Happyville [...]. The drinking water supply in Happyville is contaminated by a naturally occurring substance that each and every resident believes may be responsible for the above-average cancer rate observed there. So concerned are they that they insist you put in place a very expensive treatment system to remove the contaminant. [...]
>
> You have asked the top ten risk assessors in the world to test the contaminant for carcinogenicity [...]. These ten risk assessors tell you that while one could never prove that the substance is harmless, they would each stake their professional reputations on its being so. You have repeatedly and skillfully communicated this to the Happyville citizenry, but because of a deep-seated skepticism of all government officials, they remain completely unconvinced and truly frightened."

The mirror image of Happyville is Blissville (Viscusi, 2000). In Happyville, the risk is low but perceived as large. In Blissville, the risk is large but perceived as low. The question becomes: You are the Director of Environmental Protection both in Happyville and Blissville, in which city should you develop the higher cleanup effort?[7]

Let us formally represent this economy. Let the utility of a representative Happyville agent be

$$U(x, a, b) = u(b) - (1 - a)\delta bx - c(a),$$

where

$u(.)$	is the agent's utility from drinking water,
b	is water consumption,
a	is cleanup effort, $0 \leq a \leq 1$,
δ	is the desutility from getting a cancer,
x	is the unknown dose-response risk of carcinogenicity, $0 \leq x \leq 1$,
$c(a)$	is the cleanup cost function.

---

[7]To Viscusi, the choice seems clear-cut. More efforts have to be spent in Blissville. Indeed, if efforts are mostly spent in Happyville, this is a "statistical murder" since lives are sacrificed to focus instead on illusory fears. Yet, for many scholars, the choice is not so clear-cut (see, e.g., Pollak, 1995, 1998, and references therein). If efforts are spent in Happyville, people who were worried feel protected, and so feel better. The latter view is usually called the "populist approach" to risk regulation (Breyer, 1993, Hird 1994). Salanié and Treich (2003) compare the cleanup efforts of a paternalistic regulator to those of a populist regulator.

Assume simple functional forms

$$
\begin{aligned}
u(b) &= -(1-b)^2/2 \\
\delta &= 1 \\
c(a) &= a^2/2.
\end{aligned}
$$

Under expected utility, only the expected value of x matters. The expected value used by the regulator is denoted by r. Yet, the agent does not share the same beliefs as the regulator. The agent believes that this expected value is $s \neq r$. Happyville (resp. Blissville) is then simply characterized by a society where $s > r$ (resp. $s < r$).

For a given cleanup effort a, the agent simply chooses b to maximize his expected utility computed with s, that is

$$
E_s U(x, a, b) = -(1-b)^2/2 - (1-a)sb - c(a),
$$

so that we get

$$
b(a, s) = 1 - (1-a)s.
$$

According to the intuition, optimal water consumption b is decreasing in the perceived probability of getting a cancer s and increasing in the level of cleanup efforts a.

How does the cleanup effort depend on belief s? Our fundamental assumption in the paper will be that the regulator is paternalistic, namely his objective is to maximize the expected utility of the individual $E_r U$, i.e. the expected utility based on his own beliefs r.[8] Nevertheless the regulator must take into account the consumer's response (based on s), so that a is chosen to maximize

$$
E_r U(x, a, b(a, s)) = -((1-a)s)^2/2 - (1-a)r(1-(1-a)s) - a^2/2, \quad (3.1)
$$

the solution being

$$
a(r, s) = \frac{r - 2rs + s^2}{1 - 2rs + s^2} \in [0, 1]. \quad (3.2)
$$

This framework thus captures a complex channel for why the individual's perception s affects the regulatory choices. This channel is related to the anticipation of the faulty response of individuals, i.e. the response based on s, not on r.

From equation (3.2), it is easy to see that decision $a(r, s)$ is decreasing in s then increasing in s. Importantly, this function has its minimum at $s = r$. Thus we have $a(r, s) \geq a(r, r)$. This shows that the optimal decision

---

[8]The reader may find related theoretical discussions and counter-arguments in Hammond (1981), Sandmo (1983) and Marshall (1988).

is always larger than under identical beliefs. In other words, this shows that the difference in beliefs *always* calls for more intervention in the economy.

Further note that the difference between the optimal policy $a(r, s)$ and the policy $a(r, r)$ increases as the absolute value $|r - s|$ increases. Hence, an important lesson from that example is that the *difference* between the public and the regulator beliefs is more important than the direction of the misperception. In other words, it is not so important for the regulator to know whether consumers are optimistic or pessimistic.

To explain this result, one may give the following intuition. In Blissville, people are optimistic and consume too much water. Risk-exposure to cancer is thus too large. Public intervention is thus required since cleaning water will reduce risk-exposure. This is the direct effect of cleaning water that is beneficial here. In Happyville, the reason for why cleanup efforts increase is different. Individuals are pessimistic and do not consume water enough. Cleanup efforts then give an incentive for the population to consume more water, which is a source of welfare. Here this is the indirect effect that dominates.

These intuitive arguments suggest that the result is model-dependent, since it depends on the relative strengths of the direct and indirect effects. We check the robustness of our result in the following section. But before turning to a more general framework, let us briefly address the question of taxation, which is the focus of the merit good literature. If the regulator can set a tax t on water's consumption, the consumer optimizes over b

$$u(b) - (1 - a)bs - tb - c(a)$$

and the agent's consumption $b(a, s, t)$ is characterized by

$$u'(b(a, s, t)) = (1 - a)s + t.$$

If the redistribution of the tax is costless, the regulator's program becomes

$$\max_{a,t} \ u(b(a, s, t)) - (1 - a)b(a, s, t)r - c(a)$$

and we get

$$t = (1 - a)(r - s). \tag{3.3}$$

Hence the optimal tax is monotonic with the difference in beliefs, a key difference with the non-monotonicity of norms that we have emphasized above. This monotonicity property of the tax is emphasized in the literature (see, e.g., Besley, 1988), and corresponds to the idea that a well-chosen tax corrects erroneous beliefs, from the viewpoint of the regulator. As a consequence each agent now behaves as if he shared the same beliefs r as the regulator. This implies that the norm level can be set equal to $a(r, r)$, as if there was no risk misperceptions: the introduction of taxation allows the regulator to restore efficiency in the Happyville society.

This result, however, strongly depends on our assumption of a representative agent, as we discuss now. Suppose indeed that the agents are heterogeneous in Happyville, and in particular that different agents may have different beliefs. Then from (3.3) the regulator would have to set personalized taxes, depending on each agent's beliefs. This seems to be quite a difficult task in practice, and a natural constraint is to impose a uniform taxation. It is then easy to see that a uniform tax only allows the regulator to correct the average misperception.[9] The regulator may then try to respond to the remaining heterogeneity by adjusting the norm level. Recall moreover that the above analysis has shown that the norm level should be increased, *whatever the agent's beliefs*. It is easy to understand then that the result that the norm level increases will be robust to aggregation, and will remain unchanged if the agents have heterogeneous beliefs. The next section examines the conditions under which this result extends to more general preferences.

3.3   General Preferences

The example above has shown that risk misperceptions may call for increasing the level of norms, and that this property does not depend on whether the consumer is optimistic or pessimistic. In this section we check the robustness of this result by considering more general preferences. Let us endow the consumer with the following utility function

$$U(x, a, b) = u(a, b) - f(a, b)x.$$

As in the Happyville example, x is a positive, unknown parameter characterizing the dose-response risk. The function $f(a, b)$ thus represents the consumer's exposure to the risk x. Risk exposure is assumed to be increasing with consumption ($f_b > 0$), at an increasing rate ($f_{bb} \geq 0$), but at a decreasing rate with the norm level ($f_{ab} < 0$). $u(a, b)$ is the utility from consuming a volume b when the norm level is set to a. For regularity we assume that u is strictly concave with respect to b, and that an increase in the norm level discourages consumption ($u_{ab}(a, b) \leq 0$). This last assumption captures the idea that while a higher norm level reduces marginal risk-exposure ($f_{ab} < 0$), it also increases the water marginal cost.

As in the preceding section, let us denote s the expected value of x computed with the consumer's beliefs. Given a norm level a, the agent's con-

---

[9]We show in Salanié and Treich (2003) that if beliefs are unbiased in average, then the optimal uniform tax is zero, and thus taxation is useless.

sumption $b(a, s)$ is uniquely characterized by

$$u_b(a, b) = f_b(a, b)s \qquad (3.4)$$

Under our assumptions it is easily checked that consumption $b(a, s)$ is decreasing with the index of pessimism s. Now, if the paternalistic regulator believes that the expected value of x is equal to the value r, then he chooses a to maximize

$$E_r U(x, a, b(a, s)) = u(a, b(a, s)) - f(a, b(a, s))r.$$

In order to examine the impact of risk misperceptions on the norm level, we want to study how the maximizer $a(r, s)$ to this program varies with s. Notice first that the derivative of this objective with respect to a can be decomposed into two terms. First, there is the direct effect

$$u_a(a, b(a, s)) - f_a(a, b(a, s))r$$

that captures how a change in a modifies utility, when consumption $b(a, s)$ is kept constant. On the other hand, there is the indirect effect

$$[u_b(a, b(a, s)) - f_b(a, b(a, s))r]\frac{\partial b}{\partial a}(a, s)$$

that captures how an increase in a modifies the agent's consumption and, in turn, his utility. In the most simple cases the two effects go in opposite direction. Consider for example a pessimistic agent $(s > r)$ in the case when the water marginal cost is independent from its quality ($u_{ab} = 0$). Because his consumption $b(a, s)$ is less than $b(a, r)$, the consumer is less exposed to risk, and thus the direct effect supports a reduction in a compared to $a(r, r)$.[10] By contrast, the indirect effect supports an increase in a since the agent's consumption is too low from the regulator's viewpoint.[11]

Recall now that we want to examine the effect of a difference in beliefs on the regulatory decision. Hence we want to evaluate the effect of a variation in s on the direct and the indirect effects, that may go in opposite direction, as we have just seen. The problem looks difficult, but fortunately one may solve this comparative statics analysis by applying the following methodology.

The first step consists in proceeding to the following monotonic change of variable, from consumption b to risk-exposure β:

$$\beta \equiv f(a, b).$$

---

[10] From $u_{ab} = 0$ and $f_{ab} \leq 0$ the function $u_a - f_a r$ is increasing with b, so that the direct effect is less than $u_a(a, b(a, r)) - f_a(a, b(a, r))r$.

[11] Using the first-order condition (3.4), the indirect effect can be rewritten as $f_b[s - r]\partial b/\partial a$, in which all terms are positive.

Then define a utility function v such that

$$v(a, f(a, b)) \equiv u(a, b).$$

The model thus becomes

$$V(x, a, \beta) = v(a, \beta) - \beta x$$

in which β stands for the level of risk-exposure chosen by the agent. Thus $\beta(a, s)$ is characterized by

$$v_\beta(a, \beta(a, s)) = s \tag{3.5}$$

and must be decreasing with s. The regulator's objective becomes

$$v(a, \beta(a, s)) - \beta(a, s)r. \tag{3.6}$$

Let us differentiate this objective with respect to a:

$$v_a(a, \beta(a, s)) + [v_\beta(a, \beta(a, s)) - r]\frac{\partial \beta}{\partial a}(a, s)$$

The first term is a new direct effect, now computed keeping the risk-exposure (and not the consumption) constant. Use (3.5) to simplify the second term associated to the indirect effect of changing a, to get

$$v_a(a, \beta(a, s)) + [s - r]\frac{\partial \beta}{\partial a}(a, s).$$

Now differentiate with respect to s to get

$$v_{a\beta}(a, \beta(a, s))\frac{\partial \beta}{\partial s}(a, s) + \frac{\partial \beta}{\partial a}(a, s) + (s - r)\frac{\partial^2 \beta}{\partial a \partial s}(a, s).$$

But the first two terms exactly offset each other, thanks to the first-order condition (3.5). We thus have shown that the cross-derivative with respect to a and s of the regulator's objective (3.6) is equal to

$$(s - r)\frac{\partial^2 \beta}{\partial a \partial s}(a, s).$$

The critical condition to sign the comparative statics analysis thus depends on the sign of $\partial^2 \beta / \partial a \partial s$. If it is positive, then the cross-derivative of the objective must be positive when s is above r, and negative when s is below r. This implies that the solution $a(r, s)$ to the regulator's problem (3.6) must be increasing with s when s is above r, and decreasing with s when s is below r. Hence the optimal norm level is not monotonic; this is exactly the result we already obtained in the Happyville example.

When is $\partial^2\beta/\partial a\partial s$ positive ? We know from (3.5) that $\beta(a, s)$ is decreasing with s. Hence what we require is that an increase in the norm level makes each agent's risk-exposure less dependent on his beliefs. To make the interpretation of this property easier, let us translate it in the context of our initial model $U(x, a, b)$. Recall that $\beta(a, s) = f(a, b(a, s))$, so that

$$\frac{\partial\beta}{\partial s} = f_b\frac{\partial b}{\partial s}$$

and

$$\frac{\partial^2\beta}{\partial a\partial s} = [f_{ab} + f_{bb}\frac{\partial b}{\partial a}]\frac{\partial b}{\partial s} + f_b\frac{\partial^2 b}{\partial a\partial s}.$$

It turns out that under our assumptions, the first product is positive.[12] Hence $\partial^2\beta/\partial a\partial s$ is positive as soon as $\partial^2 b/\partial a\partial s$ is not too negative. This last condition is easier to interpret: for example, it is satisfied if an increase in a increases more the consumption of a more pessimistic consumer.

This result may give us a better intuition for why risk misperceptions may always call for increasing a. When the citizen is pessimistic $(s > r)$, water consumption is low, and therefore the direct effect is weak. In fact, the main reason for setting a above the benchmark $a(r, r)$ is the indirect effect: the regulator wants to increase water consumption. If the citizen becomes more pessimistic, water consumption is further reduced, the direct effect is also reduced, while the indirect effect becomes even more compelling. Hence a should be increased. Still this requires that the sensitivity of consumption with respect to a is not reduced too much by an increase in pessimism; or equivalently $\partial^2 b/\partial a\partial s$ is not too negative, as has been formally shown above.[13]

To sum up, in this section we have shown that risk misperceptions call for setting higher norms as soon as higher norms increase more the consumption of a more pessimistic agent. This result holds whether the agent is optimistic or pessimistic. It also holds in the presence of heterogeneous beliefs, as we have discussed at the end of Section 2; or in the presence of heterogeneous preferences.

3.4 Conclusion

There is a lot of evidence that people misperceive the risks they face and there is also a lot of evidence that regulatory programs respond heavily to public risks misperceptions (Viscusi, 1998). The objective of this paper has been to develop a theoretical linkage between these two pieces of evidence, namely risks misperceptions and over-regulation. Hence the paper's contribution may

---

[12]Indeed we have $\partial b/\partial a = (f_{ab}s - u_{ab})/(u_{bb} - f_{bb}s)$, so that $f_{ab} + f_{bb}\partial b/\partial a = (f_{ab}u_{bb} - f_{bb}u_{ab})/(u_{bb} - f_{bb}s)$, which is negative under our assumptions. Moreover $\partial b/\partial s$ is negative.

[13]A similar reasoning holds when the consumer is optimistic; then the direct effect plays the main role.

be viewed as an effort to rationalize the observed heavy-handed risk regulatory interventions advocated by many scholars (Hahn, 1996).

At this point, it is important to mention that this rationalization goes against the standard criticism of these policies, the criticism based on "populism" (Hird, 1994). A populist policy is a policy that responds to the citizens' demands, even though these demands are uninformed or misplaced. Typically, the overly expensive Superfund program in the US has been said to embody populism, namely to be designed to reassure people rather than adequately respond to the risks in presence.

Our rationalization has instead been based on paternalism, not on populism. The idea is that worried consumers are too cautious from the viewpoint of a paternalistic regulator. As a result, a more stringent policy may be efficient since it objectively increases the level of safety perceived by consumers, and therefore helps increase consumption. A typical example of such a paternalistic policy could be the European stringent beef regulation in the late 90s. This policy seems indeed to have mitigated the huge decrease in beef consumption following the "mad cow" crisis. Hence the apparently inefficient Superfund program may for instance have prevented pessimistic people from moving away from inocuous sites, thus avoiding costs which were unjustified according to the regulator.

In this paper, this latter positive effect due to the response of the consumers to the policy has been coined the indirect effect. Obviously, the presence of this effect does not mean that over-intervention is always socially efficient. This is only the case if the indirect effect offsets the negative effect of investing money where there is little need for an increase in safety in the first place, namely if it offsets the direct risk-reduction effect. This observation leads us to the other policy message emerging from our analysis. This message is that an increase in safety effort may be warranted as well in an economy where consumers are too optimistic. In this case, consumers take too much risk, and the rationale for regulatory intervention is simply the other way round, namely the willingness to reduce the level of risk in the society. This situation corresponds to a situation where the direct effect simply dominates the indirect effect. Various examples easily come mind as well: home safety prevention, compulsory seat belts, namely situations where people are prone to take too much risk, as the introductory example justifying a food tolerance standard for fat content.

We must add that we have adopted here a normative point of view. We have assumed that the regulator maximizes welfare computed with his own beliefs, not using consumers' different beliefs. This obviously raises the issue of the political feasibility of such a paternalistic policy. This policy indeed will likely be unpopular, as the paternalistic regulator's objective is, by definition, different from the citizens' one. In contrast, the populist policy maximizes citizens' perceived welfare and thus would maximize for instance the chances of the regulator to be reappointed in the short-run. A similar approach would be to consider a median voter model with beliefs heterogeneity. In this model, agents choose different consumption levels and their indirect utility function, under weak assumptions, is single peaked. The optimal norm level would then be chosen by the median of the distribution, that is by the agent having the median beliefs. Note however that a populist approach to risk regulation may have to face some ex post contestation if agents can get better information about the risks they face, e.g., by observing the early realizations of the damages. A long-term regulator should then anticipate this. His optimal strategy may in turn be more in line with the paternalist approach that we have considered in the paper. Related interesting discussions may be found in Maskin and Tirole (2004), who study the optimal incentive schemes for short-term or long-term politicians in presence of asymmetric information.

Finally, there is the classical positive question of whether safety efforts can be easily implemented. In this paper, we have abstracted from this problem. We have indeed assumed that the regulator has a perfect control over public safety investments. This amounts to assume that the regulator can perfectly monitor the firms involved in the provision of public safety, e.g., the food industry. Obviously, this is not true in practice, and, our risk regulatory model should clearly account for agency costs as well. Stimulated by some papers in this book, we even believe that this is the way to go to make the analysis of paternalistic policies more powerful. It seems, for instance, that many compulsory liability norms imposed by the state to firms permit to somehow circumvent problems posed by consumers' misperceptions.

References

Akerlof, G.A. and Dickens, W.T., 'The economic consequence of cognitive dissonance', *American Economic Review*, 72 (1982): 307-19.

Barigozzi, F. and Villeneuve, B., 'Influencing the misinformed misbehaver: An analysis of public policy towards uncertainty and external effects', (2002), *mimeo*, University of Toulouse.

Besley, T., 'A simple model for merit good arguments', *Journal of Public Economics*, 35 (1988): 371-383.

Breyer, S.G., *Breaking the Vicious Circle: Toward Effective Risk Regulation*, (Harvard University Press 1993).

Glazer, A., 'Using corrective taxes to remedy consumer misperceptions', *Journal of Public Economics*, 28 (1985): 85-94.

Gruber, J. and Koszegi, B., 'Tax incidence when individuals are time-inconsistent: the case of cigarette excise taxes', *Journal of Public Economics*, 88 (2004): 1959-1987.

Hahn, R., (Ed), *Risks, Costs, and Lives Saved - Getting Better Results from Regulation*, (Oxford University Press, 1996).

Hammond, P., 'Ex ante and ex post welfare optimality under uncertainty', *Economica*, 48 (1981): 235-50.

Harris, J.E., 'Taxing tar and nicotine', *American Economic Review*, 70 (1980): 300-311.

Hird, J., *Superfund: The Political Economy of Environmental Risk*, (John Hopkins University Press, 1994).

Kuran, T. and Sunstein, C., 'Availability cascades and risk regulation', *Stanford Law Review*, 51 (1999): 683-768.

Lichtenstein S., Slovic P., Fischhoff B., Layman B. and Combs, B., 'Judged Frequency of Lethal Events', *Journal of Experimental Psychology: Human Learning and Memory*, 6 (1978): 551-78.

Marshall, J.M., 'Welfare analysis under uncertainty', *Journal of Risk and Uncertainty*, 2 (1989): 385-403.

Maskin, E. and Tirole, J., 'The politician and the judge: Accountability in government', *American Economic Review*, 94 (2004): 1034-1054.

Musgrave, R., *The Theory of Public Finance*, (New York: McGraw-Hill, 1959).

O'Donoghue, T. and Rabin, M., 'Studying optimal paternalism, illustrated by a model of sin taxes', *American Economic Review, Papers and Proceeedings*, 93 (2003): 186-91.

Peltzman, S., 'The effects of automobile safety regulation', *Journal of Political Economy*, 83 (1975): 677-725.

Pollak, R.A., 'Regulating risks', *Journal of Economic Literature*, 33 (1995): 179-91.

Pollak, R.A., 'Imagined risks and cost-benefit analysis', *American Economic Review*, 88 (1998): 376-80.

Portney, P.R., 'Trouble in Happyville', *Journal of Policy Analysis and Management*, 11 (1992): 131-32.

Salanié, F. and Treich, N., 'Regulation in Happyville', (2003), *mimeo*, University of Toulouse.

Sandmo, A., 'Ex post welfare and the theory of merit goods', *Economica*, 50 (1983): 19-34.

Slovic, P., 'Perception of risk', *Science*, 236 (1987): 280-285.

Spence, M., 'Consumer misperceptions, product failure and producer liability', *Review of Economic Studies*, (1977): 561-572.

Tversky, A. and Kahneman, D., 'Judgment under uncertainty: Heuristics and biases', *Science*, 185 (1974): 1124-1130.

Viscusi, K.P., *Rational Risk Policy*, (Oxford University Press, 1998).

Viscusi, K.P, 'Risk Equity', *Journal of Legal Studies*, 29 (2000): 843-71.

Wirl, F., 'Paternalistic principals', *Journal of Economic Behavior and Organization*, 38 (1999): 403-19.

Chapter 4

Using Information from Insiders to Target Environmental Enforcement

Anthony Heyes
Royal Holloway College, University of London

Catherine Liston-Heyes
Royal Holloway College, University of London

4.1 Background - Targeting Regulatory Attention

Regulatory agencies in a variety of contexts are hampered by their lack of firm-specific information and shortage of enforcement resource. In this context it is natural that regulators will endeavour to use information bought to them from informed insiders to target investigative and enforcement actions. As we will evidence "whistleblower" clauses are incorporated in a number of well-known pieces of legislation that govern the operations of firms in high-risk environmental settings. Protections for insiders who disclose information to regulators are also protected (in different ways, and to differing extents) under both statute and common law.

Without offering a formal model we explore some of the issues that will determine how and when information from such sources will be useful to regulators, and how such information should be handled.

An approach routinely made in economic models of the enforcement/compliance game played between regulators and firms is to develop models in such a way that compliance inspections are essentially random. The regulated entity makes a binary compliance decision (comply/violate) or perhaps chooses a degree of non-compliance or adopts a mixed strategy. If before inspection firms are observationally-identical from the perspective of the enforcement agency, then the agency chooses a probability of inspecting a representative firm in a given period, effectively playing a mixed strategy in its game with any individual firm. Different objectives may be attributed to the regulator - maximizing some weighted or unweighted statement of social welfare, maximizing population compliance subject to a budget constraint, minimizing the social burden of achieving a certain aggregate level of pollution etc. There are many examples of such models - a survey is provided in the early parts of Heyes (2000).

From such a starting point it is natural to try think of ways in which an agency might better target or direct its investigative efforts to enhance the performance of its compliance program. In practice agencies have routinely develop practical ways to better focus their efforts in field settings. The management of this sort of targeting has only been subject to rigorous economic analysis comparatively recently.

Within a sector there are likely to be observable characteristics of firms on which inspection probability might sensibly be conditioned - such conditioning reflecting the statistical differences in propensity to violate. These might include size, age of capital stock, number of plants, nationality of ownership etc. Most theoretical models proceed on the (usually implicit) assumption that such conditioning on the basis of observables has been done, so that firms are observationally equivalent though may differ in terms of unobservable (typically cost of compliance). In some enforcement contexts issues may arise as to the legitimacy of targeting on the basis of particular characteristics.

Recognizing that most interactions between firms and regulatory agencies are repeated rather than single shot, work over the past two decades has pointed to the performance benefits that can accrue from basing enforcement strategy on the past performance of regulated entities. Of course, many legal systems base investigation and sentencing on the previous behavior of individuals, and the management of such targeting has been rigorously developed in the regulatory context. In an important and widely cited paper Harrington (1988) builds on the earlier theoretical work of Greenberg (1984) and Landsberger and Meilijson (1982). He shows that in a repeated binary compliance/enforcement game with restricted penalties the EPA maximizes the rate of steady-state compliance by operating a "state dependent" enforcement regime. in the simplest case the agency groups sources according to immediate inspection history. Group 1 contains firms found to be compliant when last inspected, group 2 those found non-compliant. Optimal policy involves levying no penalty upon a group 1 firm caught violating but a maximal penalty upon a group 2 firm caught likewise. In equilibrium a representative firm can be induced to comply a significant fraction of the time (i.e. whenever they find themselves resident in group 2) despite penalties *never actually being levied.*

State dependent models can be used to "explain" - or provides a theory consistent with - the paradox with which Harrington opens his paper. Namely, that despite the fact that (i) when the USEPA observes violations if often (almost always) chooses not to pursue the violator and, (ii) the expected penalty faced by a violator who is pursued is small compared to the cost of compliance, it remains the case that (iii) most firms comply most of the time. How does this work? A firm can be induced to comply some of the time even though the plausible limit on penalties is such that if all violations were penalized with penalties it would never do so. The (crude) state dependent regime described generates "penalty leverage". When in group 2 a source's incentive to comply is not just the maximal penalty which it avoids but also the present value of reinstatement in group 1 and the laxer treatment which this entails in the subsequent round. Whilst discussion for the purposes of illustration here has revolved around penalties, compliance-maximizing policy can be characterized by refining the crude regime to allow for differential rates of random inspection amongst group 1 and group 2 firms, and by making reinstatement to group 1 less-than-automatic.

Heyes (2002) offers a model of "filtered enforcement". That analysis pointed to the importance in setting enforcement policy (the intensity of audit programs, thoroughness of inspections, pursuant penalties etc.) of taking account of the "creation" of cases. In many or most contexts the cases of alleged pollution or excess risk-taking that end up being most thoroughly investigated by enforcement agencies will have been subject to some form of initial screen or filter, which builds a "2-stagedness" into the overall process. For example, an initial coarse screen may have been made by an inspector, who then makes a recommendation for a more thorough investigation or audit. For example, Lloyd-Bostock (1992) argues that:

> More attention should be paid to early screening or report-
> ing decisions and their role in the "creation" of cases. One such
> filter is the selection of accidents for investigation by British fac-
> tory inspectors. Large numbers of reports are scanned, and the
> overwhelming majority (about 95%) are rejected. Any subsequent
> enforcement action in response to violations, such as prosecution
> is contingent on this initial screening decision [...] yet that decision
> is made very quickly and on the basis of very scanty and imprecise
> information. (Lloyd-Bostock (1992: 7))

That first stage, coarse stage, may be something that the regulatory agency controls (e.g. as with the factory inspectors). Lloyd-Bostocks particular interest is in the psychology of routine discretion - how enforcement agents make these coarse screening decisions, and how those routine processes can impact the incentive characteristics of the whole enforcement regime.

Alternatively, it may be something outside its direct influence. For instance, in many settings much of an enforcement agency's work will be responsive or complaints-driven. In these cases the initial round of "monitoring" can be thought of as being done by members of the public, a subset of whom then report to the agency then they suspect - probably on the basis of not particularly accurate information - wrong-doing. The policy problem then becomes one of how to pursue that information.

The difficulty in setting up such a model or complaints-based enforcement is the need to provide a defensible assumption of why people complain. Naturally if one is the actual or prospective victim of damage one has the incentive to expend effort in complaint in order to seek redress (for past harm) or prevention (of future harm). Such self-motivation is straightforward to understand and articulate in a modelling context. But - in the absence of "bounties" being paid for information it is harder to see why *homo economicus* would (a) expend effort looking for and (b) go to the trouble of reporting alleged damage to an environmental "commons". As with other contribution games, the agents incentive to contribute to the public good becomes small as the agents share in the public resource becomes small.

The set-up in Heyes (2002) and other models of public citizens as complainers - providing informational raw material to the enforcement process - conceives of complaints or information coming neither from the polluter nor from the immediate victim from "concerned third parties". The (noisy) information that the concerned third party is assumed to have has regard to the environmental performance of the regulated entity.

It is natural from this discussion to think that we might wish to move on to how other groups of individuals might be "co-opted" into the regulatory process, what might motivate their participation, and what the properties of the resulting regime might look like. A natural group to think of in this sort of context are organizational insiders. Employees and others inside the firm might be expected to have privileged information about the firms activities and plans - as they relate to the natural environment - and the implications for the firm of changing those plans.

4.2 Information from the Informed Insider - "Whistleblowing"

The phenomenon of "whistleblowing" - a term that is in common parlance, a more concrete definition of which we will turn to shortly - features quite prominently in the regulatory landscape in the United States and elsewhere.

In the United States whistleblowing clauses are included in pieces of legislation such as the Occupational Health and Safety Act (OSHA) of 1970, the Clean Air Act Amendments (CAAA) of 1977 and the Financial Institutions Reform Act of 1989. Other environment-related Acts incorporating employee protection provisions are the Safe Drinking Water Act (SDWA), Toxic Substances Control Act (ToSCA) and Comprehensive Environmental Response, Compensation and Liability Act (CERCLA) - all enacted in the 1970s.

An example of the encouragement of whistleblower disclosures in a supernational setting is the "CDM-Watch" program which monitors happenings under the clean development mechanism (CDM) of the Kyoto Protocol. Their website (www.cdmwatch.org/whistleblowers) encourages: "If you have confidential information about a CDM project or related CDM matter which you think we should know about follow these steps", and goes on to provide instructions on how to render e-mail communication untraceable.

Whilst the term whistleblowing is widely-understood, a precondition for more careful analysis is a firmer definition. For our purposes the definition provided by Glazer and Glazer (1989) has three key elements, associated with the italicized words in the following. They define the whistleblower as one who acts to *prevent* harm to *others*, not him or herself, while possessing *evidence* that would convince a reasonable person. Though other definitions exist this is a frequently-cited one, and is the one that Heyes (2003) adopts in setting-up a game theoretic model of the whistleblower - firm - regulatory agency game. Hunt (1997) offers an alternative definition: "Whistleblowing is the public disclosure, by a person working within an organization, or acts, omissions, practices or policies perceived as morally wrong by that person and is a disclosure regarded as (or said to be, or treated as) wrongful by that organizations authorities".

Hunts definition is different to Glazer and Glazer's in almost every way. There is no presumption that the disclosure should *prevent* harm - it could be about passed and now extinct activities. It does not require that the damage done by the wrongdoing be to *others*. There is no presumption that the whistleblower has externally convincing evidence. Similarly a strong assumption is made about the motives of the discloser, namely that the disclosure have regard to something "morally wrong". Glazer and Glazer make no judgement on motives which seems to us more appropriate, and it is their definition that we adopt for purposes of discussion and analysis.

The biggest hurdle in setting-up an economic model of compliance based on disclosure by whistleblowers, is the need to explain why disclosure is likely to happen at all. Whistleblowing may well be a privately costly action but the benefits accrue (by definition) to others, so the behavior is not easily explained under conventional assumptions about rational, self-interested behavior. The model presented here explores alternative behavioral motivations and is in the spirit of the recent behavioral law and economics literature associated with scholars such as Cass Sunstein, Richard Zerbe, Thomas Ulen and others (see, for examples, the work contained in Sunstein (2000)).

4.2.1 What Motivates Whistleblowers?

A number of attempts have been made by scholars in fields other than economics (especially sociology, psychology) to identify a whistleblower "type" - a profile of the sorts of individuals and sorts of circumstance in which whistleblowing is likely to occur. Correlations with observables such as social class, religion and educational attainment have been shown to offer only quite modest explanatory power.

The same applies to empirical research on rescuers (of which whistleblowers are a particular sub-category) more generally. "Rescuers come from all corners and sectors of social structure, thereby calling the bluff of those believing there to be largely social determinants of moral behavior" (Bauman (1989:5)). This is reminiscent of the well-known findings of Milgram (1974), and others since, that there were no convincing social psychological correlations with disobedience.

In an interesting recent book, Alford (2001) develops a theory of whistleblowing that builds on concepts of individuality. Individuals, he concludes, are caught between loyalties to their employers, society and ego. He proposes a novel ethical category which he calls "narcissism moralized". In summary, "(n)arcissism becomes moral when the self's commitment to the highest ideals is based in the internalized image of an ideal self, so bound to its ideals that there is in the end no difference between the ideal self and ideals of the self [...] (w)histleblowers disclose because they dread living with a corrupted self more than they dread the other outcomes" (Alford (2001: 90)). This description does not, however, tell us the behavior that those highest ideals prescribe. In a legal context it does not tell us in what circumstances, if ever, violation of a law or regulation is defensible. The answers to such questions rely on individual moralities.

Hunt (1997) proposes that a "justifiable disclosure" must at minimum (a) serve some purpose in correcting or preventing harm and (b) do more harm than good. The definition of whistleblowing we adopt implies that (a) is necessarily satisfied. (b) implies some weighing up of social costs and benefits and implies, importantly, that the disclosure decision is forward-looking. Hunt notes that in practice the whistleblower may find it difficult or impossible to ascertain whether (b) is satisfied. Conceptually, though, he notes that: "[...] all the well-rehearsed arguments for and against utilitarian calculation could be invoked at this point. The moral codes of some will lead them to take the view that it reasonable to make a disclosure simply and only because 'it is the right thing to do' even if harmful consequences are known to be more likely than beneficial ones. A whistleblower in this position might feel, for example, that they are answerable to God who will judge them *only for following moral principles of honesty and fortitude, not for the consequences of the rightful act*" (Hunt (1997:2), italics added). Hunt also regards the rectification criterion as open to question: "The whistleblower, as we have seen, may not be concerned so much with the consequences of the disclosure as with simply making the truth known, because it is the truth", though the evidence remains that typically whistleblowers appeal to rectification to justify their disclosures.

In terms of unobservable characteristics of an individual psyche which might be expected to determine the propensity to disclose, psychologists offer the concepts of 'imagination for consequences' and "doubling". Doubling takes place when a part of the self develops an ability to act or think quasi-autonomously. Ethical qualms that our whole self might have can be ignored or suspended at work because a "work self" temporarily prevails. Individuals are likely to differ is in their ability to 'double' (Lifton (1986)). Doubling is a sophisticated emotional act and individuals differ in the extent to which they are able to insulate their work self from their whole being, and therefore tolerate behavior from the former which would be unconscionable to the latter.

A much popular view of whistleblowers - particularly, perhaps, in the business community - is that they are what Alford refers to as "hysterical malcontents". Such individuals motivated by a desire to punish the organization of which they are part. Some may be unhappy because of the way they are of have been treated within the organization and ready to take the opportunity to "hurt" their employer when the chance arises. Others might want to punish because of the firms planned actions or omissions in the regulatory context itself, and this sort of motivation would relate to the notion of punishment as an end in itself (see Fehr and Schmidt (1999) and Fehr and Falk (2002) for experimental evidence).

4.2.2 Economic Considerations

There is, then, a substantial lack of consensus in the non-economic literature regarding what "drives" whistleblowing.

In thinking about the way in which an enforcement regime should be set-up to process and respond to information sourced in this way some concrete assumption or assumptions need to be made about motivations. Motivation will likely determine both the *sorts of contexts* in which evidence is forthcoming and those in which it isn't, and also how the flow of reports and information from whistleblowers is likely to respond to changes in the way in which the information is handled.

Heyes (2003) develops a formal model of regulatory enforcement with information being brought to the regulator by whistleblowers according to some sort of behaviorally-based motivation function. The framework allows for systematic examination of how policy might be influenced by the motivation attributed to disclosers. He proposes a "generic" regulatory setting in which a principal may instruct an agent to engage - on behalf of the firm - in an illegal act. The agent is then given the opportunity to "blow the whistle" - to report the proposed wrong-doing to the regulator - for possible prosecution.

Importantly, there is a period between the decision to violate and violation actually occurring (and damage being done). This provides the opportunity for disclosure to *prevent harm being done* in line with Glazer and Glazer's (1989) definition. The regulatory agency receives the report and decides whether it wishes to pursue it. Pursuit means that the non-compliance is prevented and the firm is subject to a fine f for intent to violate. Indeed, it is interesting to note that any schema without some lag between the decision being taken to violate, and the actual event of violation necessarily precludes whistleblower activity. Insiders who choose to report wrongdoing after it has happened may be performing a valid public service - in, for example, facilitating compensation of victims - *but they are not blowing the whistle.*

The equilibrium of the model, along with policy recommendations, are developed under alternative assumptions regarding what motivates the prospective whistleblower.

In the most general case we would expect that any individual's decision to disclose may be sensitive to any or all of the "moving parts" of the compliance/enforcement environment - in this case c_i (the firm's compliance cost), d_i (the external damage that the violation would impose), π (the probability that a report from a whistleblower will be pursued and f (the penalty imposed for intent to violate) - in addition to individual emotional/moral/behavioral characteristics.

Heyes posits a general "motivation function" that describes the probability that an individual will disclose when faced with a particular set of these variables and parameters,

$$\rho_\Delta(c_i, d_i, \pi, f),$$

where the subscript Δ denotes may denote particular behavioral assumptions in play.

In the previous Section we presented a short summary of some parts of the relevant literatures (from psychology, sociology, etc.) on *why* people become whistleblowers. A wider reading suggests three preponderant alternative views about why employees blow the whistle on errant employers: (a) conscience-cleansing (a moral inability to be complicit with wrongdoing), (b) social motivation (based on a calculation of social costs and benefits) and (c) the desire to punish the wrongdoer.

Because of the difficulty of arriving at a consensus view we conduct our analysis under three different assumptions, proposing simple decision rules which operationalize each of these.

Without going through detailed derivation we can think here about how such behavioral motivations can be operationalized.

For summary purposes, we will focus here on the first and third.

The prospective violator, recall, takes a decision to impose external environmental damage of value d in order to save private costs c_i. An individual can take a view on the morality of that decision - and decide whether or not his conscience allows him to "live" with that decision - without reference to the enforcement environment. The impact of non-compliance is fixed at d, so we can regard c_i as a measure of the defensibility of the firm's decision, then assume that an individual has some threshold of defensibility beyond which his conscience compels him to speak out.

Choosing to not comply if c is very small (even zero) would be a particularly indefensible thing for a firm to choose to do. It would - under the current vision of what drives behavior - be a decision particularly likely to prompt disclosure. As c gets larger it is more defensible that the firm would opt to violate, though individuals would be expected to differ in their "forgiveness". Heyes operationalizes this by making *Assumption* α, which states that individual i will disclose a planned violation if and only if $c_i < \mu_i$.

The μ_i term can be thought of as capturing an element of individual i's moral code. In particular, it is some sort of "threshold of conscienability". The probability that planned violation by firm i will be disclosed to the regulator can then be described as some function of the defensibility of the act, $\rho_\alpha(c_i)$, with other arguments surpressed. If μ_i's vary across individuals according to some single-peaked distribution then ρ_α will reflect the cumulative of that distribution - an upward sloping ogive.

Other things equal, then, under this sort of moral defensibility assumption regarding the behavioral basis for whistleblowing, a firm with a low cost of compliance who chooses to violate is *more* likely to be reported than a high cost firm. This is simply because such a decision is - it is contended - less morally defensible in that case and therefore it is more likely that an employees "threshold of conscienability" will be transgressed.

The firm, then, chooses to comply upfront if

$$c_i \leq \rho_\alpha(c_i).\pi.(c_i + f) \qquad (4.1)$$

The left-hand side is the cost of compliance, the right-hand side is the expected cost of non-compliance. Compliance incentives - the net expected benefits from compliance - decrease monotonically with c_i and the firm will comply voluntarily if its cost is less than some critical value $\widehat{c}(\pi, f|\alpha)$ implicitly defined by setting the above expression to equality.

The policy questions are two-fold: (a) How responsive should the regulatory agency be to reports from whistleblowers? And (b) how should firms caught through such disclosures be punished?

Assume that the regulator acts to minimise social loss, defined as the sum of expected compliance costs and external damage. The regulator is assumed to want to minimise social loss:

$$SL(\pi, f|\alpha) = \int_0^{\widehat{c}} c_i g(c) + \int_{\widehat{c}}^{\infty} [\rho_\alpha(c_i)\pi c_i + (1 - \rho_\alpha(c_i)\pi)d] \, g(c) dc \qquad (4.2)$$

Firms with compliance costs below $\hat{c}$ comply voluntarily. Firms with costs above $\hat{c}$ will comply only if coerced. This occurs with probability $\rho_\alpha(c_i)\pi$. Otherwise the firm will be left to execute its plans for violation, imposing external damage d.

The policy problem faced by the agency is how to set the enforcement variables, π and f. How often to follow-up on disclosures made to them by whistleblowers, and how penal to be on firms caught in this way. In one of the Propositions in his paper, Heyes (2003) establishes that:

Proposition 4.1 *If whistleblowing is motivated by conscience cleansing, then optimal policy will be characterized by (a) a maximal penalty and (b) an inspection intensity set less than maximally. This applies even though inspection is costless.*

Significantly, alternative behavioral assumptions generate qualitatively quite different policy prescriptions.

Consider, for example, the case in which a certain proportion of staff might be "disgruntled employees", unhappy for reasons unconnected with the firms planned non-compliance with the regulation, but opportunistic in blowing the whistle when so doing creates sufficiently substantial discomfort (i.e. cost) for their employer. This would correspond with a populist view of whistleblowers - perhaps amongst the business community - as corporate vandals. What Alford (2001:18) refers to as the popular conception of the whistleblower as "corporate malcontent".

In an inarguably rather simplistic way we can operationalize such a notion of punishment-motivated disclosure by making *Assumption* γ: An employee discloses planned violation if and only if the expected cost impact upon the firm is sufficiently large: $\pi.(c_i + f) > \delta_i$. The left hand side here captures the cost to the firm of having the whistle blown on their proposed wrongdoing - with probability π the report is followed up by the regulator in which case the firm has to comply (at cost c_i) and pay the fine for intent to violate (f). Again we propose that the corporate vandal might have some sort of motivational threshold δ_i - they want to hurt their firm and will report if so doing hurts enough.

Again, these thresholds are likely to vary across a population of employees (or the attitude of a particular employee may vary over time). An implication of this sort of assumption is that the probability that planned violation will be disclosed is *increasing in the cost of compliance*. Other things equal a firm that would find being brought into compliance particularly expensive or inconvenience is more likely to be reported by an individual subject to this sort of behavioral motivation, because it is the imposition of that expense or inconvenience that motivates the complaint.

It is also now increasing in the two enforcement parameters. Prospective whistleblowers want to punish, and so increased expected penalty increases the likelihood of report.

Without presenting the manipulation we can report that Heyes arrives at the following result:

Proposition 4.2 *If whistleblowing is punishment motivated, for any given probability that disclosure will be pursued the optimal penalty for planned violation may be less than maximal.*

This is not necessarily too surprising. The information being brought forward and offered to the regulator now is "less valuable" in some sense. In the earlier case it tended to be the low cost firms that were disclosed - precisely the ones which a welfare-motivated agency would like to visit and coerce into compliance. Here the pattern of complaints is skewed the other way - the reports are likely to be about firms that are disproportionately at the high cost end of the distribution. These are precisely the firms at which the social gains from coercing compliance are likely to be low, or where compliance may actually be welfare-reducing.

4.2.3 Empirical Evidence on Motivation

Given that our belief about what "drives" observers of wrongdoing to become disclosers or whistleblowers, robust empirical evidence would be particularly valuable. There is no agreed model of whistleblowing amongst academicians in this area. Gundlach et al (2003) provides a nice overview of the empirical evidence currently available, and the interested reader is directed there.

It is an occupational hazard of model-building, of course, that the categorizations adopted for the purposes of analytics do not always correspond cleanly with the outputs of empirical and case study work. One or two examples are worth highlighting.

Gundlach et al (2004) test a model of the whistleblowing process. Their results showed that when organizational wrongdoing was attributable to a controllable cause there was a positive influence on propensity to disclose. The set-up used in the analysis here regards the external damage as always controllable but *at a cost*. Controllability is, then, a continuous rather than binary variable. The conscience cleansing version of the model presented here implies that increases in that variable will increase the likelihood of disclosure. These authors go on to emphasize the role that *anger* plays in stimulating disclosure, where a more controllable (preventable) incident stirs up greater anger amongst observers.

The prosocial perspective emphasizes that whistleblowing is a form of altruistic behavior, performed with the intention of promoting welfare. For this reason researchers have sought to investigate relationships between disclosure and measures ordinarily thought to correlate with pro-social behavior, for example the subjects level of cognitive moral development. With regard to cost-benefit analysis Miceli and Near (1985: 542) "see some evidence of a subjectively rational decision process, [...] whereby observers of wrongdoing weight costs and benefits of taking action."

4.3   Conclusions

In formulating policy about how to respond to information brought forward from informed insiders ("whistleblowers") it is critical to think about what the motivation for disclosure are likely to be. A rich non-economics literature points to several potential behavioral bases. The types of models envisaged here fall into the category of "behavioral law and economics", a comparatively new field (see Sunstein 2000a, 2000b, 2003 for examples and surveys).

Injecting scope for whistleblowing into a "generic" model of compliance/enforcement and manipulating it under alternative behavioral assumption scan generate some interesting results. Worryingly, in terms of policy formulation, policy depends qualitatively upon the assumption made.

A few general points are worth underlining. First, the value of the information that whistleblowers bring to the enforcement agency - and what the agency will wish to do with that information - depends upon the motives assumed to whistleblowers. If the motive is either conscience cleansing (or welfarist) then whistleblowers will be more likely to report a planned act of violation at a firm with low compliance costs. These are the cases in which the agency would find it beneficial to coerce compliance. If, on the other hand, the motivation is punishment or "spite" then other things equal a case is more likely to be disclosed at a firm where compliance costs are high, precisely those cases where coerced compliance is of least (or even negative) social value.

Second, in adjusting the enforcement instruments - π and f - attention has to be paid to the change induced in the flow of disclosures. Again, the quantitative and qualitative response will depend upon whistleblower motives. Under conscience cleansing the propensity for individuals to disclose isn't sensitive to the enforcement regime. Under other assumptions they plausibly are, as we have shown.

The model contained in Heyes (2003) and outlined here has been set-up in an unashamedly "behavioral" way - in effect trying to move complex and ill-understood social psychological motives into simple, operational decision rules. It is natural to ask (if you are economist) whether all or any of these heuristics for behavior could be derived from self-interested behavior in, for example, a model of repeat interaction. Whether this is theoretically possible remains for future research. But experimental and other evidence of the past decade should caution us against thinking that behavior apparently driven by emotion, or concepts of fairness, altruism, rebuke or conscience can necessarily be subsumed as special cases of self-interest, except in a trivial sense.

References

Alford, C.F., *Whistleblowers*, (Ithaca: Cornell University Press, 2001).

Bauman, Z., *Modernity and Ambivalence*, (Ithaca: Cornell University Press, 1989).

Fehr, E. and Falk, A., 'Psychological Foundations of Incentives', *European Economic Review*, 46(4-5) (2002): 687-724.

Fehr, E. and Schmidt, K.M., 'A Theory of Fairness, Competition and Cooperation', *Quarterly Journal of Economics*, 114(3) (1999): 817-68.

Glazer, M.P. and Glazer, P., *The Whistleblowers*, (New York: Basic Books, 1989).

Greenberg, J., 'Avoiding Tax Avoidance', *Journal of Economic Theory*, 32(1) (1984): 1-13.

Gundlach, M., Douglas S. and Martinko, M., 'The Decision to Blow the Whistle: A Social Information Processing Framework' *Academy of Management Review*, 28(1) (2003): 107-34.

Gundlach, M., Douglas S. and Martinko, M., 'Testing a Cognitive-Emotional Model of Whistleblowing Decisions', (2004), *Academy of Management Conference Paper*.

Harrington, W., 'Enforcement Leverage when Penalties are Restricted', *Journal of Public Economics*, 37(1) (1988): 29-53.

Heyes, A., 'Implementing Environmental Regulation: Enforcement and Compliance', *Journal of Regulatory Economics*, 17(2) (2000): 107-29.

Heyes, A., 'A Theory of Filtered Enforcement', *Journal of Environmental Economics & Management*, 43(1) (2002): 34-46.

Heyes, A., 'Whistleblowers', Royal Holloway, (2003), University of London *Working Paper*, #03-10: London.

Hunt, G., 'Whistleblowing', *The Encyclopedia of Applied Ethics*, (New York: Academic Press, 1997).

Landsberger M. and Meilijson, I., 'Incentive Generating State Dependent Penalty Systems: The Case of Income Tax Evasion', *Journal of Public Economics*, 19(3) (1982): 333-52.

Lifton, R.J., *The Nazi Doctors: Medical Killing and the Psychology of Genocide*, (Cambridge MA: Harvard University Press, 1986).

Lloyd-Bostock, S., 'The Psychology of Routine Discretion', *Law & Policy*, 14(1) (1992): 45-76.

Near J.P. and Miceli, M.P., 'Organizational Dissidence: The Case of Whistleblowing', *Journal of Business Ethics*, 4 (1985): 1-16.

Miethe, T., *Whistleblowing at Work: Tough Choices in Exposing Fraud, Waste and Abuseon the Job*, (Boulder, Colorado: Westview Press, 1999).

Milgram, S., *Obedience to Authority*, (New York: Harper & Row, 1974).

Sunstein, C.R., *Behavioral Law & Economics*, Cambridge Series on Judgement and Decisionmaking, (Cambridge University Press, 2000).

Sunstein, C.R., 'Moral Heuristics', John M. Olin Law & Economics, *Working Paper*, #180, (2003), University of Chicago Law School.

Sunstein, C.R., Schkade D. and Kahneman, D., 'Do People Want Optimal Deterrence?', *Journal of Legal Studies*, 29(1) (2000): 237-253.

Chapter 5

Environmental Protection, Consumer Awareness, Product Characteristics, and Market Power[*]

Marcel Boyer
Université de Montréal

Philippe Mahenc
Université de Perpignan

Michel Moreaux
Université de Toulouse

5.1   Introduction

Textbooks generally claim that a monopoly is more environmental friendly than a competitive industry (see, for example, Kolstad (2000)). As far as polluting emissions are positively related to production, by restricting output to extract more surplus from consumers, the monopolist tends to reduce emissions. The present paper departs from this fairly simple idea by showing that a monopolist can increase emissions while restricting output. We argue that, in the presence of environmentally aware consumers, a monopolist may internalize, at least in part, the damage caused by the pollution she emits. However, it is only the damage incurred by the consumers who actually purchase the good that is of concern to the monopolist. Thus, the externality is only partially internalized under the sole pressure of market forces. As a result, the monopolist produces too little and pollutes too much.

When markets are not perfectly competitive, environmental safety may be a fairly complex issue to address. This was pointed out by Buchanan (1969)

---

[*]We thank Bernard Caillaud, Philippe Bontems and an anonymous referee for their comments. Financial support from CIRANO (Canada), CNRS (France) and INRA (France) is gratefully acknowledged. We remain of course solely responsible for the content of this paper and of its shortcomings.

and then emphasized by Barnett (1980) and Baumol and Oates (1988). The reason is that the exercise of market power already imposes on society the cost of output and price distortions. If in addition those distortions are combined with the generation of pollution, the resulting social damage yields a further problem of efficiency. The present paper is related to the literature on environmental policy in a context of imperfect competition as developed among others by Levin (1985), Conrad and Wang (1993), Carraro, Katsoulakos and Xepapadeas (1996), and Innes and Bial (2002). The oligopolistic paradigm is quite realistic for addressing the environmental question in such markets as electric and other public utilities (see Baron (1985)), coal mining, chemicals, motor vehicles, among others.

Furthermore, the analysis of a polluting monopolist has been confined to contexts in the monopolist is not directly concerned by the consumers' valuation of a cleaner or safer environment. Recently, several articles have appeared which consider or show that consumers are willing to pay higher prices for products that generate less environmental harm; see for example Carraro and Soubeyran (1996), Cason and Gangadharan (2000), Foulon, Lanoie and Laplante (2001), and Bansal and Gangopadhyay (2003). Due to this environmental awareness, pollution by a firm may shift its demand downward. In such a context, a producer enjoying market power has an indirect incentive to reduce pollutant emissions if consumers, aware of the public bad nature of pollution generated by the production of the goods they consume, modify downward their consumption plans. The consumers' purchasing behavior communicates to the producer their preferences concerning the private good and the associated public bad. This is likely to make pollution more costly for a producer endowed with market power than for a perfectly competitive producer. The present paper shows that the emergence of consumers' environmental awareness plays a crucial role in the monopolist's internalization of the externality due to pollution.

To get further insight along this line, we investigate the behavior of a polluting monopolist facing no threat of entry in a market *à la* Hotelling. Consumers appreciate the good supplied by the monopolist but they doubly suffer from pollution: production causes a global damage affecting consumers and non-consumers alike, and consumption causes a specific damage affecting consumers only. The monopolist makes three decisions concerning respectively the product characteristics or variant, the pollution intensity, and the price. Depending on the monopolist's choices, the market is fully or partially covered. The market is fully covered when everyone on the Hotelling interval is a consumer of the good even if pollution generates a utility loss from both production (general) and consumption (specific).

We compare two contexts in terms of product variant, intensity of pollution and market coverage. The first context is the standard unregulated monopolist choosing the profit maximizing product variant, pollution intensity and market coverage or price. The second one is the monopolist subject to environmental regulation, under which the level of pollution intensity, or the production technology, which is here completely determined by the pollution intensity, is chosen by a regulator while the product variant and the market coverage are chosen by the monopolist. This modeling strategy is certainly not the most general conceivable but it has the advantage of being quite explicit in the variables under the control of the firm or the regulator and to be prone to more general albeit tractable formulations.[1]

We show that the monopolist chooses the same variant, whether she is regulated or not. The private and the social incentives to choose the variant that maximizes the global consumers' surplus coincide. The unregulated monopolist proposes the socially most appealing variant of the product in order to extract the largest possible surplus from consumers. However, once the variant has been chosen, the private and the social incentives for production and pollution levels may not coincide.

Confronted with environmentally aware consumers, the monopolist anticipates how her pricing and polluting behavior affects the purchasing decisions. Unlike a price-taking competitive producer, the unregulated monopolist has the power to make consumers pay for pollution abatement. We show here that the monopolist pollutes less when she is confronted with consumers that are more environmentally aware. However, if the market is not fully covered, she may then serve more or less consumers but always at a higher price relative to what would prevail in the absence of consumption-specific damage. If the market is fully covered, she raises her price as consumers are more environmentally aware if and only if her chosen pollution intensity level is relatively elastic with respect to the consumption-specific damage level. Hence, the unregulated monopolist internalizes part of the externality associated with pollution, namely that part associated with the consumption-specific damage.

Nevertheless, it is the socially efficient global damage from pollution that is of concern to the environmental regulator. As a result, the unregulated monopolist generates pollution up to the level at which the marginal benefit in terms of reduced production costs equals the marginal consumption-specific damage. The regulator on the other hand chooses a pollution intensity level such that the marginal benefit, again in terms of reduced production costs, equals the marginal social global damage.

Whatever the market coverage, the unregulated monopolist pollutes more but produces no more than the monopolist subject to an environmental regulator. If the efficient market coverage is partial, the unregulated monopolist produces strictly less, pollutes strictly more, and charges a lower price than the monopolist subject to environmental regulation. If the unregulated monopolist were covering the whole market, she still would do it when subject to environmental regulation. If she were not covering the whole market, she

---

[1]We develop some of those more general formulations in our companion paper Boyer, Mahenc and Moreaux (2004).

would increase production when subject to the regulatory pollution intensity standard.

A noteworthy conclusion is that, as a result of environmental regulation, the monopolist always raises the price of her product but never reduces production. This is a striking result: in the presence of consumers who are environmentally aware and of producers who have market power (monopoly in the present case), the implementation of a socially optimal pollution intensity standard leads to both higher prices and larger production. Our analysis identifies two reasons why. First, a stricter standard of pollution intensity increases the consumers' surplus since the latter are environmentally aware, hence there is a larger part of this surplus that is likely to be captured by the firms exercising their market power (the monopolist here) through higher prices and more consumers served. Second, the equilibrium price reflects the increase in marginal production costs due to the stricter standard of pollution intensity.

5.2 The Model

Consider an industry in which the range of potential product varieties is represented by a Hotelling interval $[0, 1]$. There is a single private good, characterized by its variant $a \in [0, 1]$, produced by a protected monopolist (no threat of entry). As a by-product of the private good, the monopolist produces a bad that is nonexcludable and nonrival in consumption,[2] such as a greenhouse gas: all consumers are subject to the environmental harm and a consumer's consumption of the public bad imposes no costs or benefits on its consumption by others. Distinction will be made between the global damage and the consumption-specific damage caused by emissions. Emissions are transformed to ambient concentrations of pollution generating a global damage affecting all individuals whether they consume the product or not and, moreover, emissions cause a consumption-specific damage or risk such as exposure to a toxic substance, which affects only those who consume the product.[3]

Let e be the intensity of pollution defined as the amount of pollution per unit of the good produced. Total emissions are then $E = eq$ where q is the quantity produced. The marginal cost of producing the good is represented by $c(e)$. The function $c(e)$ is assumed to be convex in e and to reach a minimum at $\bar{e}$, that is, $c''(e) > 0$ and $c'(\bar{e}) = 0$.[4] Let $d(E)$ denote the individual

---

[2] A nonrival bad is not depletable in the sense of Baumol and Oates (1988).

[3] Tietenberg (2000) claims that: "Some 55 000 of the potential substances that could prove toxic are in active use" (p. 493). Pesticides and other chemicals as well as food additives may cause chronic illnesses not only for those directly and indirectly in contact with them but also for the general public, albeit with a smaller incidence.

[4] This is a reasonable assumption to make. It says that once the pollution level $\bar{e}$ is reached, there is no more net benefits to be captured, the firm itself suffering from its own pollution.

damage due to pollution; it could be interpreted either as the individually perceived cost of ambient pollution, or as the expected personal cost of an environmental accident whose probability of occurrence increases with E, or as the expected personal cost of an environmental accident whose damage, if the accident occurs, is an increasing function of E.[5] The function $d(E)$ can also be viewed as the consumers' willingness to pay for a clean and safe environment. We will refer to E as pollution in the present paper. We will reiterate in the conclusion the relevance of our results in the context of major industrial risk.

Consumers are represented by their most preferred product variant, and so are located in the Hotelling interval. We will assume that they are uniformly distributed over $[0,1]$ with a density of 1. There is a "preference gap" between a consumer x (located at x) and the supplied variant a; we assume that this gap is measured by the linear function $t\,|x-a|$ where t is a positive constant.[6] All consumers have the same gross reservation value r for the product. Let $\beta > 0$ denote a parameter measuring the (constant) marginal consumption-specific damage of the product. The consumer characterized by the most preferred variant x (that is, located at x in the characteristics space) derives the indirect utility

$$u(E,a,x) \equiv \begin{cases} r - (1+\beta)d(E) - t\,|x-a| - p, \\[1em] \qquad \text{if he buys the product variant } a \text{ offered at price } p, \\[1em] -d\,(E), \quad \text{if he does not buy.} \end{cases}$$

$$(5.1)$$

Hence, those who do consume and those who do not consume have different willingness to pay for reducing the pollution generated by production, namely $(1+\beta)d(E)$ for the former and $d(E)$ for the latter.[7] Hence, $r - (1+\beta)d(E)$ denote the consumers' net willingness to pay (NWP) for the product, that is, net of their willingness to pay for a cleaner and/or safer environment $(1+\beta)d(E)$, while the non-consumers' willingness to pay for a cleaner and/or safer environment is $d(E)$. Each consumer buys one unit of the product if and only if it is offered to him at a full price (the product price plus the "preference

---

[5]From Boyer and Dionne (1983), we know that a risk averse agent will prefer a reduction in the magnitude of loss to a reduction in the probability of loss when both generate the same reduction in expected loss. The reason is that the former is a mean preserving transformation (negative mean-preserving spread) of the latter.

[6]As suggested by a referee, one could also consider that the pollution intensity level e affects the preference gap factor, $t = t(e)$; specific formulations could take the form $k(e)t$ or simply et. We do not pursue this alternative modeling strategy in this paper but we intend to do it in a sequel paper.

[7]In the context of product safety, it is usual to suppose that consumers of dangerous products have a NWT for improved safety which is larger than the NWP of non-consumers for similar improved safety; see for instance Daughety and Reinganum (2003).

gap" cost) which is less than his NWP differential between consuming the product and non-consuming it, that is $r - \beta d(E)$. We will assume a specific form for $d(E)$, namely $d(E) \equiv E$ in order to ease the presentation. Hence, $(1 + \beta)$ represents how much the individual consumer of the product would be willing to pay for one unit reduction in the pollution generated while the individual non-consumer would be willing to pay 1 for the same unit of pollution reduction.

Let us now derive the demand curve. Without loss of generality, we can assume that $a \leq \frac{1}{2}$. A consumer buys the product if he derives more utility in consuming than in not consuming. If the consumer located at $x = 1$, who suffers the largest "preference gap" cost, derives a positive surplus by purchasing from the monopolist, then the market is covered ($q = 1$, $E = e$), that is the price satisfies $p \leq r - \beta e - t(1 - a)$. For $p > r - \beta e - t(1 - a)$ some consumers are worse off buying.[8] As long as there is a single consumer who is indifferent between buying or not, his location x (to the right of $\frac{1}{2}$) verifies:

$$r - \beta E = r - \beta e x = p + t \left| x - a \right|. \tag{5.2}$$

This is the case when $r - \beta e - t(1 - a) \leq p < r - a(t + 2\beta e)$. The solution of (5.2) is then given by

$$x\left(p, e, a\right) \equiv \frac{r - p + ta}{\beta e + t}. \tag{5.3}$$

In such a case, the potential but unserved consumers are located on the right hand side of the market only. For $p = r - a(t + 2\beta e)$, equation (5.2) has a lower and an upper root, respectively 0 and

$$q(p, e) \equiv 2 \, \frac{r - p}{2\beta e + t}. \tag{5.4}$$

For higher levels of p, that if $r - a(t + 2\beta e) < p \leq r$, the market coverage is given by $q(p, e)$ and unserved consumers can now be found on both sides of the market. In that case, the market coverage is symmetric with respect to the product variant a: it extends from $a - q(p, e)/2$ to $a + q(p, e)/2$ and thus the level of sales no longer depends on a (See Figure 5.1).

It follows that the demand function is given by

$$D\left(p, e, a\right) = \begin{cases} 1 & \text{if } 0 \leq p \leq r - \beta e - t(1 - a), \\ x\left(p, e, a\right) & \text{if } r - \beta e - t(1 - a) \leq p \leq r - a(t + 2\beta e), \\ q(p, e) & \text{if } r - a(t + 2\beta e) \leq p \leq r, \\ 0 & \text{if } r \leq p. \end{cases}$$
$$\tag{5.5}$$

---

[8]They are nevertheless affected by pollution E.

Hence,

$$D_p\,(p,e,a) = \begin{cases} 0 & \text{if } r \le p \text{ or } 0 \le p \le r - \beta e - t(1-a), \\ -1/(\beta e + t) & \text{if } r - \beta e - t(1-a) < p < r - a(t + 2\beta e), \\ -2/(2\beta e + t) & \text{if } r - a(t + 2\beta e) < p \le r. \end{cases}$$

$$(5.6)$$

5.3 The Protected Monopolist

The monopolist makes three decisions: the product characteristics or variant a, the pollution intensity e, and the price p. We assume that those decisions are made in a two stage set-up: first the product variant a and pollution intensity e are chosen simultaneously in stage 1 and then the price p in stage 2. There are many justifications for such a modeling strategy. The product characteristics (variant) choice and the technological (pollution intensity) choice are long term decisions involving important sunk costs once incurred. Price on the other hand may be considered as quite flexible. In a perfect information context with no uncertainty (our case), all decisions would be made simultaneously since there is no development of any kind between the different stages and moreover, decisions once taken will not be revised. In such a world, flexibility has literally no (real options) value and irreversibility has no cost. Considering sequential decisions is tantamount to imposing a sequence of decisions under certainty to mimic the sequence of decisions under uncertainty or imperfect information when information on market evolution or changes is gathered over time.[9] The price will be determined as a function of product variant a and pollution intensity e. Given the pricing decision function, the choice of a and e can be characterized.

5.3.1 The Pricing Decision

Let $\pi(p, e, a)$ denote the profit for the monopolist:

$$\pi(p,e,a) \equiv (p - c(e))D(p,e,a). \tag{5.7}$$

From (5.5), the profit function is continuous in p. Moreover:

---

[9]Consider the following framework. Demand (in our case the value of r) is uncertain. The firm observes signals as time goes by which reduces the uncertainty about r. Since the decisions on product variant and technology take time (it takes time to determine the proper characteristics of the product and to install the chosen technology, that is, product variant and production/pollution technology are somewhat irreversible), the firm must decide on product variant and pollution intensity when it is relatively uninformed about r. Later, the firm observes a signal on r, so that when it makes its decision on price, it is better informed if not perfectly so. The analysis of such a context is done in our companion paper Boyer, Mahenc and Moreaux (2004).

Lemma 5.1 *Given a and e, the profit function $\pi(p, e, a)$ is strictly concave in p and piecewise differentiable.*

Proof: Let $\pi_p^-(p^0, e, a)$ and $\pi_p^+(p^0, e, a))$ be the left-hand and right-hand partial derivatives of the profit function with respect to p at $p = p^0$. From the demand function given in (5.5), the function $\pi(p, e, a)$ is strictly concave in p on each interval where it is differentiable. Thus, it remains to show that the function $\pi(p, e, a)$ is strictly concave in the neighborhood of each point where it is not differentiable. Clearly, $\pi(p, e, a)$ is strictly increasing in the interval $[0, r - \beta e - t(1 - a)]$ and $\pi_p^-(r - \beta e - t(1 - a), \cdot) = 1$. The derivative of $\pi(p, e, a)$ is $\pi_p(p, e, a) = D(p, e, a) + (p - c(e))D_p(p, e, a)$, where $D_p(p, e, a)$ is given by (5.6). First, $\pi_p^+(r - \beta e - t(1 - a), e, a) < 1$ since, for all $p \in (r - \beta e - t(1 - a), r - a(t + 2\beta e))$, $D(p, e, a) < 1$ and $D_p(p, e, a) < 0$. Hence, $\pi(p, e, a)$ is strictly concave in the neighborhood of $r - \beta e - t(1 - a)$. Second, from (5.6), $\pi_p^-(r - a(t + 2\beta e), e, a) > \pi_p^+(r - a(t + 2\beta e), e, a)$. Hence $\pi(p, e, a)$ is strictly concave in the neighborhood of $r - a(t + 2\beta e)$. ∎

Hence there is a unique profit maximizing price for the monopolist, denoted by $\hat{p}(e, a)$. The different cases are depicted in Figures 5.2A-5.2D. Defining the critical location points (critical variants) a_1, a_2, a_3 as follows:

$$a_1 \equiv \frac{2\beta e + 2t - (r - c(e))}{t}, \quad a_2 \equiv \frac{r - c(e)}{3t + 4\beta e} > 0, \quad a_3 \equiv \frac{r - c(e)}{2(t + 2\beta e)} > 0,$$

$$(5.8)$$

we obtain the following expressions for the profit maximizing price.

Proposition 5.1

1. *If $2\beta e + 2t \le r - c(e)$, that is, $a_1 < 0$, the profit maximizing price is*

$$\hat{p}(e, a) = r - \beta e - t(1 - a) \quad (5.9)$$

   and the market is fully covered (Figure 5.2A).

2. *If $2\beta e + \frac{3}{2}t \le r - c(e) < 2\beta e + 2t$, that is, $a_1 > 0$ and $a_3 > a_2 > \frac{1}{2}$, then:*

   (a) *for $a \in [0, a_1)$, the profit maximizing price is*

$$\hat{p}(e, a) = (r + ta + c(e)) / 2 \quad (5.10)$$

   and the market is well covered on the left side (the consumer at $x = 0$ strictly prefers to buy) but not on the right side (Figure 5.2B);

   (b) *for $a \in [a_1, \frac{1}{2}]$, the profit maximizing price is given by (5.9) and the market is fully covered (Figure 5.2A).*

3. *If* $2\beta + te \leq r - c(e) < 2\beta e + \frac{3}{2}t$, *that is,* $a_1 > \frac{1}{2}$, $a_2 < \frac{1}{2}$, *and* $a_3 > \frac{1}{2}$, *then:*

   (a) *for* $a \in [0, a_2)$, *the profit maximizing price is given by* (5.10) *and the market is well covered on the left side but not on the right side* (Figure 5.2B);

   (b) *for* $a \in \left[a_2, \frac{1}{2}\right]$, *the profit maximizing price is*

   $$\widehat{p}(e, a) = r - a(t + 2\beta e) \tag{5.11}$$

   and the market is just barely covered on the left (the consumer at $x = 0$ *is indifferent between buying or not)*[10] *but not on the right side, except at* $a = \frac{1}{2}$ *(Figure 5.2C).*[11]

4. *If* $0 \leq r - c(e) < 2\beta e + t$, *that is,* $a_1 > \frac{1}{2}$ *and* $a_2 < a_3 < \frac{1}{2}$, *then:*

   (a) *for* $a \in [0, a_2)$, *the profit maximizing price is given by* (5.10) *and the market is well covered on the left side but not on the right side* (Figure 5.2B);

   (b) *for* $a \in [a_2, a_3)$, *the profit maximizing price is given by* (5.11) *and the market just barely covered on the left side but not on the right side* (Figure 5.2C);

   (c) *for* $a \in \left[a_3, \frac{1}{2}\right]$, *the profit maximizing price is*

   $$\widehat{p}(e, a) = (r + c(e))/2 \tag{5.12}$$

   and the market is covered neither on the left side nor on the right side (Figure 5.2D).

Note that in (5.9) and (5.11), the price is independent of the marginal cost $c(e)$ while in (5.10) and (5.12), the price is independent of the marginal consumption-specific damage β.

From the above, we can derive the effects (partial derivatives) on price $\widehat{p}$ of pollution intensity e, product variant a, consumption-specific damage β, and preference heterogeneity factor t; they are summarized in the following proposition.

Proposition 5.2 *The monopolist's price* $\widehat{p}(e, a)$ *is:*

- *a decreasing function of e in all cases;*

---

[10] The consumer at $x = 0$ is indifferent because, for $\widehat{p}(e, a) = r - a(t + 2\beta e)$ and $x(p, a, e)$ given by (5.3), we have $\widehat{p}(e, a) + ta = r - \beta ex$.

[11] The consumer at $x = 1$ buys the good when $a = \frac{1}{2}$ because, for $\widehat{p}(e, \frac{1}{2}) = r - \frac{1}{2}t - \beta e$, we have $r - \beta e = \widehat{p}(e, \frac{1}{2}) + \frac{1}{2}t$.

- *a non-monotonic function of a: the price is* increasing *in a when either the market is fully covered (expression (5.9)) or covered on the left but not on the right (expression (5.10)),* decreasing *in a when the market is just barely covered on the left side but not on the right side (expression (5.11)), and* independent *of a when the market is covered neither on the left side nor on the right side (expression (5.12));*

- *non-increasing with respect to consumption-specific damage β (corresponding to the partial derivative $\frac{\partial \widehat{p}}{\partial \beta}$, hence for given e and a): the monopolist's price* decreases *with β when either the market is fully covered or the market is just barely covered on the left side and not on the right side (expression (5.9) and (5.11)); otherwise the price is* independent *of β (expression (5.10) and (5.12)).*

- *a non-monotonic function of t (corresponding to the partial derivative $\frac{\partial \widehat{p}}{\partial t}$, hence for given e and a): the price is* decreasing *in t when either the market is fully covered (expression (5.9)) or just barely covered on the left but not on the right (expression (5.11)),* increasing *in t when the market is well covered on the left side but not on the right side (expression (5.10)), and the price is* independent *of t when the market is covered neither on the left side nor on the right side (expression (5.12));*

When pollution intensity is lower (larger unit pollution abatement), there are two forces that complement one another to yield a price increase: first, as consumers are more willing to pay for the product, the monopolist can extract more surplus from them, and second, marginal production costs are higher. Proposition 5.1 shows that the reasons why the monopolist's price $\widehat{p}(e, a)$ is decreasing in e differ in a subtle way according to whether the whole market is covered or not. When the whole market is covered (cases 1 and 2b in Proposition 5.1), the price is given by (5.9) with $\frac{\partial \widehat{p}(e, a)}{\partial e} = -\beta$ and therefore the monopolist increases her price to extract the additional consumer's surplus generated by the reduction in pollution. On the other hand, when the market is uncovered at least on one side (cases 2a, 3a, 4a and 4c in Proposition 5.1), the price is given by (5.10) or (5.12) with $\frac{\partial \widehat{p}(e, a)}{\partial e} = \frac{1}{2}c'(e)$ and therefore the monopolist raises her price as a reaction to the increase in marginal production costs.

Two conflicting forces explain the non monotonicity of the price as the variant a moves toward the market center. One deals with the "preference gap" cost, the other with the gain of market coverage (assuming it is not complete) as price decreases. The monopolist can set a higher price (hence capture a larger part of the surplus) if the "preference gap" cost decreases and the "preference gap" cost decreases as a moves toward the market center. Such is the first force favoring a positive relationship between the price and the product variant as the latter moves toward the market center.[12] When the market is just barely covered on the left side and not covered on the right side (cases 3b and 4b in Proposition 5.1), then the monopolist finds it profitable to lower her price as a moves to the right in order to keep selling to the consumers on the left and gaining more consumers on the right. When a reaches a_3, then the two forces balance each other and the monopolist keeps her price constant for product variants $a > a_3$ (case 4c of Proposition 5.1). The price function is illustrated in Figure 5.3.

When the market is fully covered, the lower the consumption-specific damage factor β, the higher the price charged by the monopolist since there is a larger consumer's surplus to be captured. The monopolist captures the whole surplus of end-point consumers ($x = 0$ and $x = 1$) when $a = \frac{1}{2}$. Interestingly enough, the price remains unchanged for variations in the consumption-specific damage factor β when the market is uncovered on both sides. In this case, the price elasticity of demand is, from (5.4) and (5.6), given by $\frac{p}{r - p}$ and does not directly depend on β (price p is considered as given). A larger β shifts demand downward in such a way that for each given price, the price elasticity remains the same. Consequently, the monopolist does no longer take into account the consumption-specific damage when choosing her price.

For the same reason, the consumers' preference heterogeneity parameter t has no influence on the monopolist's price when the market is uncovered on both sides. On the other hand, when the market is fully covered, the monopolist's price decreases as t increases. The reason is that there is less consumers' surplus to extract when preferences are more heterogeneous.

5.3.2 Simultaneous Choice of the Product Variant and the Pollution Intensity

As we mentioned before, the product characteristics (location) and the technology characteristics (pollution intensity) are both relatively inflexible once chosen and are therefore considered here as long run variables. They are chosen in a first stage followed in the second stage by the pricing decision characterized above.

The Choice of Product Variant

Recognizing that whatever the values of a and e chosen, the price p will be chosen to maximize profit, the profit function can now be written in reduced form as

$$\widehat{\pi}(a, e) \equiv \pi(a, e, \widehat{p}(a, e))$$

---

[12]The rate at which $\hat{p}(e, a)$ increases may be t (cases 1 and 2b of proposition 5.1) or $\frac{1}{2}t$ (cases 2b, 3a and 4a of proposition 5.1).

From Proposition 5.1, we obtain that the reduced-form profit function for stage 1, namely $\hat{\pi}(a, e)$, can take four different forms, where a_1, a_2 and a_3 are given by (5.8).

1. $\hat{\pi}(a, e) = r - \beta e - t(1 - a) - c(e)$ when
   either $2\beta e + 2t \leq r - c(e)$
   or $\{2\beta e + \frac{3}{2}t \leq r - c(e) < 2\beta e + 2t \text{ and } a \in [a_1, \frac{1}{2}]\}$;

2. $\hat{\pi}(a, e) = [(r + ta + c(e))/2 - c(e)]D((r + ta + c(e))/2, e, a)$ when
   either $\{2\beta e + \frac{3}{2}t \leq r - c(e) < 2\beta e + 2t \text{ and } a \in [0, a_1)\}$,
   or $\{0 \leq r - c(e) < 2\beta e + \frac{3}{2}t \text{ and } a \in [0, a_2)\}$

3. $\hat{\pi}(a, e) = [r - a(t + 2\beta e) - c(e)]q(r - a(t + 2\beta e), e)$ when
   either $\{2\beta e + t \leq r - c(e) < 2\beta e + \frac{3}{2}t \text{ and } a \in [a_2, \frac{1}{2})\}$
   or $\{0 \leq r - c(e) < 2\beta e + t \text{ and } a \in [a_2, a_3)\}$;

4. $\hat{\pi}(a, e) = [(r + c(e))/2 - c(e)]q((r + c(e))/2, e) = [(r - c(e))/2]q((r + c(e))/2, e)$ when
   $$\{0 \leq r - c(e) < 2\beta e + t \text{ and } a \in [a_3, \frac{1}{2})\}.$$

To characterize the optimal variant $\hat{a}$, we can perform a case-by-case analysis, following Proposition 5.1. In what follows, the value of e is considered as given while p is given by $\hat{p}(a, e)$. We will argue that $\hat{a} = \frac{1}{2}$.

1. If $2\beta e + 2t \leq r - c(e)$, then $\hat{\pi}(a, e) = r - \beta e - t(1 - a) - c(e)$. Hence the monopolist is better off choosing $\hat{a} = \frac{1}{2}$ for all e.

2. If $2\beta e + \frac{3}{2}t \leq r - c(e) < 2\beta e + 2t$, then the monopolist strictly prefers $\hat{a} = \frac{1}{2}$ also. Indeed:

   (a) For $a \in [0, a_1)$, we have $\hat{\pi}(a, e) = (\hat{p}(e, a) - c(e))x(\hat{p}(e, a), e, a)$ strictly increasing in a from (5.3).

   (b) For $a \in [a_1, \frac{1}{2}]$, the market is fully covered and $\hat{\pi}(a, e) = r - \beta e - t(1 - a) - c(e)$ is strictly increasing in a.

3. If $2\beta e + t \leq r - c(e) < 2\beta e + \frac{3}{2}t$, the monopolist is indifferent between all product variants in $[a_2, \frac{1}{2}]$. Indeed:

   (a) For $a \in [0, a_2)$, we have $\hat{\pi}(a, e)$ strictly increasing in a for the same reason as in case 2a above.

   (b) For $a \in [a_2, \frac{1}{2}]$, the profit $\hat{\pi}(a, e)$ is given by $(\hat{p}(e, a) - c(e))x(\hat{p}(e, a), e)$ is strictly increasing in a.

4. If $0 \leq r - c(e) < 2\beta e + t$, then, whatever e, we have $\hat{\pi}(a, e)$ strictly increasing in a for $a \in [0, a_2)$, increasing in a for $a \in [a_2, \frac{1}{2}]$, and constant for variants in $[a_3, \frac{1}{2}]$.

To determine a unique profit maximizing product variant for the monopolist, we must introduce either a variant cost or a refinement concept to identify the most likely product variant among all those which maximize profit. Introducing a variant cost in the space of characteristics may be somewhat arbitrary unless we can derive it from empirical observations which can only be specific to the industry or product class considered.[13] In the context we have considered so far in this paper, the preferred route is clearly to introduce a refinement concept. It is reasonable to assume that when the monopolist is indifferent between a set of product variants, she chooses the one which maximizes global consumer surplus, that is the surplus of all consumers, actual and potential. Given that the "preference gap" cost is linear, this means that the monopolist will choose a product variant as close as possible to $\frac{1}{2}$, the center of the market. Hence,

Proposition 5.3 *The monopolist always chooses a product variant at the market center.*

The monopolist always choose the product variant that is the most appealing to consumers, that is, the variant which maximizes the interest of potential consumers in the product. In other words, she chooses the variant which minimizes the total "preference gap" cost over all potential consumers.

The Choice of the Pollution Intensity

Maximizing her profit given a and e, the monopolist chooses the price $\hat{p}(e, a)$ such that $\pi_p(\hat{p}(e, a), e, a) = 0$.[14] It follows that

$$D(\hat{p}(e, a), e, a) = -(\hat{p}(e, a) - c(e))D_p(\hat{p}(e, a), e, a). \tag{5.13}$$

Now the variant $\hat{a} = 1/2$ is chosen independently of the pollution intensity e and we can characterize the monopolist's choice of e as maximizing the reduced-form profit function

$$\hat{\pi}(e, 1/2) \equiv (\hat{p}(e, 1/2) - c(e))D(\hat{p}(e, 1/2), e, 1/2)$$

with respect to e. The following analysis shows that the profit function $\hat{\pi}(e, 1/2)$ attains a unique maximum which is denoted by $\hat{e}$.

From the analysis leading to Proposition 5.2, we know that the monopolist will, conditional on e, either covers the whole market or leaves some consumers unserved on both sides of the market.[15] From Proposition 5.1, we know that

---

[13]One could claim that given that $a = \frac{1}{2}$ is appealing to more people, it would be reasonable to expect that it is more expensive to design.

[14]The profit function $\pi(\hat{p}(e, a), e, a)$ is not always differentiable with respect to p at its maximum. At such point, the left derivative is positive and the right derivative is negative.

[15]When $a = \frac{1}{2}$, both sides of the market are symmetric.

the market will be fully covered when the monopolist chooses $\hat{a} = \frac{1}{2}$ and pollution intensity e in the closed interval

$$\mathcal{E}_M \equiv \{e \in \mathbb{R}^+ \mid r - c(e) - t - 2\beta e \geq 0\},$$

while choosing the profit maximizing price $\hat{p}(e, 1/2)$ in stage 2. Let e_1 and e_2 be respectively the left endpoint and the right endpoint of $\mathcal{E}_M$,[16] that is, the minimal and maximal pollution intensities for which the market is fully covered.

For matter of simplicity and to concentrate on the more interesting cases, we will make four specific assumptions, which could clearly be relaxed at the cost of a more lengthy and complex analytical treatment. The first assumption says that the production costs under no pollution would be high enough that the monopolist would not cover the whole market:

Assumption 5.1 $r - c(0) - t < 0.$

It follows from Proposition 5.1(4c) that $\hat{p}(0, 1/2) = (r + c(0))/2$. We will also assume that the monopolist's profit increases with e for $e \leq e_1$, namely:

Assumption 5.2 $\pi_e(e, 1/2) > 0$ for all $e \leq e_1$.

Hence, the profit maximizing pollution intensity will either be in $\mathcal{E}_M$ or to the right of e_2. To concentrate on the more interesting and relevant cases, we will assume that for $\beta = 0$ (no consumption-specific damage or no environmentally aware consumers), the monopolist, choosing the pollution intensity $\bar{e}$ at which $c'(e) = 0$, would not cover the whole market (while charging a price $\hat{p}(\bar{e}, 1/2)$ given by (5.12)), namely:

Assumption 5.3 $e_2 < \bar{e}.$

Finally, as a sufficient condition for the second-order conditions to be satisfied (below), we will assume that the cost function is sufficiently convex in the following sense:

Assumption 5.4 For all e, $c'(e)^2 < (r - c(e))c''(e).$

We first consider the maximization of $\hat{\pi}(e, 1/2)$ in $\mathcal{E}_M$. For all e inside $\mathcal{E}_M$, the market is fully covered and therefore the reduced-form profit function is $\hat{\pi}(e, 1/2) = r - \beta e - t/2 - c(e)$, which is concave in e due to the convexity of $c(e)$. Hence, the function attains a unique local maximum in $\mathcal{E}_M$, which we shall denote by $\tilde{e}$. If $\tilde{e}$ is in the interior of $\mathcal{E}_M$, then

$$\hat{\pi}_e(\tilde{e}, 1/2) = -\beta - c'(\tilde{e}) = 0. \tag{5.14}$$

---

[16]The set $\mathcal{E}_M$ expands with reservation value r and shrinks with the preference-gap cost factor t and with the marginal consumption-specific damage β.

Let us now consider the maximization of $\widehat{\pi}(e,1/2)$ outside $\mathcal{E}_M$. The market is then covered neither on the left side nor on the right side. Given $\widehat{a}=1/2$, the monopolist's profit is

$$\widehat{\pi}(e,1/2) = (\widehat{p}(e,1/2) - c(e))q(\widehat{p}(e,1/2),e)$$

where $\widehat{p}(e,1/2) = (r+c(e))/2$ (case 4c of Proposition 5.1). We get

$$\widehat{\pi}_e(e,1/2) = (\widehat{p}(e,1/2) - c(e))q_e(\widehat{p}(e,1/2),e) - q(\widehat{p}(e,1/2),e)c'(e)$$

$$+\pi_p(\widehat{p}(e,1/2),e,1/2)\widehat{p}_e(e,1/2).$$
(5.15)

From the envelope theorem, the indirect effect of e on $\widehat{\pi}(e,1/2)$ through the change in price is zero since the optimal price $\widehat{p}(e,1/2)$ satisfies $\pi_p(\widehat{p}(e,1/2),e,1/2) = 0$. But there is a conflict between the two effects captured by the first two terms of (5.15). The first term is a demand effect: From (5.4), increasing the pollution intensity level shifts demand downward and reduces profit. The second term is a cost effect: Increasing the pollution intensity level reduces the production cost and increases profit. Substituting $D(\widehat{p}(e,\widehat{a}),e,a) = q(\widehat{p}(e,1/2),e)$ in (5.13), we obtain: $\widehat{p}(e,1/2) - c(e) = -\left(q(\cdot)/q_p(\cdot)\right)$. The derivative $\widehat{\pi}_e(e,1/2)$ can then be written as, using $\widehat{p}$ for $\widehat{p}(e,1/2)$ when no confusion is possible:

$$\widehat{\pi}_e(e,1/2) = q(\widehat{p},e)\left(-\frac{q_e(\widehat{p},e)}{q_p(\widehat{p},e)} - c'(e)\right).$$
(5.16)

Using (5.4) and (5.6), we obtain:

$$\widehat{\pi}_e(e,1/2) = q(\widehat{p},e)\left(-\beta q(\widehat{p},e) - c'(e)\right).$$
(5.17)

Hence, the first order condition for profit maximization outside $\mathcal{E}_M$ yields

$$-\beta q(\widehat{p},e) - c'(e) = 0,$$
(5.18)

(from which, $\widehat{e} < \overline{e}$) and the second order condition requires

$$\widehat{\pi}_{ee}(e,1/2) = -2\beta q(\widehat{p},e)q_e(\widehat{p},e) - q_e(\widehat{p},e)c'(e) - q(\widehat{p},e)c''(e) < 0. \quad (5.19)$$

The monopolist's production level as a function of e is given by

$$q(e) \equiv q(\widehat{p}(e,1/2),e) = q\left(\frac{r+c(e)}{2},e\right) = \frac{r-c(e)}{2\beta e + t}.$$
(5.20)

Hence,

$$q_e(\widehat{p}(e,1/2),e) = \frac{-c'(e) - 2\beta q(e)}{2\beta e + t};$$
(5.21)

substituting the first-order condition (5.18) into equation (5.21) yields

$$\frac{dq((\widehat{p}(e,1/2),e)}{de}\bigg|_{e=\widehat{e}} = q_e(\widehat{e}) = -\frac{\beta q(\widehat{e})}{2\beta\widehat{e}+t} < 0. \tag{5.22}$$

Condition (5.22) holds for $e = \widehat{e}$ but not in general.

Using (5.20) and (5.21), the second order condition given by (5.19) is equivalent to

$$(2\beta q(\widehat{p},e) + c'(e))^2 - (r - c(e))c''(e) < 0, \tag{5.23}$$

which is satisfied under Assumption 5.4. Furthermore, we have $\widehat{\pi}_e^-(e_2,1/2) = -\beta - c'(e_2) = \widehat{\pi}_e^+(e_2,1/2)$ from (5.17), which means that the profit function is differentiable at e_2. Hence, under Assumptions 5.1 to 5.4, (5.14) and (5.18), one of the following two cases appears at the right endpoint e_2 of $\mathcal{E}_M$:

1. either we have $-c'(e_2) \leq \beta$, in which case $\widehat{\pi}(e,1/2)$ is maximized at $\widetilde{e}$ in $\mathcal{E}_M$ and therefore $\widehat{e} = \widetilde{e}$ and $-c'(\widehat{e}) = \beta$;

2. or we have $-c'(e_2) > \beta$ and $\widehat{\pi}(e,1/2)$ is maximized at $\widehat{e}$ above e_2 (but below $\overline{e}$), where, from (5.18), $-c'(\widehat{e}) = \beta q(\widehat{p},\widehat{e})$.

Proposition 5.4A *The profit maximizing intensity of pollution $\widehat{e}$ chosen by the monopolist satisfies*

$$\beta D(\widehat{p},\widehat{e},1/2) = -c'(\widehat{e}). \tag{5.24}$$

The monopolist chooses a level of pollution intensity e that is lower than the level a competitive producer would choose.

The monopolist chooses to generate a level of pollution at which the marginal benefit of pollution in terms of a reduced production cost, $-c'(\widehat{e})$, equals the marginal consumption-specific damage from pollution incurred by all served consumers, $\beta D(\widehat{p},\widehat{e},1/2)$. This strongly contrasts with the behavior of a price-taking competitive producer who would generate a level of pollution $e^c = \overline{e}$ at which the marginal benefit of pollution in terms of a reduced production cost $c'(e^c)$ is zero. Unlike the competitive producer, the monopolist takes into account the consumers' willingness to pay for a clean and/or safe environment. She recognizes the effect of pollution on the NWP and of the NWP on her profit maximizing price, to the extent that the detrimental effect of pollution has a consumption-specific component. The monopolist internalizes the externality, at least in part, because she behaves strategically with respect to consumers and properly takes into account their reaction to changes both in the pollution intensity level and the price level.

Proposition 5.4B

- *If $-c'(e_2) \leq \beta$, then $D(\widehat{p},\widehat{e},1/2) = 1$.*

- If $-c'(e_2) > \beta$, then $D(\widehat{p}, \widehat{e}, 1/2) = q(\widehat{p}(\widehat{e}, 1/2), \widehat{e}) < 1$.

An interpretation of condition $-c'(e_2) \leq \beta$ is that the marginal benefit of pollution (in reducing production costs) is lower than the marginal consumption-specific damage when the market is not fully covered, since $-c'(e) < -c'(e_2)$ for all $e \in [e_2, \overline{e})$. In this case, the monopolist is better off reducing pollution intensity until the whole market is covered: the consumers' relatively large willingness to pay to reduce pollution is large enough to dominate the negative impact of the increase in production costs. On the other hand, when $-c'(e_2) > \beta$, the marginal consumption-specific damage β is smaller than the marginal benefit of pollution $-c'(e)$ when $e < e_2$, that is, when the market is fully covered. In this case, the monopolist is better off increasing pollution intensity above e_2 even though she losses some consumers (among the more reluctant to buy her product).

When the market is fully covered (production or market coverage is constant at 1) for $e = \widehat{e}$, the intensity of pollution $\widehat{e}$ also represents the monopolist's total pollution which thus increases with $\widehat{e}$. When the market is not fully covered, the monopolist's overall level of pollution $E(\widehat{e}) = \widehat{e}q(\widehat{p}, \widehat{e})$ increases with pollution intensity $\widehat{e}$ since, using (5.22),

$$
\begin{aligned}
E_e(\widehat{e}) &= q(\widehat{p}, \widehat{e}) + \widehat{e}q_e(\widehat{p}, \widehat{e}) \qquad\qquad\qquad (5.25)\\
&= \frac{t + \beta\widehat{e}}{2\beta\widehat{e} + t} q(\widehat{p}, \widehat{e}) > 0.
\end{aligned}
$$

As previously seen, the total derivative $q_e(\widehat{p}, e)$ evaluated at $e = \widehat{e}$ is negative, showing that a higher intensity of pollution $\widehat{e}$ has both a positive direct effect on total pollution, captured by the first term in the right hand side of (5.25), and an indirect negative effect through the decrease in the monopolist's production level. The direct effect of e on total pollution dominates the indirect effect. As a result, a lower intensity of pollution in the neighborhood of $\widehat{e}$ results in a wider market coverage (if not already fully covered), a lower global level of pollution, and a higher price since $\widehat{p}(e, 1/2) = \dfrac{r + c(e)}{2}$ decreases with e in the interval $[e_2, \overline{e})$. Hence:

Proposition 5.4C *Evaluated at $e = \widehat{e}$, the relation between the global level of pollution E and the pollution intensity level e is positive.*

Indeed,

$$
\left.\frac{dE}{de}\right|_{e=\widehat{e}} = \left.\frac{d[eq(\widehat{p}(e, 1/2), e)]}{de}\right|_{e=\widehat{e}} < 0.
$$

The proposition below summarizes the results for the unregulated monopolist.

Proposition 5.5 *The unregulated monopolist behaves as follows.*

1. *If at the right endpoint of $\mathcal{E}_M$, we have $-c'(e_2) \leq \beta$, then the unregulated monopolist covers the whole market by choosing a product variant at the center of the Hotelling market, a pollution intensity level $\hat{e}$ satisfying $\beta = -c'(\hat{e})$, and a price $\hat{p}(\hat{e}, 1/2) = r - \beta\hat{e} - t/2$. Hence $\hat{p}_e(\hat{e}, 1/2) < 0$ and production is constant at 1, in which case the global level of pollution $E(\hat{e})$ is equal to $\hat{e}$.*

2. *If at the right endpoint of $\mathcal{E}_M$, we have $-c'(e_2) > \beta$, then the unregulated monopolist leaves consumers unserved on both sides of the market by choosing a product variant at the center of the market, a pollution intensity level $\hat{e} < \bar{e}$ satisfying $\beta q(\hat{p}(\hat{e}, 1/2), \hat{e}) = -c'(\hat{e})$, where*
$$q(\hat{p}(\hat{e}, 1/2), \hat{e}) = \frac{r - c(\hat{e})}{t + 2\beta\hat{e}} = -\frac{c'(\hat{e})}{\beta}, \text{ and a price } \hat{p}(\hat{e}, 1/2) = \frac{r + c(\hat{e})}{2}.$$
   Hence, $\hat{p}_e(\hat{e}, 1/2) < 0$ and $\hat{q}_e(\hat{p}(\hat{e}, 1/2), \hat{e}, 1/2) < 0$; moreover, the global level of pollution $E(e) = eq(\hat{p}(e, 1/2), e)$ is increasing in e at $e = \hat{e}$.

The monopolist's price always decreases in e, albeit for different reasons according to whether the monopolist fully covers the market or not. First, when the monopolist covers the whole market, a reduction in pollution intensity induces her to capture an additional surplus from consumers since they are willing to pay more for a cleaner product. Thus, the monopolist raises her price. Second, when the monopolist finds it more profitable not to cover the whole market, the reduction in pollution intensity, raising marginal production costs, induces the monopolist to raise also her price and, interestingly enough, to produce more.

As mentioned by an anonymous referee, the present result that the monopolist raises her price as pollution intensity decreases, is not specific to the horizontal differentiation in consumers' taste. This result may be true, for instance, in the following vertical differentiation context which is reminiscent of Mussa and Rosen (1978) or Gabszewicz and Thisse (1979):[17] consumers have heterogeneous reservation values for the product given by r/e, where r is uniformly distributed over $[0, R]$; they derive a surplus $r/e - p$ from purchasing the good at price p; hence, the demand function is given by $1 - ep/R$. Moreover, the interested reader can check that the result that the monopolist's price declines with pollution intensity also holds with structures of the demand function that are more general than the linear ones, to the extent that demand decreases with e.

---

[17] We are grateful to Philippe Bontems for suggesting this example.

5.3.3 The Impact of the Consumption-Specific Damage Factor β

Given the parameters of the problem at hand, the monopolist's choices of product characteristics, pollution intensity and price are the result of three forces: market power, the positive benefit of pollution intensity in terms of cost reduction, and the negative impact of pollution on the consumers' willingness to pay for the product. To measure the impact of the consumption-specific damage factor β on the monopolist's choice of pollution intensity and price, let us define the following elasticity η of pollution intensity with respect to β: $\eta \equiv \frac{\beta}{\hat{e}}\frac{d\hat{e}}{d\beta}$. We show below that this elasticity is negative regardless of the market coverage. Thus, the higher the consumption-specific damage (or the more environmentally aware consumers are), the lower the pollution intensity chosen by the unregulated monopolist. Moreover, we will show that the absolute value of η determines whether the consumption-specific damage has a positive or a negative effect on the monopolist's price, that is, whether $\frac{d\hat{p}}{d\beta}$ is positive or negative. Using Proposition 5.5, we obtain

Proposition 5.6

- *The pollution intensity level $\hat{e}$ decreases with β.*

- *If $-c'(e_2) \leq \beta$, the market is fully covered and the monopolist's price increases with β if $\eta < -1$ while it decreases with β if $\eta > -1$.*

- *If $-c'(e_2) > \beta$, the market is partially covered and the monopolist's price increases with β while the production level increases with β if $\eta < -2$ while it decreases with β if $\eta > -2$.*

Proof: As long as the market is fully covered, then from (5.9), we have

$$\frac{d\hat{p}(\hat{e}, 1/2)}{d\beta} = -\hat{e} - \beta\frac{d\hat{e}}{d\beta} = -\hat{e}(1 + \eta).$$

Taking then the total differential of (5.14) yields $d\hat{\pi}_e(\hat{e}, 1/2) = -d\beta - c''(\hat{e})d\hat{e} = 0$, that is

$$\frac{d\hat{e}}{d\beta} = -\frac{1}{c''(\hat{e})} < 0; \tag{5.26}$$

hence the elasticity η is negative and the first part of the proposition follows if the market is fully covered. On the other hand, when some consumers on both sides are unserved, then, from (5.12), we have

$$\frac{d\hat{p}(\hat{e}, 1/2)}{d\beta} = \frac{1}{2}c'(\hat{e})\frac{\hat{e}}{d\beta} = \frac{1}{2}c'(\hat{e})\frac{\hat{e}}{\beta}\eta,$$

whose sign depends on η. To obtain the expression for η in this case, let us substitute $q(\hat{p}, e)$ from (5.20) in equation (5.24), which yields $\beta \dfrac{r - c(e)}{2\beta e + t} + c'(e) = 0$, and take the total differential to obtain

$$\frac{d\hat{e}}{d\beta} = -\frac{r - c(\hat{e}) + 2\hat{e}c'(\hat{e})}{\beta c'(\hat{e}) + c''(\hat{e})(t + 2\beta\hat{e})}.$$

From (5.20), we get that $q(\hat{p}(\hat{e}, 1/2), \hat{e}) = -\dfrac{c'(\hat{e})}{\beta}$. Using this equation yields

$$\frac{d\hat{e}}{d\beta} = \frac{tq(\hat{p}(\hat{e}, 1/2), \hat{e})^2}{c'(\hat{e})^2 - (r - c(\hat{e}))c''(\hat{e})} < 0 \text{ by Assumption 5.4;}$$

hence $\eta < 0$ and $\dfrac{d\hat{p}(\hat{e}, 1/2)}{d\beta} > 0$. Moreover, from (5.20) and (5.22), we obtain

$$\frac{dq(\hat{p}, \hat{e})}{d\beta} = -2\hat{e}\frac{r - c(\hat{e})}{(2\beta\hat{e} + t)^2} - \frac{\beta q(\hat{e})}{2\beta\hat{e} + t}\frac{d\hat{e}}{d\beta}$$

$$= \frac{\hat{e}q(\hat{e})}{2\beta\hat{e} + t}(-2 - \eta);$$

hence the remainder of the proposition follows. ∎

Increases in the consumption-specific damage factor β generates two conflicting effects on the monopolist's price. The direct effect is that the monopolist charges a lower price as there is less surplus to extract from consumers. The indirect effect is that the monopolist chooses a lower level for pollution intensity, which increases both consumers' surplus and production costs. The increase in consumers' surplus in turn relaxes the downward pressure on price.

Consider first that the market is fully covered. The indirect effect dominates the direct effect on price provided that η is larger than one in absolute value ($\hat{e}$ is β-elastic): the monopolist raises her price as consumption-specific damage increases. By contrast, when η is less than one in absolute value ($\hat{e}$ is β-inelastic), the monopolist reduces her price as consumption-specific damage increases.

Consider now the case where the market is uncovered on both sides. As previously seen, the price elasticity of demand does not depend on β. Consequently, the direct effect on price of an increase in β vanishes and the indirect effect is the only active one. Thus, confronted with an increase in β, the monopolist reduces pollution intensity, raises her price because she incurs higher production costs, and at the same time increases production and market coverage provided that $\hat{e}$ is sufficiently β-elastic, namely η is larger than two in absolute value.

5.4 The Regulated Choice of the Pollution Intensity

Suppose that a regulator can determine the emission intensity level e, leaving the choice of a and $p(a, e)$ to the monopolist. This is a second-best problem which concentrates on the efficiency of the pollution abatement control as the sole mean of command. Environmental regulation is done in most countries through a separate regulatory authority, which operates more or less independently from other regulatory authorities, such as those dealing with antitrust, competition policy, copyrights and patents, or occupational health and safety standards. In order to see the specific effects of a stand alone environmental regulatory authority in the present context, we assume that none of the other regulatory controls are present and that the environmental regulator's sole instrument is indeed the pollution intensity parameter e. Note however that, in the present context, it amounts to a complete control on the production technology.[18]

The choice of e is made in stage one by the regulator while a in stage one and p in stage two remain under the control of the monopolist. Let e^* denote the pollution intensity standard chosen by the regulator. From Proposition 5.3, we know that the monopolist chooses a product variant at the market center regardless of e: $\hat{a}(e^*) = 1/2$. The social welfare is defined as the sum of consumers' and monopolist's surplus less environmental damages. Given a price p, the profit is $p - c(e)$ per consumer or per unit consumed and the net surplus of a consumer located at x is given by $r - (1 + \beta) d(E) - t |x - 1/2| - p$ if he purchases the good at price p from the monopolist located at the market center, and $-d(E)$ otherwise. The regulator's objective is to maximize the sum of profit and net consumer surplus over all those consuming the product, that is, $r - \beta d(E) - t |x - 1/2| - c(e)$, less the global damage from pollution $d(E)$, which affects both served and unserved consumers.

Suppose the level of production is q and let $\mathcal{A}(q, e)$ denote the set of consumers (consumer locations) with a positive surplus given the product variant $\hat{a} = 1/2$ and that level of output q:

$$\mathcal{A}(q, e) \equiv \{x \in [0, 1] \ : \ r - \beta e q - t |x - 1/2| - c(e) \geq 0\}.$$

---

[18]Alternatively, the regulatory control could be modeled as affecting directly the characteristics of the product, hence the choise of location a in the present case. In some applied cases, it is not the technology but the product characteristics that are controlled by the regulator. Indeed, our preliminary results, not reproted here, show that the regulator may choose a location $a*$ away from the market center in order to make the product either less accessible or less desirable for some consumers, namely those located closer to one of the end points of the market line. See Boyer, Mahenc, Moreaux (2004) for more on this regulatory policy for environmental protection.

The social welfare function $W(q, e)$ can then be written as follows:

$$W(q, e) = \int_{A(q,e)} [r - \beta eq - t\,|x - 1/2| - c(e)]\,dx - eq$$
$$= (r - c(e) - \beta eq - e)q - tq^2/4 \tag{5.27}$$

Given a pollution intensity standard e, the monopolist located at the market center chooses in stage 2 the price $\widehat{p}(e, 1/2)$ characterized in Proposition 5.1. When choosing e, the regulator anticipates that the monopolist will charge $\widehat{p}(e, 1/2)$ and produce up to a level that satisfies the demand expressed at that price. From (5.5), we get:

$$D(\widehat{p}(e, 1/2), e, 1/2) = \begin{cases} 1, & \text{if } 2\beta e + t \le r - c(e), \\ q(\widehat{p}(e, 1/2), e) & \text{if } 0 \le r - c(e) \le t + 2\beta e. \end{cases} \tag{5.28}$$

Social welfare $W(D(\widehat{p}(e, 1/2), e, 1/2), e)$ can thus be written as

$$\int_{A(D(\widehat{p}(e,1/2),e,1/2),e)} [r - \beta e D(\widehat{p}(e, 1/2), e, 1/2) - t\,|x - 1/2| - c(e)]\,dx$$
$$-eD(\widehat{p}(e, 1/2), e, 1/2)$$

From now on, we will use the simplified notations:

$$W(e) \equiv W(D(\widehat{p}(e, 1/2), e, 1/2), e),$$

$$\pi(e) \equiv \widehat{\pi}(1/2, e) = (\widehat{p}(e, 1/2) - c(e))\, D(\widehat{p}(e, 1/2), e, 1/2),$$

$$q(e) \equiv q(\widehat{p}(e, 1/2), e) = \frac{r - c(e)}{t + 2\beta e}.$$

Hence the following expressions for social welfare:

$$W(e) = \begin{cases} r - c(e) - (1 + \beta)e - t/4 = \pi(e) + t/4 - e, \\ \qquad \text{if } t + 2\beta e \le r - c(e) \\ \qquad \text{(the market is fully covered);} \\ \\ (r - c(e) - \beta e q(e) - e)q(e) - tq(e)^2/4 = \pi(e) + tq(e)^2/4 - eq(e), \\ \qquad \text{if } 0 \le r - c(e) \le t + 2\beta e \\ \qquad \text{(the market is uncovered on both sides).} \end{cases} \tag{5.29}$$

Clearly, the private and the social incentive to curb pollution emissions have no reason to coincide. The environmental regulator is responsible for achieving an optimal balance between the cost of abatement (the monopolist's benefit of pollution) and the global damage from pollution, while it is only the consumption-specific damage that is of concern to the unregulated monopolist.

The derivative W_e is given by:

$$W_e(e) = \begin{cases} \pi_e(e) - 1, & \text{if } t + 2\beta e \leq r - c(e); \\ \\ \pi_e(e) + tq(e)q_e(e)/2 - q(e) - eq_e(e), & \text{if } 0 \leq r - c(e) \leq t + 2\beta e. \end{cases}$$
$$(5.30)$$

Moreover, using Proposition 5.4, the derivative of welfare W with respect to the pollution intensity e, evaluated at the pollution intensity chosen by the monopolist $\hat{e}$, is:

$$W_e(\hat{e}) = \begin{cases} -1, & \text{if } t + 2\beta\hat{e} \leq r - c(\hat{e}); \\ \\ -\dfrac{q(\hat{e})}{t + 2\beta\hat{e}}(\beta\hat{e} + t + \beta tq(\hat{e})/2) < 0, & \text{if } 0 \leq r - c(e) \leq t + 2\beta e. \end{cases}$$
$$(5.31)$$

A straightforward consequence of (5.31) is that the welfare maximizing pollution intensity level satisfies: $e^* < \hat{e}$. To determine whether this local maximum is a global maximum, we must study the behavior of $W_e(e)$ in the neighborhood of e_1 and e_2. For matter of simplicity, let us assume that $W(e)$ is increasing in $[0, e_1]$, so that the optimum is above e_1; this allows us to concentrate on e_2. From (5.31), we have $W_e^-(e_2) = \pi_e^-(e_2) - 1$ and $W_e^+(e_2) = \pi_e^+(e_2) - 1 + q_e(e_2)(t/2 - e_2)$. Remembering that $\hat{\pi}_e^-(e_2, 1/2) = \hat{\pi}_e^+(e_2, 1/2)$, we obtain that the welfare function $W(e)$ is not differentiable at e_2 and may be concave or not, depending on the sign of $q_e(e_2)(t/2 - e_2)$. A sufficient condition for $W(e)$ to be concave at e_2 is that the left hand derivative be larger than or equal to the right hand derivative or, equivalently, that $q_e(e_2)(t/2 - e_2) \leq 0$ where $q_e(\cdot)$ is given by (5.21). We will make the following assumption:

Assumption 5.5 $W_e^-(e_2) \geq W_e^+(e_2)$.

Assumption 5.5 is verified when, either $-c'(e_2)/2 \leq \beta$ and $t/2 \geq e_2$, that is, consumers' preferences are sufficiently heterogeneous (t large) when the consumption-specific damage they incur is rather high, or $-c'(e_2)/2 \geq \beta$ and $t/2 \leq e_2$, that is, consumers' preferences are sufficiently homogeneous (t low) when the consumption-specific damage they incur is rather low.

We can now compare the market coverages of both monopolists, regulated and unregulated. This is done in the next proposition. Under Assumption 5.5, three cases must be considered depending on how significant is the marginal benefit of pollution at e_2, that is $-c'(e_2)$, relative to the marginal damages, both consumption-specific and global.

Proposition 5.7 *Under Assumption 5.5,*

1. *if* $-c'(e_2) < \beta$, *then* $e^* < \hat{e} < e_2$ *and* $q(\hat{e}) = q(e^*) = 1$,

2. *if* $\beta \leq -c'(e_2) \leq 1 + \beta$ *then* $e^* < e_2 < \hat{e}$ *and* $q(\hat{e}) < q(e^*) = 1$,

3. *if* $1 + \beta < -c'(e_2)$ *then* $e_2 < e^* < \hat{e}$ *and* $q(\hat{e}) < q(e^*) < 1$.

In case 1, the marginal benefit of pollution is lower than the marginal consumption-specific damage when the market is not fully covered. The unregulated monopolist is better off choosing $\hat{e} < e_2$ and so covers the whole market. From (5.31), $W_e^-(e_2) < 0$ and Assumption 5.5 is sufficient to ensure the concavity of $W(e)$. Assumption 5.5 is equivalent here to $t/2 \geq e_2$ since $q_e(e_2) \leq 0$, that is, consumers' preferences must be sufficiently heterogeneous. In this case, the regulator chooses a pollution intensity standard $e^* < \hat{e}$ such that $1 + \beta = -c'(e^*)$. This is the Samuelson condition for the optimal provision of the public bad: the efficient standard of pollution intensity requires that the monopolist's private marginal cost saving from polluting is equal to the marginal global damage, hence is equal to the sum over all consumers of their willingness to pay for reducing the pollution generated by production. Then, the market is fully covered regardless of whether the monopolist is regulated or not.

In case 2, the marginal consumption-specific damage is lower than the marginal benefit of pollution when the market is fully covered, namely $\beta \leq -c'(e_2)$, and the marginal benefit of pollution is lower than the marginal global damage when the market is not fully covered, namely $W_e^-(e_2) \leq 0$ or, equivalently, $-c'(e_2) \leq 1 + \beta$. From Proposition 5.5, the unregulated monopolist leaves consumers unserved on both sides of the market by choosing $\hat{e} \geq e_2$. Furthermore, Assumption 5.5 ensures the concavity of $W(e)$: the regulator chooses $e^* < e_2 < \hat{e}$ and the market is then fully covered.

In case 3, the marginal damages, both consumption-specific and global, are lower than the marginal benefit of pollution when the market is fully covered. Then the regulator chooses $e^* \in [e_2, \hat{e})$ and the market coverage is partial if $W_e^+(e_2) > 0$. Under the latter condition, the regulator chooses e^* solving equation

$$\pi_e(e) + tq(e)q_e(e)/2 = q(e) + eq_e(e) = E_e(e), \tag{5.32}$$

which equates the marginal social value of the pollution intensity standard to the marginal global damage.

Moreover, it follows from Lemma 5.2 and Proposition 5.5 that more consumers are served at a higher price by the monopolist subject to environmental regulation than by the unregulated monopolist.

Proposition 5.8 *The unregulated monopolist pollutes more and produces as much or less than the regulated monopolist whatever the market coverage. If* *the efficient market coverage is partial, the unregulated monopolist produces strictly less, pollutes strictly more, and charges a lower price than the monopolist subject to environmental regulation. If the unregulated monopolist was covering the whole market, she still does it when subject to environmental regulation. If she was not covering the whole market, then she increases production when subject to the regulatory pollution intensity standard.*

This proposition sheds light on the social inefficiencies of a polluting monopolist confronted to environmentally aware consumers. The unregulated monopolist always pollutes too much from the environmental regulator's viewpoint. Moreover, the unregulated monopolist would always raise her price if she were asked to reduce pollution intensity, either to benefit from the resulting increase in consumers' surplus or as a reaction to the resulting increase in her production costs or both. By contrast, it may happen that the unregulated monopolist is better off covering the whole market when social efficiency requires to do so: such is the case when the marginal benefit from pollution is low relative to the marginal consumption-specific damage. Otherwise, according to the regulator's benchmark, the unregulated monopolist produces too little, in which case she would actually extend the market coverage and simultaneously raise price to meet the regulator's requirement.

The main technical difficulties with the present analysis emerge from the fact that the demand function is not everywhere differentiable. By contrast, the case in which the efficient market coverage is partial – namely case 3 in Proposition 5.7 – turns out to be "well-behaved" in the sense that the demand function is everywhere differentiable. Then, the model has strong affinities to Spence (1975) who addresses the problem of regulating the quality choice of a monopolist. The latter is shown to undersupply quality relative to the social optimum when the consumer's marginal valuation of quality decreases with quantity. Our result is somewhat related to Spence's result: the unregulated monopolist pollutes more than the regulated monopolist. Interestingly enough, it can be checked from (5.4) that the inverse demand curve is given by $P(q,e) = r - (2\beta e + t)q/2$. Thus $P_{qe} < 0$: paraphrasing Spence, the consumer's marginal valuation of environmental quality decreases with quantity. This provides new insight on the consequence of reducing pollution intensity on the monopoly price. The first effect is that consumers are willing to pay more for the product. This upward-shift in demand induces the monopolist to extract more surplus from the consumers. The second effect is that production costs are higher, which gives the monopolist a further incentive to raise price.

5.5 Conclusion

In Hotelling's (1929) horizontal differentiation context, the emergence of consumers concerned with an individual damage due to pollution compels the monopolist to pollute less. Clearly, this is beneficial from the environmental regulator viewpoint. Depending on the β-elasticity of the pollution intensity at its optimal level, the monopolist serves fewer or more consumers while raising her price relative to what would prevail with environmentally unaware consumers.

Nevertheless, the unregulated monopolist's choice of pollution intensity is biased upward with respect to the environmental regulator's target because the monopolist disregards the global damage and only takes into account the consumption-specific damage when maximizing profit; thus, the pollution externality fails to be fully internalized.

A notable result is that the monopolist's price declines with pollution intensity. This illustrates that the presence of environmentally aware consumers is likely to exacerbate the conflicts between an environmental regulator and an economic regulator which is responsible for controlling market power, in particular the pricing policy of a monopolist, such as a public utility commission. In the present context, if the monopolist is asked by the environmental regulator to reduce pollution intensity, then she will have two incentives to raise her price: first, a higher price allows the monopolist to extract more surplus from consumers since they are willing to pay more for a cleaner product; second, a cleaner product entails higher production costs for the monopolist, hence induces her to raise price. However, the potential increase in price clearly complicates the task of the economic regulator. However, further research will be needed to study more generally the influence of demand elasticity on pollution intensity distortion when consumers are environmentally aware. Given the increasing environmental awareness among consumers and policy makers, this topic should be high on the agenda of academic researchers.

References

Bansal, S. and Gangopadhyay, S., 'Tax/Subsidy Policies in the Presence of Environmentally Aware Consumers', *Journal of Environmental Economics and Management*, 45 (2003): 333-355.

Barnett, A., 'The Pigouvian Tax Rule under Monopoly', *American Economic Review*, 70 (1980): 1037-1041.

Baron, D.P., 'Regulation of Prices and Pollution under Incomplete Information', *Journal of Public Economics*, 28 (1985): 211-231.

Baumol, J. and Oates, W.E., *The Theory of Environmental Policy*, Second Edition, (Cambridge: Cambridge University Press, 1975).

Boyer, M. and Dionne, G., 'Riscophobie et étalement à moyenne constante : Analyse et applications,' *Actualité économique / Revue d'analyse économique*, 59 (1983): 208-229.

Boyer, M., Mahenc P. and Moreaux, M., 'The role of environmental awareness in environmental protection,' (2004), *Working Paper*, Université de Perpignan.

Buchanan, J.M., 'External Diseconomies, Corrective Taxes and Market Structures', *American Economic Review*, 59 (1969): 174-177.

Carraro, C. and Soubeyran, A., 'Environmental Taxation, Market Share and Profits in Oligopoly' in Carraro, C., Y. Katsoulacos and A. Xepapadeas (eds) *Environmental Policy and Market Structure*, (Dordrecht: Kluwer Academic Publishers, 1996).

Carraro, C., Katsoulacos Y. and Xepapadeas A., eds., *Environmental Policy and Market Structure*, (Dordrecht: Kluwer Academic Publishers, 1996).

Cason T. N. and Gangadharan, L., 'Environmental Labelling and Incomplete Consumer Information in Laboratory Experiments', *Journal of Environmental Economics and Management*, 43 (2002): 113-134.

Conrad, K. and Wang, J., 'The Effect of Emission Taxes and Abatement Subsidies on Market Structure', *International Journal of Industrial Organization*, 11 (1993): 499-518.

Foulon J., Lanoie P. and Laplante, B., 'Incentives for Pollution Control; Regulation or Information?', *Journal of Environmental Economics and Management*, 44 (2002): 169-187.

Gabszewicz, J. and Thisse, J.F., 'Price Competition, Quality and Income Disparities', *Journal of Economic Theory* 20 (1979): 340-359.

Hotelling H., 'Stability in Competition', *Economic Journal*, 39 (1929): 41-57.

Innes R. and Bial, J.J., 'Inducing Innovation in the Environmental Technology of Oligopolistic Firms', *The Journal of Industrial Economics*, 51 (2003): 265-287.

Kolstad, C.D., *Environmental Economics*, (Oxford University Press, 2000).

Levin, D., 'Taxation within Cournot Oligopoly', *Journal of Public Economics*, 27 (1985): 281-290.

Mussa, M. and Rosen, S., 'Monopoly and product Quality', *Journal of Economic Theory*, 18 (1978): 301-317.

Spence A.M., 'Monopoly, Quality and Regulation', *The Bell Journal of Economics*, 6 (1975): 417-429.

Tietenberg, T., *Environmental and Natural Resource Economics* (5th edition), (Addison Wesley Longman, 2000).

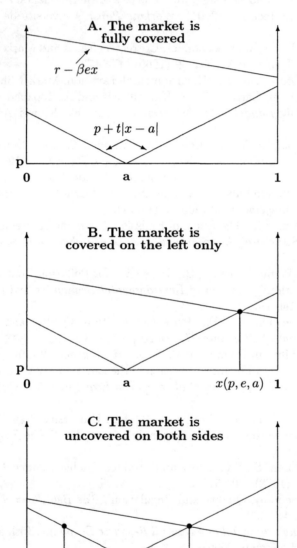

Figure 5.1: The market coverage

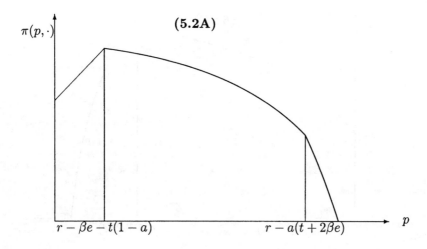

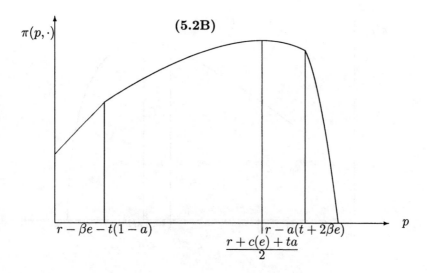

Figure 5.2: The profit function

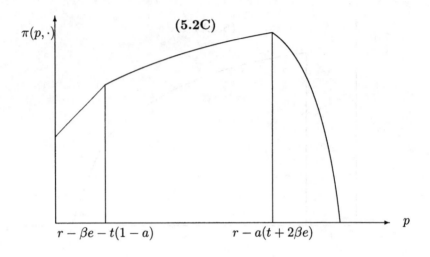

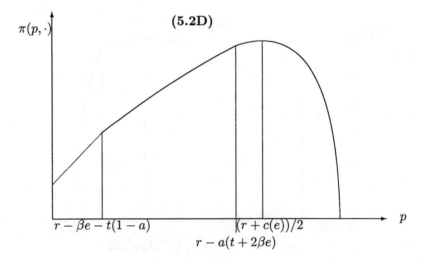

Figure 5.2: The profit function

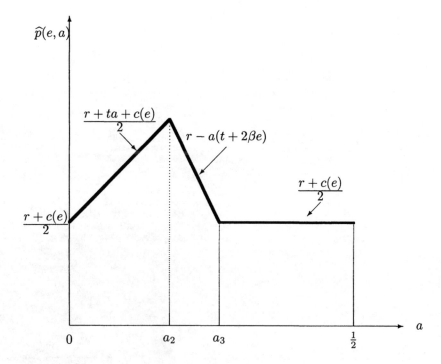

Figure 5.3: The price function $\widehat{p}(e,a)$

Part 2

Advances in Legal Design

Chapter 6

Optimal Punishment for Repeat Offenders when the Government Can and Cannot Commit to Sanctions[*]

Winand Emons
University of Bern

6.1 Introduction

Most legal systems punish repeat offenders more severely for the same offense than non-repeat offenders. Second-time offenders, for example, receive more severe punishment than first-time offenders. Penalty escalation characterizes traditional crimes such as theft and murder, but also violations of environmental and labor regulations, tax evasion, etc. This principle of escalating sanctions based on offense history is so widely accepted that it is embedded in many penal codes and sentencing guidelines.

For the rather well developed law and economics literature on optimal law enforcement escalating sanction schemes are still a puzzle.[1] This literature looks for an efficiency-based rationale for such a practice. Does a sanction scheme that maximizes welfare indeed have the property of sanctions increasing with offense history? So far the results have been mixed. At the very best the literature, which we describe at the end of this introduction, has shown that under special circumstances escalating penalty schemes may be optimal.

The purpose of this paper is to shed some light on this puzzle. We consider agents who may commit a crime twice. The act is inefficient; the agents

[*]I thank Simon Anderson, Mike Baye, Gary Biglaiser, Bob Cooter, Nuno Garoupa, Tom Gilligan, Manfred Holler, Tracy Lewis, Thomas Liebi, David Martimort, Mitch Polinsky, Eric Rasmusen, Bill Rogerson, François Salanié, David Sappington, Suzanne Scotchmer, Steve Shavell, Curtis Taylor, Asher Wolinski, and an anonymous referee for helpful comments. The hospitality of the department of economics at Purdue University is gratefully acknowledged.

[1]See, e.g., Garoupa (1997) or Polinsky and Shavell (2000a) for surveys of this literature.

are thus to be deterred. The agents are wealth constrained so that increasing the fine for the first offense means a reduction in the possible sanction for the second offense and vice versa. The agents may follow history dependent strategies, i.e., commit the crime a second time if and only if they were (were not) apprehended the first time. The government seeks to minimize the probability of apprehension.

First we assume, as is typical in the literature on optimal law enforcement, that the government can commit to sanction schemes. This means the government can use any set of threats to penalize wrongdoers. Our basic result is that the optimal sanction scheme is decreasing rather than increasing in the number of offenses. Indeed, in our framework it is optimal to set the sanction for the first offense equal to the entire wealth of the agents while the sanction for the second offense equals zero. The key intuition is as follows: A money penalty imposed for the second offense reduces the amount a person can pay for the first offense, since the wealth available to pay penalties is assumed to be fixed over the two periods. For that reason, a higher probability event – namely, a first offense that is detected – will be more effective use of the scarce money penalty resource than a lower probability event – namely, a second detected offense.

Why is the probability of detection lower for the second rather than for the first crime? An agent faces the *possibility* of being sanctioned for the second crime if and only if she has already been sanctioned for the first time. Moreover, suppose the first act went undetected and the agent commits the second crime; then there is the possibility that she is apprehended for the second crime for which she is charged, however, the first-time sanction since she has no criminal record. Accordingly, whatever strategy the agent opts for, she is more likely to pay the sanction for the first rather than for the second crime. Shifting scarce wealth from the second to the first sanction, therefore, increases deterrence.

Then we give up the assumption that the government can commit to whatever sanction scheme. We consider the analysis of optimal sanctions without the possibility to fully commit important because judges often have a lot of discretion as to the size of the penalty: they may, for example, reduce sanctions to account for the financial possibilities, the education, the family background, etc. of the wrongdoer. Accordingly, we allow only for sanctions that the government actually wishes to implement should a crime have occurred.[2]

Ruling out full commitment changes the optimal enforcement schemes. Suppose, for example, the government does not care about the sanction as is typically assumed in the literature. Then it will not enforce the penalty if a crime has happened given that there is, say, a small cost of doing so. The rational criminal will anticipate the ex post enforcement behavior of

---

[2]In many countries the President may pardon wrongdoers which, essentially, means that sanctions can be reduced.

the government. Therefore, she will commit the crime because the threat of being sanctioned is not credible. Once we drop the commitment assumption, the typical deterrence equilibria of the law enforcement literature between potential wrongdoers and the government are based on empty threats. In the jargon of game theory, the equilibria are not subgame perfect or time consistent.

Our decreasing sanction scheme from the first part raises of course the issue of time consistency. Will the government really charge the agent the entire wealth when she was apprehended for the first crime, knowing that then she will commit the second act for sure? Isn't it better for the government to renege and charge little for the first act so that the agent still has sufficient wealth to pay a sanction that deters the second crime? Given that the first act has been committed anyway, that way the government can at least deter the second act.

To analyze this problem we consider a rent-seeking government. The sanctions paid by the criminals enter the government's welfare function. Our government, therefore, has an ex post incentive to collect fines. The government can commit to a probability of apprehension but not to sanctions. Our basic result is that if the agent's benefit and/or the harm from the crime are small enough, then the scheme where the sanction for the first crime is the entire wealth and the sanction for the second crime is zero is indeed subgame perfect.

To see this, consider the government after the agent has been apprehended for the first crime. If it sticks to our decreasing sanction scheme, it appropriates the entire wealth yet incurs the harm of the second crime. Thus, the lower the harm of the second crime, the more attractive is this option.

The alternative is to set the sanction for the second crime to a level that deters the act. With this option the government does not incur the harm of the second crime, yet forgoes the sanction for the second crime because it is deterred. If the benefit from the crime goes up, the optimal probability of apprehension increases, yet by more than the benefit; accordingly, the actual sanction necessary to deter the second crime falls. Since a low sanction for the second crime means a high amount the government can charge for the first crime, a high benefit of the second crime makes this option attractive. Therefore, only for low benefits the government sticks to the decreasing sanction scheme.

If the benefit and/or the harm of the second crime are large, our decreasing sanction scheme is no longer subgame perfect. The government prefers to deter the second crime should the first crime have occurred. Accordingly, only sanction schemes where each sanction by itself deters the corresponding crime are time consistent. In this case the optimal subgame perfect sanction scheme entails equal sanctions in both periods. Enforcement costs are higher than with the decreasing sanction scheme.

Let us now discuss the related literature. In Rubinstein (1979) even if an agent abides by the law, she may commit the act accidentally. The government wishes to punish deliberate offenses but not accidental ones. Rubinstein shows that in the infinitely repeated game an equilibrium exists where the government does not punish agents with a "reasonable" criminal record and the agents refrain from deliberate offenses.

Rubinstein (1980) considers a setup where an agent can commit two crimes. A high penalty for the second crime is exogenously given. The sanction for the first crime may be lower than the sanction for the second crime. Rubinstein shows that for any set of parameters there exists a utility function such that deterrence is higher if the sanction for the first crime is lower than the sanction for the second crime. Rubinstein does not allow for the second sanction to be lower than the first one.

Landsberger and Meilijson (1982) develop a dynamic model with repeat offenses. Their concern is how prior offenses should affect the probability of detection rather than the level of punishments.

In Polinsky and Rubinfeld (1991) agents receive an acceptable as well as an illicit gain from the criminal activity. The government cannot observe the illicit gains. Repeat offenses are, however, a signal of a high illicit gain. For certain parameter values of the model it may be optimal to punish repeat offenders more severely.

In Burnovski and Safra (1994) agents decide ex ante on the optimal number of crimes. They show that if the probability of detection is sufficiently small, reducing the sanction on subsequent crimes while increasing the penalty on previous crimes decreases the overall criminal activity. This paper is similar in spirit to ours. The main differences are: In their framework agents cannot choose strategies that depend on history, in our setup they can. Moreover, we derive the optimal policy that minimizes enforcement costs and we address the problem of subgame perfection.

In Polinsky and Shavell (1998) agents live for two periods and can commit a crime twice. The sanctions depend on the agent's age and her criminal record. They show that the following policy may be optimal: Young first-time offenders and old second-time offenders are penalized with the maximum sanction. Old first-time offenders may be treated leniently. Accordingly, this result does not say that repeat offenders are punished more severely; old first-time offenders may be punished less severely than old repeat- and young first-time offenders.

Chu, Hu, and Huang (2000) consider like Rubinstein (1979) a legal system that may also convict innocent offenders. The government takes the possibility of erroneous conviction as a social cost into account. The optimal penalty scheme punishes repeat offenders (slightly) more than first-time offenders. Reducing the penalty for first-time and increasing it slightly for repeat offenders has no effect on deterrence. The cost of erroneous convictions is, however, reduced because the probability of repeated erroneous conviction is lower than for first-time mistakes.

Dana (2001) argues that contrary to the assumptions in the literature, probabilities of detection increase for repeat offenders. As a result, the optimal deterrence model dictates declining, rather than escalating, penalties for repeat offenders. Taking the salience and optimism biases from behavioral economics into account makes the case for declining penalties even stronger.

Baik and Kim (2001) extend Polinsky and Rubinfeld (1991) by introducing the possibility of social learning of illicit gains between the two periods. If social learning is more important than the inherent characteristics in inducing offenses, it may be optimal to punish first-time offenders as severely as repeat offenders.

In Emons (2003b) agents have the choice between being criminals or being law abiding. If they choose the criminal career, they commit the act twice; there is thus a barrier to exit. If they choose to be law abiding, they may still commit the act accidentally. If the benefit from the crime is small, the optimal sanction scheme is decreasing in the number of offenses. By contrast, if the benefit is large, the sanction for the first offense is zero while the sanction for the second offense is the agents' entire wealth.

The only paper we are aware of that deals with the problem of subgame perfect sanctions is Boadway and Keen (1998). They consider a government choosing a capital income tax rate and an enforcement policy. The government can commit to the enforcement policy but not to the tax rate. Ex ante the government wishes to announce a low tax rate to induce savings; ex post, when savings have been made, it will renege and apply a high tax rate. Boadway and Keen show that by committing to a lax enforcement policy the government can alleviate the welfare loss implied by its inability to commit to the tax rate.

In the next section we describe the model. In Section 6.3 we derive the optimal sanctions for a government that can commit and in Section 6.4 for a government that cannot commit. Section 6.5 concludes.

6.2   The Model

Consider a set of individuals who live for two periods.[3] In each period the agents can engage in an illegal activity, such as speeding, polluting the environment, conspiring to raise prices, or evading taxes. If an agent commits the act in either period, she receives a monetary benefit $b > 0$. We consider crimes without social gains. Using the term of Polinsky and Rubinfeld (1991),

---

[3]The set of agents has mass 1.

b is the illicit gain and the crime creates no acceptable gain.[4] The act causes a monetary harm h to society which is borne by the government. Since the damage $h > 0$, the act is not socially desirable. The individuals are thus to be deterred from the activity.[5]

To do so the government chooses monetary sanctions. The government observes whether the crime is the first or the second one. The government uses fines $s_1, s_2 \geq 0$ where s_1 applies to first-time and s_2 to second-time observed offenders.[6] Moreover, the government chooses a probability of apprehension p. This probability is the same for first- and second-time offenses.[7] Since apprehension is costly, the government wishes to minimize p.

In the following section we assume that the government can choose any set of sanctions $s_1, s_2 \geq 0$. In Section 6.4 the government can no longer fully commit to sanctions. It can commit to a maximum sanction but the government can choose a lower sanction from the one announced at the outset once a crime occurred. Typically, a judge always finds good reasons to reduce sanctions. By contrast, the probability of apprehension is irrevocably fixed before the agents take their actions. The government cannot easily change the amounts spent on, say, training the police or the tax authorities. Accordingly, in Section 6.4 we assume that the government can commit to p while it cannot commit to sanctions.[8]

In the law enforcement literature the optimal policy is derived by maximizing the sum of the offenders' benefits minus the harm caused by the offenses minus law enforcement expenditures. Sanctions do not enter the benevolent

---

[4]See also Chu, Hu, and Huang (2000) for an analysis of crimes without social gains. They argue that the gains to the offender are not considered because the crime is not socially acceptable or because the gains of offenders, such as theft or other zero-sum crimes, offset with the victims' losses.

[5]We assume that the benefits and the harms are the same for both crimes. If, e.g., the benefit of the second crime is much higher than the benefit of the first one, this might provide a rationale for escalating penalties.

[6]To have a game with complete information which we can solve by backward induction, we assume that the government also observes the period in which the agent commits the crime. However, we do not allow sanctions to depend on the agent's age. This assumption may be justified by equity reasons in the sense that the fine may not change when the agent is 46 rather than 39. Note that if fines also depend on age, the results will be different: if, e.g., an old offender is apprehended for a crime, be it the first or the second one, then the government will seize her entire assets. The analysis of optimal sanctions when fines depend on the number of crimes and on the age of the wrongdoer is an interesting topic for future research. See Polinsky and Shavell (1998) for a set-up where fines also depend on age.

[7]We thus rule out the case where agents with a criminal record are more closely monitored than agents without a record. See Landsberger and Meilijson (1982) for an analysis of optimal detection probabilities.

[8]Boadway and Keen (1998) use the same commitment structure when studying the time consistency problem in the taxation of capital income.

government's objective function because they are a mere transfer of money.[9] Within this framework the literature derives the results on optimal fines and optimal probabilities of apprehension. See, e.g., Garoupa (1997) or Polinsky and Shavell (2000a).

Nevertheless, these results hold true if and only if the government can fully commit to the probability of apprehension *and* to the announced sanction. To see this, suppose the government incurs a small cost $c > 0$ of cashing in on the fine. Suppose the agent has been apprehended for the crime and then the government strategically decides whether or not to impose the sanction. With such a sequencing, the rational government will not impose the fine: it does not care about the fine anyway and it can safe the cost c. Anticipating this ex post behavior of the government, the threat of being sanctioned is not credible and the agent will commit the act in the first place. To put it in the language of game theory: the equilibrium in the game between the offender and the government is not subgame perfect.

If we want to take the issue of subgame perfection (or time consistency) seriously, we must give the government an incentive to actually collect the fines. We do so by including the sanctions in the government's payoffs.[10] Our government thus maximizes revenues from sanctions minus the harms minus the enforcement expenditure and has thus an incentive to collect the fine should a crime have occurred. To save on notation we take the probability of detection p as a measure of the enforcement expenditure.

This approach can be justified in several ways. Garoupa and Klerman (2002) take the public choice perspective of a self-interested, rent-seeking government which maximizes revenues minus the harm borne by the government minus expenditure on law enforcement.[11] Polinsky and Shavell (2000b) consider the standard benevolent welfare function and add a term reflecting individuals' fairness-related utility. If this fairness-related utility equals the actual sanction, their government maximizes the same welfare function as ours.[12] Finally, if it is costly for the government to raise taxes due to the distortions they create, it has strong incentives to raise money from offenders.

Individuals are risk neutral and maximize expected income. They have initial wealth $W > 0$. Think of W as the value of the privately owned house

---

[9]In the explicit formulation welfare is the criminal's utility (benefit minus expected sanction) plus the government's utility (expected sanction minus harm) minus enforcement costs.

[10]In terms of the explicit welfare function given in the preceding footnote, we simply exclude the criminal's utility (benefit minus expected sanction).

[11]Dittmann (2001) uses a similar approach.

[12]In Rubinstein (1979) the government's payoffs also depend on whether or not it punishes the offender. Unlike the other papers, Rubinstein's government is worse off if it punishes the offender, independently of whether the act was committed intentionally or not.

or assets with a long maturity.[13] The agents hold on to their wealth over both periods unless the government interferes with sanctions. Any additional income they receive in both periods, be it through legal or illegal activities, is consumed immediately. Accordingly, all the government can confiscate is W. If the fine exceeds the agent's wealth, she goes bankrupt and the government seizes the remaining assets. This implies that the fines s_1 and s_2 have to satisfy the "budget constraint" $s_1 + s_2 \leq W$.[14]

To save on notation we set the interest rate zero. An agent can choose between the following strategies:

1. She can choose not to commit the act at all. Call this strategy (0,0) which gives rise to utility $U(0,0) = W$. This is the strategy we want to implement.

2. She can commit the act in period 1 and not in period 2. We call this strategy (1,0); here we have $U(1,0) = W + b - ps_1$. The act generates benefit b; with probability p the agent is apprehended and pays the sanction s_1.

3. The agent can commit the crime in period 2 but not in period 1. Call this strategy (0,1) generating utility $U(0,1) = W + b - ps_1$. With strategy (0,1) the agent has the same utility as with strategy (1,0) because the government observes only one offense.

4. Furthermore, the agent can commit the act in both periods which we denote by (1,1) and $U(1,1) = W + b - ps_1 + b - p((1-p)s_1 + ps_2)$. The second crime is detected with probability p. With probability p the agent has a criminal record in the second period and thus is fined s_2; with probability $(1-p)$ she has no record and pays s_1 if apprehended.

5. Finally, the agent can choose two history dependent strategies.[15]

   - First, she commits the act in period 1. If she is not apprehended, she also commits the act in period 2; however, if she is apprehended in period 1, she does not commit the act in period 2. Call this strategy (1,(1|no record;0|otherwise)) with $U(1, (1|\text{no record}; 0|\text{otherwise})) = W + b - ps_1 + (1-p)(b - ps_1)$. Since the agent

---

[13]The policy of the Swiss competition authority is not to use fines that drive the wrong-doer into bankruptcy. Accordingly, in this case W is the amount the firm can just afford to pay.

[14]This assumption distinguishes our approach from Polinsky and Shavell (1998) who work with a maximum per period sanction s_m. Accordingly, they may set $s_1 = s_2 = s_m$, which is typically the optimal enforcement scheme. In their framework s_m is like a per period income which cannot be transferred into the next period. Burnovski and Safra (1994) use the same budget constraint as we do.

[15]These history dependent strategies distinguish our paper from Burnovski and Safra (1994) where individuals decide ex ante simply on the number of crimes.

stops her criminal activities if she is apprehended once, she is never sanctioned with s_2.

- Second, she commits the act in period 1. If she is not apprehended, she does not commit the act in period 2; yet, if she is apprehended in period 1, she commits the act in period 2. Call this strategy $(1,(0|\text{no record};1|\text{otherwise}))$ with $U(1,(0|\text{no record};1|\text{otherwise})) = W+b-ps_1+p(b-ps_2)$. It turns out that this strategy defines the agents' binding incentive constraint for the optimal sanctions.

Before we start deriving optimal sanctions, we have to ensure that the government indeed wants complete deterrence. This is achieved by assuming $1 < 2h - W$. If the government completely deters, there is neither harm nor revenue and the maximum possible expenditure for deterrence is 1 (recall that we take the probability of apprehension as a measure for enforcement cost). If the government does not deter at all, enforcement costs are zero, the government incurs the harm twice, and the maximal revenue it can obtain is the agents' wealth W. Therefore, if the harm is large enough, the rent-seeking government wants complete deterrence.

Let us now derive the sanctions that give the agents proper incentives not to engage in the activity in both periods. We first derive the cost-minimizing sanction scheme that achieves perfect deterrence ignoring the government's commitment problem. This is the standard approach found in the literature. The literature does not further discuss why the authorities are able to commit. One argument in favor of commitment is that the government plays repeated games with potential wrongdoers and, therefore, wants to build up a reputation of being tough. Moreover, laws may be written such that the judge has little to no discretion as to the size of the penalty.[16]

The analysis of the commitment scenario follows Emons (2003a). We will then consider the government's incentives to actually implement this penalty scheme without commitment in section 6.4. This section is based on Emons (2004).

6.3 Optimal Fines if the Government can Commit

Agents are assumed to have enough wealth so that deterrence is always possible, i.e., $2b < W$. The agent does not follow strategy $(1,0)$, if $U(1,0) \leq U(0,0)$,

---

[16]The Three "Strikes and You're Out" Law, California Penal Code Section 667 (b), is an attempt to do just this. In a similar spirit, New York's Rockefeller Drug Laws require that judges impose a mandatory minimum sentence of 15 years to life upon conviction for selling more than two ounces or possessing more than four ounces of a narcotic substance.

she does not follow strategy (0,1), if $U(0,1) \leq U(0,0)$, etc. Straightforward calculations confirm that the agent does not engage in strategies (1,0), (0,1), and (1,(1|no record;0|otherwise)), if

$$s_1 \geq b/p; \tag{6.1}$$

she does not pick strategy $(1,1)$, if

$$s_2 \geq (2b/p^2) - s_1((2/p) - 1); \tag{6.2}$$

and she does not pick strategy (1,(0|no record;1|otherwise)), if

$$s_2 \geq (b(1+p)/p^2) - s_1/p. \tag{6.3}$$

Accordingly, with all sanction schemes (s_1, s_2) to the right of the bold line in Figures 6.1 and 6.2, the agent has proper incentives and commits no crime. For example, the equal sanction scheme $s_1 = s_2 = b/p$ induces no crimes.

Let us now minimize the enforcement costs, as given by p, while providing incentives not to commit any crime.[17] We will minimize p taking the incentive constraint (6.3) into account. Then we show that the optimal $\hat{p}$ also satisfies the incentive constraints (6.1) and (6.2).

Obviously, Becker's (1968) maximum fine result applies here, meaning that in order to minimize p the government will use the agent's entire wealth for sanctions.[18] Therefore, plugging the budget constraint $s_1 + s_2 = W$ into (6.3) and differentiating the equality yields

$$dp/ds_1 = (p - p^2)/(b - s_1 - 2p(W - s_1)) < 0$$

for $b < s_1 \leq W$. Consequently,

$$\hat{s}_1 = W, \ \hat{s}_2 = 0, \ \text{and} \ \hat{p} = b/(W - b).$$

Since $b/p < 2b/p(1 - p) < b(1 + p)/p \ \forall p \in (0,1)$, the incentive constraints (6.1) and (6.2) are also satisfied. Accordingly, we have:

Proposition 6.1 *With commitment to sanctions the optimal sanction scheme is given by $s_1^* = W$, $s_2^* = 0$ and $p^* = b/(W - b)$.*

We thus find that the cost minimizing sanction scheme sets $\hat{s}_1 = W$ and $\hat{s}_2 = 0$. First time offenders are punished with the maximal possible sanction while second time offenders are not punished at all. The sanction s_1 is so high that it not only deters first-time offenses but also second-time offenses even though they come for free.

---

[17]Since in our setup the harm of the crime exceeds its acceptable benefit, maximizing social welfare boils down to minimizing enforcement costs.
[18]If $s_1 + s_2 < W$, sanctions can be raised and p lowered so as to keep deterrence constant.

The intuition for this result follows immediately from the incentive constraint (3). The agent pays the sanction s_1 with probability p and the sanction s_2 only with probability p^2. Stated differently: The agent is charged s_2 with probability p if and only if she has paid already s_1. Since paying the fine s_1 is more likely than paying s_2, shifting resources from s_2 to s_1 increases deterrence for given p. Consequently, p is minimized by putting all the scarce resources into s_1.

It is somewhat surprising that the strategy $(1,(0|\text{no record};1|\text{other-wise}))$ and not the strategy $(1,(1|\text{no record};0|\text{otherwise}))$ defines the binding incentive constraint in the optimal penalty structure. Given that the optimal penalties are decreasing, an agent who was not apprehended for the first crime has a strong incentive not to commit the act a second time: if she is apprehended she pays the high sanction s_1. If the agent was, however, apprehended for the first crime, the second crime comes for free. The sanction s_1 has to be high enough so that she doesn't commit the first crime in the first place.

6.4 Optimal Fines if the Government cannot Commit

Let us now analyze under which conditions the sanction scheme $\hat{s}_1 = W$, $\hat{s}_2 = 0$ together with the minimal enforcement probability $\hat{p} = b/(W - b)$ is subgame perfect. This means: Does the government really want to implement these sanctions once the agent has committed the first offense? To do so, consider the subgame starting when the agent has been apprehended for the first crime.

If the government sticks to the penalty scheme $\hat{s}_1 = W$, $\hat{s}_2 = 0$, the agent will commit the second offense for sure because it comes for free. The government's payoff then amounts to $W - 2h - \hat{p}$. It incurs the harm twice and seizes the agent's entire wealth with s_1.

The alternative is to lower s_1 and at the same time increase s_2 such that the agent does not commit the second act. Clearly, the rent-seeking government will set $s_2 = b/\hat{p}$, the minimal sanction achieving deterrence. The government goes for the minimal sanction guaranteeing deterrence because, by its very nature, the government will not get this money; that way, s_1 is as large as possible. Using $\hat{p} = b/(W - b)$, we find $s_2 = W - b$ and $s_1 = b$. If the government follows this strategy, its payoffs are $-h + b - \hat{p}$. It incurs the harm from the first crime, collects $s_1 = b$ and there is no more crime.

Comparing the two payoffs, obviously the government prefers to stick to $\hat{s}_1 = W$, $\hat{s}_2 = 0$ if $W - h \geq b$. The government gets the entire wealth less the harm by sticking to the optimal incentive scheme whereas it gets $s_1 = b$ if it chooses to deter the second offense. Therefore, we may conclude that

$s_1^* = W$, $s_2^* = 0$ is subgame perfect if the agent's benefit b and/or the harm are not too large. See Figure 6.1.

Let us now determine the optimal subgame perfect sanction scheme together with the probability of detection p if $W - h < b$. Consider again the government deciding on sanctions after the agent has been apprehended for the first act. If the government wants to deter the second act, it will set $s_2 = b/p$. It chooses the minimal sanction ensuring deterrence because it will not get the money. In this way it can collect the maximum amount $s_1 = W - b/p$ for the first act from the agent.

By contrast, the government may wish to induce the second crime. It does so by setting $s_2 < b/p$. The government gets s_2 only with probability p; it collects s_1 for sure because we are in the node where the agent has just been apprehended for the first crime. Since $W = s_1 + s_2$, the revenue maximizing government sets $s_1 = W$ and $s_2 = 0$ if it wants to induce the second crime. This generates a payoff of $W - 2h - p$ for the government.

The government prefers the strategy of inducing the second crime to optimally deterring the second crime if $W - 2h - p > W - h - b/p - p \Leftrightarrow b/p > h$. Deterring the second crime has the cost of the foregone revenue $s_2 = b/p$; inducing the second crime has the cost of the harm h.

The left-hand side of the inequality $b/p > h$ is a decreasing function of p. Accordingly, if it is not satisfied for the minimal probability of apprehension inducing no crimes $\hat{p} = b/(W - b)$, it does not hold for any p deterring both crimes. Therefore, if $b/\hat{p} < h \Leftrightarrow W - h < b$, the government prefers to deter the second crime and does so optimally by setting $s_1^* = s_2^* = W/2$ and $p^* = 2b/W$.[19] See Figure 6.2.

A low probability of apprehension increases b/p, the sanction which is necessary to deter the second crime. Deterring a second crime thus becomes unattractive. By choosing a low p, the government commits not to raise s_2 to a level which deters. This result is similar in spirit to Boadway and Keen (1998) where the government commits to a lax enforcement in order not to raise tax rates after savings decisions have been made.

We summarize the preceding observations with the following proposition.

Proposition 6.2 *Without commitment not to lower sanctions if $W - h \geq b$, the optimal subgame perfect sanction scheme is given by $s_1^* = W$, $s_2^* = 0$ and $p^* = b/(W - b)$; if $W - h < b$, the optimal subgame perfect sanction scheme is given by $s_1^* = s_2^* = W/2$ and $p^* = 2b/W$.*

The government is better off in the first case where it uses the decreasing sanction scheme. In both cases crime is completely deterred. With the decreasing sanction scheme the probability of detection and hence enforcement cost is lower than in the second case of constant sanctions.

---

[19]If an old agent is apprehended for the first crime, the government would like to raise s_1 to W. Nevertheless, it cannot do so because it committed to maximum sanctions.

6.5   Conclusions

The purpose of this paper is to help understand the difficulties the law and economics literature has in explaining escalating penalties. If a higher sanction for the second crime means a lower sanction for the first crime and vice versa, cost minimizing deterrence in the case of commitment is decreasing, rather than increasing, in the number of offenses.[20] Since an agent can only be a repeat offender if she has been a first-time offender, there is no second offense if we completely deter the first one. This effect seems to be quite robust and should also apply to non-monetary sanctions. Accordingly, if one wants to give a rationale for the widely prevailing escalating penalties, one has to go beyond the simple deterrence model à la Becker.

Section 6.4 is an attempt to do just this. There we analyze subgame perfect sanction schemes, i.e., sanctions which the government indeed wants to implement should a crime have occurred. We consider the problem of time consistency important because judges tend to have a lot of discretion as to the size of the penalty. They anticipate that a high penalty now may reduce the potential for future sanctions. Rational criminals will anticipate this and thus not be deterred by empty threats. A rent-seeking government will stick to the optimal decreasing sanction scheme if it gets more money by allowing the second crime and cashing in the agent's entire wealth with the first sanction than by deterring the second crime. In the opposite case the government prefers to deter the second crime. It does so with equal sanctions for both crimes.

Accordingly, the constraint of time consistency bites. If the government can commit to penalties, decreasing sanctions are always optimal; if the government cannot commit to penalties, decreasing sanctions may still be optimal but so may be equal sanctions. We have not explained escalating sanctions based on offense history which are embedded in many penal codes and sentencing guidelines. Explaining escalating sanctions seems to be not an easy task for the law enforcement literature; see our discussion in the Introduction.[21] Nevertheless, in our set-up the commitment issue ruled out decreasing sanction schemes in some cases. It thus seems that the problem of time consistency is a fruitful track for future research to understand escalating sanction

---

[20]Similar results hold in repeated moral hazard situations. For example, if agents decide strategically over time on how carefully to treat a consumer durable, optimal incentive compatible warranties tend to increase, rather than decrease, with the product's age. See Emons (1989).

[21]The explanations of Rubinstein (1979) and Polinsky and Rubinfeld (1991) seem to be the most reasonable ones. Both models are based on adverse selection. Repeat offenses are a strong signal that the wrongdoer is a hard-core criminal whom the government wants to punish heavily; the government does not want to punish accidental crimes. By contrast, we look at the pure moral hazard problem where the government wants to deter crimes. In this class of models the results of the literature are less convincing and subgame perfection has the potential to add new insights.

schemes. It is, for example, of interest how the optimal sanction scheme looks like when fines may depend on the number of crimes and on the age of the wrongdoer.

References

Baik, K. H. and Kim, I.G., 'Optimal Punishment when Individuals may learn deviant Values', *International Review of Law and Economics*, 21 (2001): 271-285.

Becker, G., 'Crime and Punishment: An Economic Approach', *Journal of Political Economy*, 76 (1968): 169-217.

Boadway, R. and Keen, M., 'Evasion and Time Consistency in the Taxation of Capital Income', *International Economic Review*, 39(2) (1998): 461-476.

Burnovski, M. and Safra, Z., 'Deterrence Effects of Sequential Punishment Policies: Should Repeat Offenders be more Severely Punished', *International Review of Law and Economics*, 14 (1994): 341-350.

Chu, C. Y. Cyrus, Sheng-cheng Hu, and Ting-yuan Huang, 'Punishing Repeat Offenders more Severely', *International Review of Law and Economics*, 20 (2000): 127-140.

Dana, D. A., 'Rethinking the Puzzle of Escalating Penalties for Repeat Offenders', *Yale Law Journal*, 110 (2001): 733-783.

Dittmann, I., 'The Optimal Use of Fines and Imprisonment if Governments Don't Maximize Welfare', *Discussion Paper*, Humboldt-Universität Berlin, 2001. (*papers.ssrn.com/sol3/papers.cfm?abstract_id* = 274449).

Emons, W., 'On the Limitation of Warranty Duration', *Journal of Industrial Economics*, 37 (1989): 287-302.

Emons, W., 'A Note on the Optimal Punishment for Repeat Offenders', *International Review of Law and Economics*, 23 (2003a): 253-259.

Emons, W., 'Escalating Penalties for Repeat Offenders', University of Bern (2003b), *www.vwi.unibe.ch/theory/papers/emons/esc_p.pdf*.

Emons, W., 'Subgame Perfect Punishment for Repeat Offenders', *Economic Inquiry*, 42 (2004): 496-502.

Garoupa, N., 'The Theory of Optimal Law Enforcement', *Journal of Economic Surveys*, 11 (1997): 267-295.

Garoupa, N. and Klerman, D., 'Optimal Law Enforcement with a Rent-Seeking Government', *American Law and Economics Review*, 4(1) (2002): 116-140.

Landsberger, M. and Meilijson, I., 'Incentive Generating State Dependent Penalty System, The Case of Income Tax Evasion', *Journal of Public Economics*, 19 (1982): 333-352.

Polinsky, M. and Rubinfeld, D., 'A Model of Fines for Repeat Offenders', *Journal of Public Economics*, 46 (1991): 291-306.

Polinsky, M. and Shavell, S., 'On Offense History and the Theory of Deterrence', *International Review of Law and Economics*, 18 (1998): 305-324.

Polinsky, M. and Shavell, S., 'The Economic Theory of Public Enforcement of Law', *Journal of Economic Literature*, 38 (2000a): 45-76.

Polinsky, M. and Shavell, S., 'The Fairness of Sanctions: Some Implications for optimal Enforcement Policy', *American Law and Economics Review*, 2(2) (2000b): 223-237.

Rubinstein, A., 'An Optimal Conviction Policy for Offenses that May Have Been Committed by Accident', 406-413, Applied Game Theory, S. Brams, A. Schotter, and G. Schwödiauer (Eds.), (Würzburg: Physica-Verlag, 1979).

Rubinstein, A., 'On an Anomaly of the Deterrent Effect of Punishment', *Economics Letters*, 6 (1980): 89-94.

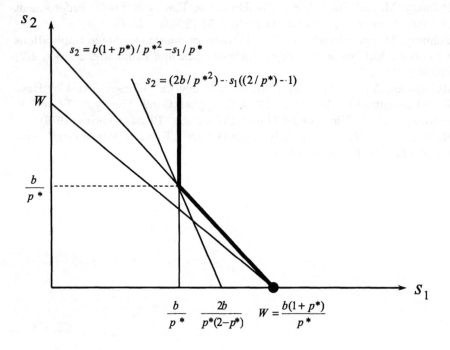

Figure 6.1: The set of incentive compatible sanctions and the optimal sanction scheme $(s_1^*, s_2^*) = (W, 0)$ and $p^* = b/(W - b)$

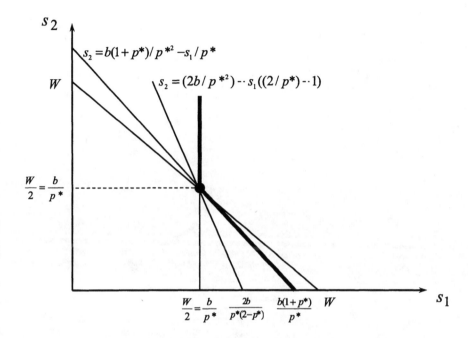

Figure 6.2: The set of incentive compatible sanctions and the optimal sanction scheme $(s_1^*, s_2^*) = (W/2, W/2)$ and $p^* = 2b/W$

Chapter 7

Nash Implementable Liability Rules for Judgement-Proof Injurers[*]

Patrick González
Université Laval

7.1 Introduction

In this paper, I provide a complete characterization of Nash implementable allocations of spending in prevention by judgement-proof injurers (hereafter *players*). This characterization is used to identify the rule that allows for the maximum total spending in prevention. Models of liability rules for two players go back to Brown (1973). Liability rules for more than two heterogeneous players can be found in Shavell (1987) and Emons and Sobel (1991). In a series of papers, Kornhauser and Revesz (1989, 1990, 1994) have pointed out that the characterization of liability rules for many judgement-proof injurers, that is injurers with a limited liability, is problematic. Limited liability is a real concern in liability cases involving life or environmental matters where the magnitude of damages can quickly skyrocket well beyond the actual capacity of paying of injurers. A liability rule designed to provide incentives to players to undertake due care must take into account these constraints.

Kornhauser and Revesz' analysis is restricted to two-player situations. They consider the equilibria induced by *ad hoc* liability rules that have been proposed in law and economics or that are actually used in real life legal disputes. By contrast, the number of players here is arbitrary and I cast the problem as one of mechanism design so that the whole set of feasible rules can be analyzed in a single step.

My analysis is related to that of Bergstrom, Blum and Varian (1986, 1992) who study the voluntary private provision of a public good by players with

---

[*]I thank Michel Truchon, Yves Richelle, Claude Fluet, Michel Poitevin, Karine Gobert, David Martimort and an anonymous referee for their helpful comments. All errors are mine. This paper was completed while visiting the *LERNA* in Toulouse.

different wealth endowments. These authors show how the voluntary provision to a public good is affected by a redistribution of wealth when the redistribution modifies the subset of net contributors. In the present context, liability plays the role assigned to wealth in the public good problem.

My main result states that to achieve the maximum total spending in prevention, it is strictly efficient to make all players strictly liable except one that I call the "deep-pocket" or the "victim". This player is identified as the one who is the most responsive to monetary incentives under the strict liability rule. He is to be subjected to the negligence rule: he shall evade liability only if he has undertaken a due amount of spending in effort. Actually, the money gathered from the strictly liable players is used to provide additional monetary incentives to the deep-pocket. Under that regime, all spending in effort are undertaken by the deep-pocket only.

This is a striking result. The typical analysis of liability rules deals with the problem of disciplining a single player and leads to the conclusion that the negligence rule is strictly better than the strict liability rule when the player has a low solvency. The analysis here shows that when a *group* of potential players is involved, the negligence rule should be applied only on a single subsidized player; all the others should be subjected to the strict liability rule.

Who shall be that subsidized player is problematic: on one hand, since monetary incentives are scarce, we would like him to be highly responsive under the negligence rule. Players who are the more responsive under the strict liability rule are also those who are the most responsive under the negligence rule. Yet, these players are also the ones who have the highest solvency, hence the ones who would provide more monetary incentives under the strict liability regime. I show that the first effect will dominate favoring the deep-pocket.

The rest of this paper is structured as follows. In the next section, I present the formal model. In Section 7.3, I show that implementation in dominant strategy is not possible with judgement-proof players. This justify the focus on Nash implementation. In Section 7.4, I recast the classical result of the dominance of the negligence rule over the strict liability rule to discipline a judgement-proof player. In Section 7.5, I generalize this result to the multi-player setting by characterizing the set of allocations of spending that can be achieved under Nash implementation. With this characterization at hand, I identify the rule that provides the maximum incentives to spend in prevention. This is done in the last section which also provides two interpretations of this rule depending on whether the subsidized party is perceived as a "victim" or as a "deep-pocket".

7.2 The Model

There are N players indexed with i. These players spend in prevention to reduce the probability of an accident. If an accident occurs, the courts applies a liability rule that specifies the different compensating damages to be paid or received by the players.

All players are assumed to have quasi-linear preferences. Player i's utility in the no accident state is V_i. The indexes are chosen so that

$$V_1 \leq V_2 \leq \ldots \leq V_N.$$

I assume that the last inequality is strict so that $V_N > V_i$ for $i < N$. Player i's utility in the accident state is a measure of the value of his seizable assets. It is denoted U_i. I assume that $U_i \geq 0$ and strictly so for at least one player j. The cost of an accident is thus $C_i = V_i - U_i$. I assume that an accident is bad for all players ($C_i \geq 0$) and strictly so for at least one player. Because of player j above, it follows that $V_N \geq V_j \geq U_j > 0$; hence $V_N > 0$.[1]

After an accident, player i must pay compensatory damages L_i. This raises his cost of an accident to $C_i + L_i$. This payment could actually lead to a reduction of that cost if $L_i < 0$; that is, if i is a "victim" that gets compensated.

Player i's strategy amounts to choose a level of spending $X_i \geq 0$ in prevention to decrease the likelihood of the accident state. Let $X = [X_1, \ldots, X_N]$ denotes the strategy profile. I use the standard game-theoretic notation for alternate profiles

$$X_{-i} = [X_1, \ldots, X_{i-1}, X_{i+1}, \ldots, X_N],$$
$$[X_{-i}, \chi] = [X_1, \ldots, X_{i-1}, \chi, X_{i+1}, \ldots, X_N].$$

Beside, $x = \sum X_i$ is the total sum of spending in accident prevention and $x_{-i} = x - X_i$ is the share of this sum supported by the players other than i. A similar notation is used for the variables U_i, V_i, C_i and L_i. For instance, it was assumed above that $V \geq U \geq \mathbf{0}$ (the null vector) and that $u > 0$ and $c > 0$.

The higher x, the lower the probability $P(x)$ of an accident. Hence the spending of all players are perfect substitutes in the prevention technology. To account for decreasing marginal returns in prevention, P is assumed to be differentiable, strictly decreasing, strictly convex and to satisfy the standard Inada conditions: $\lim_{x \to 0} P'(x) = -\infty$, $\lim_{x \to \infty} P'(x) = 0$ and $\lim_{x \to \infty} P(x) = 0$.

Define player i's (expected) private cost as

$$\phi_i(X, L_i) = P(x)(C_i + L_i) + X_i.$$

---

[1] The same implication results from $V_N > V_i \geq U_i \geq 0$ for $i < N$.

Private cost strictly increases with L_i. Prevention is a public good whose provision may be problematic: Notwithstanding its effect on the allocation of liabilities, a raise by player i of his contribution X_i reduces not only his (expected) private cost of an accident – his private benefit – but it reduces that of the other players as well – an external effect.

I assume that the accident state entails an additional external cost a to society. Expected social cost is defined as the expected sum of private costs plus a and minus the sum of transfers

$$P(x)(a - l) + \sum \phi_i(X_i, L_i) = P(x)(a + c) + x.$$

Notice that, given x, social cost is independent of X.

Consider the function $P(\chi)K + \chi$ where K is a parameter. This function is strictly quasiconvex: If $K > 0$, it is strictly convex; if $K \leq 0$, it is strictly increasing. In both cases, strict quasiconvexity follows. Private and social costs are special cases of this function with respectively $K = C_i + L_i$ and $K = a + c$. We shall repeatedly encounter the case where $K = V_i$ so it is worthwhile to define

$$\psi_i(\chi) = P(\chi)V_i + \chi,$$

which is also strictly quasiconvex by the same argument. Strict quasiconvexity ensures a unique minimum; hence define

$$\psi_i^* = \min_{\chi \geq 0} \psi_i(\chi) = \psi_i(\xi_i).$$

The value ξ_i is player i's *maximum level of spending*. Indeed, should player i expect to be fully liable ($L_i = U_i$) and to spend alone in prevention ($x_{-i} = 0$), his cost of an accident would be $C_i = V_i$ and he we would choose to spend ξ_i to minimize his private cost. A straightforward application of the envelope theorem establishes that ξ_i increases with V_i, hence with i: player N has the highest maximum level of spending. Furthermore, since $V_N > 0$ and $P'(0) \to -\infty$, that level is strictly positive: $\xi_N > 0$. I call player N the *deep-pocket* or the *victim* (both interpretations are discussed in the conclusion).

Applying again the envelope theorem to the minimization of social cost above, we see that, as the external cost a changes, any level $x \geq 0$ may be rationalized as socially efficient.

A *liability rule* is a function R that maps the courts' available information into a vector of liabilities L to be imposed to the players in the accident state. In this paper, I assume that the courts have ex post perfect information but I shall only make explicit the liability rule dependence on X by writing $L = R(X)$. A liability rule is *separable* (with respect to X) if L_i depends on X through X_i alone. It is *admissible* if, for all $X \geq \mathbf{0}$, it satisfies the limited liability constraints

$$R(X) \leq U,$$

and budget balance

$$\sum R_i(X) \geq 0,$$

so that the courts are not a net contributor.[2]

All the relevant information (a, P, U, etc) is common knowledge among the players when they choose their strategy profile X. In particular, they commonly know which rule R will be applied in the accident state. A liability rule is then a *mechanism* that structures the prevention game through the payoff functions $-\phi_i(\cdot, R_i(\cdot))$.

In this paper, I characterize the admissible rule that provides the best incentives to minimize expected social cost. The choice of an optimal mechanism depends on the solution concept assumed to give a good description of how the game will be played. Among the solution concepts encountered in the literature, those of *(weakly) dominant strategy equilibrium* and of *Nash equilibrium* are the most common.

An allocation X may be implemented in dominant strategies (DS) if there exists a rule such that X_i is a (weakly) dominant strategy for each player i. An allocation X is Nash implementable (NI) if there exists a rule such that X_i is a best reply to X_{-i} for each player i. An admissible NI allocation is a NI allocation that can be implemented with an admissible rule.

7.3 Implementation in Dominant Strategy

If the limited liability constraints are discarded, any allocation X may be implemented in DS as follows. Let f_i be any function such that

$$f_i(\chi) \geq \phi_i([\mathbf{0}_{-i}, \chi], 0),$$

and that reaches its minimum at X_i. Define the rule

$$R_i(X) = \frac{f_i(X_i) - X_i}{P(x)} - C_i,$$

so that player i's private cost becomes

$$\phi_i(\chi, R_i([X_{-i}, \chi])) = f_i(\chi).$$

By construction, whatever the value x_{-i}, spending X_i in prevention minimizes his private cost. Notice that budget balance is ensured since

$$R_i([X_{-i}, \chi]) \geq \frac{P(\chi) - P(x_{-i} + \chi)}{P(x_{-i} + \chi)} C_i \geq 0.$$

---

[2]When this inequality is strict, the money collected is used to restore the resource and/or is redistributed among the general public.

However, this rule does not satisfy the liability constraints. If player i believes that the other players will invest a lot so that the probability of an accident becomes small, he knows that in any event his liability will be bounded at U_i. Since he does not expect the accident state to occur anyway, he will invest zero. But players are similar and investing a lot and not investing can't concurrently be dominant strategies. This result is generalized in the next proposition (all proofs are in the appendix).

Proposition 7.1 *The set of allocations that can be implemented in dominant strategies by an admissible rule is empty.*

The scope of Proposition 7.1 goes beyond stating that it is difficult to handle the crowding out problem (giving player i incentives to invest reduces the same incentives for the other players). It emphasizes that there is also a coordination problem since no equilibrium in dominant strategies would exist even if no liability was imposed ($R \equiv \mathbf{0}$).

Proposition 7.1 relies a lot on the Inada conditions imposed on the prevention technology. Arguably, in less stringent environments, one could find allocations implementable in DS with an admissible rule. But since existence of such allocations is not guaranteed, DS is not an attractive concept to study liability rules. On the other hand, existence is not an issue with Nash implementation. Besides, the set of NI allocations obviously includes that of allocations implementable in DS. Identifying the set of NI allocations is thus an important step to devise a sensible liability rule (see Footnote 3).

7.4 The Strict Liability and Negligence Rules

Suppose that R implements X as a Nash equilibrium:

$$X_i \in \arg\min_{\chi \geq 0} \phi_i([X_{-i}, \chi], R_i([X_{-i}, \chi])), \quad \forall i. \tag{7.1}$$

Then we can always define the separable rule

$$R_i'(\chi) = R_i([X_{-i}, \chi]),$$

that implements X as well. Furthermore, if R is admissible, so is R'. Hence, without loss of generality, we can focus on separable rules.[3]

---

[3] To go back to the issue of robustness provided by DS, if a separable rule R' induces X as a Nash equilibrium, then investing X_i is optimal for i given x_{-i} but not if i expects $x'_{-i} \neq x_{-i}$. Proposition 7.3 states that, generally, there does not exist an admissible rule that works for *all* possible deviation x'_{-i} but the rule could be made robust against many. That is, we can enrich R' by defining a non-separable rule R'' for which investing X_i is optimal against many possible deviations x'_{-i}. Nevertheless, $R''([X_{-i}, X_i]) = R'(X_i)$ is still a necessary condition for R'' to be implementable.

I begin the analysis of admissible separable rules by fixing X_{-i} and the liabilities $L_j = R_j(X_j)$ for $j \neq i$. This leaves X_i and the function R_i to be specified. To ensure that best replies are well defined, I shall restrict R_i to be lower semi-continuous. Besides, admissibility imposes that

$$-l_{-i} \leq R_i(X_i) \leq U_i. \tag{7.2}$$

A classical result in law and economics is the weak dominance of the negligence rule over the strict liability rule under limited liability (see Shavell, 1986). These two rules differ in the definition of the event in which player i is liable. Under the negligence rule, player i's liability is conditional on the event that he has spent less than some standard of care X_i (see below). Under the strict liability rule, player i is fully liable for the damage in any event. When the value of the damage is greater than the value of his assets, player i's liability binds at U_i. Because I am interested in cases where there is under-provision of spending in prevention, that is in cases where player i's liability is likely to bind, I associate the strict liability rule with the constant rule $R_i(X_i) \equiv U_i$ that specifies the same maximum payment regardless of X_i. Player i's lost in the accident state is then raised to $C_i + L_i = V_i$ and his expected cost becomes

$$\phi_i([X_{-i}, X_i], U_i) = \psi_i(x_{-i} + X_i) - x_{-i}. \tag{7.3}$$

That cost is minimized in

$$X_i^* = \max\{0, \xi_i - x_{-i}\},$$

to

$$\phi_i^* = \phi_i([X_{-i}, X_i^*], U_i),$$
$$= \begin{cases} \psi_i(x_{-i}) - x_{-i} & \text{if } x_{-i} > \xi_i, \\ \psi_i^* - x_{-i} & \text{if } x_{-i} \leq \xi_i. \end{cases}$$

That is, either the other players spend more than player i's maximum level of spending ($x_{-i} > \xi_i$) so that i spends zero and bears the expected cost $P(x_{-i})V_i = \psi(x_{-i}) - x_{-i}$, or player i is willing to contribute $X_i^* = \xi_i - x_{-i} \geq 0$ to raise x to ξ_i and bears the expected cost $P(\xi_i)V_i + X_i^* = \psi_i^* - x_{-i}$. If we define

$$\xi_i^* = \max\{x_{-i}, \xi_i\},$$

then we may write

$$X_i^* = \xi_i^* - x_{-i} \quad \text{and} \quad \phi_i^* = \psi_i(\xi_i^*) - x_{-i}.$$

Consider now the constant rule where player i always receives $-l_{-i}$ in the accident state so that his expected cost is $\phi_i(X, -l_{-i})$. Let χ_i^* minimize this cost and define the lower contour set

$$\mathbf{X}_i = \{\chi \geq 0 : \phi_i([X_{-i}, \chi], -l_i) \leq \phi_i^*\}.$$

By (7.2), player i's cost is reduced under this ruled so that $\mathbf{X}_i$ is not empty. Since ϕ_i is quasiconvex and continuous in X_i, the set $\mathbf{X}_i$ is a closed interval

$$\mathbf{X}_i = [X_i^{\min}, X_i^{\max}]$$

where both ends are implicitly defined as the solutions of

$$\phi_i([X_{-i}, \chi], -l_i) = \phi_i^*. \tag{7.4}$$

I define the negligence rule as

$$R_i(\chi) = \begin{cases} -l_{-i} & \text{if } (\chi - X_i)(X_i - \chi_i^*) \geq 0, \\ U_i & \text{else.} \end{cases} \tag{7.5}$$

Hence, when a high level of care is expected $(X_i > \chi_i^*)$, player i receives l_{-i} if he has spent at least X_i and pays U_i otherwise.[4]

Proposition 7.2 establishes the weak optimality of the negligence rule since it can implement any level o spending that could be implemented with any other rule.

Proposition 7.2

1. X_i can be implemented with a rule bounded by (7.2) if and only if $X_i \in \mathbf{X}_i$.

2. Any $X_i \in \mathbf{X}_i$ may be implemented with the negligence rule.

By construction, player i's expected cost is minimized in X_i with the negligence rule. Hence, his liability is given by $L_i = R_i(X_i) = -l_{-i}$. (Notice though that when $-l_{-i} = U_i$, the negligence rule and the strict liability rule are confounded.) Proposition 7.2 is illustrated in Figure 7.1. There, the two constant rules where player i always pays U_i or $-l_i$ induce two U-shaped expected cost functions $\phi_i([X_{-i}, \chi], U_i)$ and $\phi_i([X_{-i}, \chi], -l_i)$. Any other admissible rule $\tilde{R}$ induces an expected cost function between those two. Hence, the minimum cost with such a rule is necessarily reached within a set $\mathbf{X}_i$ delimited by the lower contour set of $\phi_i([X_{-i}, \chi], -l_i)$ at ϕ_i^*. By construction,

---

[4] Again, this rule is somewhat different than the classical negligence rule because it always specifies the maximum amount the player can pay in the event of default, regardless of the actual damage, and the maximum reward otherwise. Again, this simplification is made because I am interested in cases where damages are large so that the liability constraints bind.

the negligence rule induces an expected cost function that is discontinuous at X_i (the thick line). When $X_i \geq X_i^*$, it follows $\phi_i([X_{-i}, \chi], U_i)$ for $\chi < X_i$ and $\phi_i([X_{-i}, \chi], -l_i)$ thereafter. It is minimized in X_i, a value that may be set anywhere in $\mathbf{X}_i$.

7.5   Nash Implementation

Up to now, X_{-i} and L_{-i} were assumed fixed. I now consider the case where all players choose their level of spending simultaneously given a liability rule R. Proposition 7.2 states that for any player i, and given X_{-i} and L_{-i}, there is no loss in generality in imposing the negligence rule to that player. Hence, the only unknowns to be specified are the allocation X of standards and the liabilities L. Hence, in what follows, I resume the description of a liability rule by the vector L where it is understood that, given X and L, R is given by (7.5).

With many players, there is a dilution of incentives. Suppose that all players are liable under the strict (constant) liability rule; $L = U$. In that case, each player minimizes (7.3)

$$\min_{\chi \geq 0} \psi_i(x_{-i} + \chi) - x_{-i}$$

Let M be the subset of m players who spend in prevention in a Nash equilibrium. Since player N's maximum level of spending ξ_N is strictly positive, we know that $m \geq 1$ because if nobody else would spend, player N would. Then

$$x = \sum_{i \in M} X_i = \sum_{i \in M} (\xi_i - x_{-i}),$$

$$= \sum_{i \in M} (\xi_i - x + X_i),$$

$$= \sum_{i \in M} \xi_i - (m-1)x,$$

$$= \frac{1}{m} \sum_{i \in M} \xi_i. \tag{7.6}$$

The only way (7.6) may hold is if $M \equiv \{N\}$, $m = 1$ and $x = \xi_N$; that is, if player N is the only one who spends in prevention. In what follows, I note $X^i(x)$ such allocation where player i alone spends the total amount x in prevention:

$$X^i(x) = [0, \ldots, 0, x, 0, \ldots, 0],$$

with x in the ith position. Hence, under the strict liability rule, the spending in prevention by player N crowds out the incentives for the other players to spend as well. This rule generates excess liability since $l = u > 0$. This suggests a better rule where that money could be used to provide additional incentives to player N. I will show that such a rule is indeed optimal.

To get this result, we need to define the function $F : \mathbb{R}_+^N \to \mathbb{R}_+$,

$$F(X) = \sum \psi_i(\xi_i^*).$$

Notice that F is a continuous function. Because all players have quasi-linear preferences, we get the following characterization of admissible NI allocations.

Proposition 7.3 *An allocation X is NI with an admissible rule if and only if*

$$P(x)c + Nx \leq F(X). \tag{7.7}$$

Besides, if (7.7) holds, then

$$L_i^+ = U_i - \frac{\psi_i(x) - \psi_i(\xi_i^*)}{P(x)}$$

is admissible and implements X as a Nash equilibrium.

Proposition 7.3 gives a characterization of admissible NI allocations that does not depend on the rule actually used to implement them. As we shall see, not all levels of spending x can be achieved with an admissible NI allocation and two allocations that provide the same level of total spending x, hence the same level of expected social cost, may not be both admissible NI allocations because the distribution of spending among players matters for implementation. Inequality (7.7) makes a clear distinction between the total level of spending on the l.h.s. and the distribution of spending on the r.h.s. In particular, it is clear that given x, (7.7) is easier to satisfy when F is maximized.

The liabilities L^+ are set so that (7.1) holds (X_i is a best reply for each player). The rule induced by L^+ is admissible by construction:

- If $\xi_i \geq x_{-i}$, then $\psi_i(\xi_i^*) = \psi_i^*$ and $L_i^+ \leq U_i$ since ψ_i has ψ_i^* for minimum.

- If $\xi_i < x_{-i}$, then $\psi_i(\xi_i^*) = \psi_i(x_{-i})$. Both x and x_{-i} are two values to the right of the minimum ξ_i of ψ_i. Since that function is quasiconvex, the difference $\psi_i(x) - \psi_i(x_{-i})$ is positive. It follows that $L_i^+ \leq U_i$.

Multiplying L_i^+ by $P(x)$ and summing over i then yields

$$P(x)l^+ = F(X) - (P(x)c + Nx),$$

so that budget balance holds if (7.7) holds.

7.6 Providing the Maximum Incentives

Recall that, depending on the external cost a, any x may be rationalized as a socially efficient amount of spending in prevention when designing a liability scheme. In this section, I use Proposition 7.3 to determine the maximum level of spending x^* that can be achieved with a NI admissible allocation and the optimal liability rule that implements this allocation.

As a corollary of Proposition 7.3, notice that for $X^N(\xi_N)$, inequality (7.7) becomes

$$P(\xi_N)c + N\xi_N \leq P(\xi_N)v + N\xi_N,$$
$$-P(\xi_N)u \leq 0, \tag{7.8}$$

which is true so that $X^N(\xi_N)$ is NI with an admissible rule (the strict liability rule for every player) as it has already been suggested at the beginning of section 7.5. It follows that $x^* \geq \xi_N$. Since both sides of (7.7) are continuous functions of X, it is clear that x^* is reached when (7.7) holds with equality and F is maximized given x. As the next lemma shows, F is maximized in $X^N(x)$ when x is sufficiently large.

Lemma 7.1 *For $x > \xi_N$, $X^N(x)$ uniquely maximizes F subject to $\sum X_i = x$.*

If (7.8) did hold with equality, we would have found x^* but it does not since $u > 0$. Hence $x^* > \xi_N$ and we may use Lemma 7.1 in (7.7) to reach x^*:

$$P(x^*)c + Nx^* = F(X^N(x^*)),$$
$$P(x^*)(C_N + c_{-N}) + Nx^* = \psi_N(\xi_N) + \sum_{i<N} \psi_i(x^*),$$
$$P(x^*)(C_N + v_{-N} - u_{-N}) + Nx^* = \psi_N^* + P(x^*)v_{-N} + (N-1)x^*,$$
$$P(x^*)(C_N - u_{-N}) + x^* = \psi_N^*,$$
$$\phi_N(X^N(x^*), -u_{-N}) = \phi_N^*. \tag{7.9}$$

Comparing (7.9) with (7.4), we see that x^* is implemented with the liabilities

$$L^* = (U_1, U_2, \ldots, U_{N-1}, -u_N);$$

that is, by providing player N with the maximum level of (admissible) incentives $l_{-N} = u_{-N}$ under the negligence rule and by setting a maximal standard at the top of $\mathbf{X}_N$ with these incentives. This important result is formalized in the next proposition.

Proposition 7.4 *Let x^* solve (7.9). Then, within the class of liabilities rules defined by (7.5), $X^N(x^*)$ and L^* uniquely implement x^*.*

Since $-l_i = U_i$ for all $i < N$, the optimal multiplayer liability rule puts every player under a strict liability regime except player N who stays under the negligence rule and who actually receives the money collected from the other players when an accident happens.

It is easy to understand Proposition 7.4 if we relate it to the classical problem of the private provision of a public good first analyzed by Warr (1983) and Bergstrom, Blum and Varian (1986). These authors show that the amount of public good provided is independent of the distribution of income unless the set of contributors is affected by the distribution. The optimal multiplayer rule achieves this by concentrating all the ex post wealth (the compensatory damages from the liable agents) into the hands of a single player. That player is then disciplined through the negligence rule. Because there is a single player who spends in prevention, there is no dilution of incentives and a maximum of spending is undertaken.

Under this rule, the deep pocket is expected to undertake all spending. Again, the comparison with the public good problem helps to understand the result: if all wealth is to be given to a single player to spend on a public good and if any player would spend less than the socially optimal amount, then it makes sense to give the wealth to the player who values the most the public good. If we concentrate all incentives upon a single player i which we submit to the negligence rule, then the cost of an accident for this player becomes

$$C_i + L_i = V_i - U_i + (-u_{-i}) = V_i - u.$$

Hence, the most responsive player under that rule is the "deep pocket" for whom the cost of an accident under strict liability (V_i) is the greatest.

I have suggested that there are two interpretations of player N as the "deep pocket" or as the "victim". By definition,

$$V_i \equiv C_i + U_i.$$

Hence, the cost of an accident and the ex post liability of a player are jointly identified in this model. The "deep-pocket" interpretation is natural when there is little variation in the C_is relatively to the U_is. Then, all players would be similarly careless in absence of a liability regime but player N is highly motivated to produce the required amount of care once his assets U_i are at stake. He is then chosen because he is the most responsive to monetary incentives under the negligence rule.

When there is a lot of variation in the C_is relatively to the U_is, interpreting player N as a "victim" is more natural. Then, all players have roughly the same ability to pay ex post but player N has an higher ex ante incentive to spend in prevention because of his higher cost of an accident.

Appendix

The proofs of the propositions and Lemma 7.1 follow.

Proof of Proposition 7.1 First, define

$$z(K) = \min_{\chi \geq 0} P(\chi)K + \chi.$$

Because $P'(0) \to -\infty$, $z(K) = 0$ implies that $K \leq 0$.

Suppose that X can be implemented in DS with R. Then either $x = 0$ (so that $X = \mathbf{0}$) or $x > 0$.

Suppose that $x = 0$ and consider a rule R that implements $X = \mathbf{0}$. For $X_i = 0$ to be dominant when $x_{-i} = 0$, it must be that

$$z(C_i + R_i(\mathbf{0})) = 0,$$

which implies

$$C_i + R_i(\mathbf{0}) \leq 0, \tag{A.7.1}$$

as above. Summing (A.7.1) over i yields

$$l = \sum R_i(\mathbf{0}) \leq -c < 0;$$

a violation of budget balance.

Suppose that $x > 0$. Then there exists a player i for which setting $X_i > 0$ is a weakly dominant strategy at least as good as investing nothing:

$$P(x)(C_i + R_i([X_{-i}, X_i])) + X_i \leq P(x_{-i})(C_i + R_i([X_{-i}, 0])), \quad \forall X_{-i}.$$

Since R_i is bounded above by U_i, as $x_{-i} \to \infty$, the probability of an accident vanishes on both sides and this inequality yields $X_i \leq 0$; a contradiction. ∎

Proof of Proposition 7.2

1. To prove sufficiency, consider any lower semi-continuous rule R_i bounded by (7.2). Notice that ϕ_i is increasing in its second argument. Hence, for any best reply X_i to any lower semi-continuous rule R_i, one has

$$\phi_i(X, -l_i) \leq \phi_i(X, R_i(X_i)) \leq \phi_i([X_{-i}, \chi], R_i(\chi)) \leq \phi_i([X_{-i}, \chi], U_i),$$

for any $\chi \geq 0$. In particular, for $\chi = X_i^*$,

$$\phi_i(X, -l_i) \leq \phi_i([X_{-i}, X_i^*], U_i) = \phi_i^*,$$

so that $X_i \in \mathbf{X}_i$. To prove necessity, consider point 2 below and the fact that the negligence rule is a lower semi-continuous rule.

2. Assume that we want to implement $X_i \in \mathbf{X}_i$. We verify that the negligence rule incites player i to invest X_i.

If $X_i = \chi_i^*$, then $R_i(\chi) \equiv -l_i$ and cost are minimized by setting $\chi = \chi_i^* = X_i$. If $X_i > \chi_i^*$, then

$$R_i(\chi) = \begin{cases} -l_i & \text{if } \chi \geq X_i, \\ U_i & \text{else.} \end{cases}$$

Playing $\chi < X_i$ yields at least ϕ_i^* while playing $\chi \geq X_i$ minimizes cost to

$$\min_{\chi \geq X_i} \phi_i([X_{-i}, \chi], -l_i) = \phi_i(X, -l_i) \leq \phi_i^*.$$

The minimum is in X_i because the unconstrained solution is $\chi_i^* < X_i$ and $\phi_i([X_{-i}, \chi], -l_i)$ is quasiconvex. If $X_i < \chi_i^*$, a similar argument applies and cost are also minimized in X_i. ∎

Proof of Proposition 7.3 Suppose that X is NI and that L implements X. Then, for all i, $X_i \in \mathbf{X}_i$ so that

$$\phi_i(X_i, L_i) \leq \phi_i^*,$$
$$P(x)(C_i + L_i) + X_i \leq \psi_i(\xi_i^*) - x_{-i},$$
$$P(x)(C_i + L_i) + x \leq \psi_i(\xi_i^*). \tag{A.7.2}$$

Clearly, if (A.7.2) holds for all i then L implements X as a Nash equilibrium. Summing (A.7.2) over i yields

$$P(x)(c + l) + Nx \leq F(X).$$

From (A.7.2), it is clear that if $L \leq U$ implements X, so does any $L' \leq L$. Hence if $l > 0$, we can always find $L' \leq U$ that implements X as well and such that $l' = 0$. Hence (7.7) holds.

For the sufficiency part: given X such that (7.7) holds, L^+ solve (A.7.2) with equality so that X is NI. The discussion in the text establishes that $L^+ \leq U$. Besides, since (7.7) holds, $l^+ \geq 0$ and L^+ is admissible. ∎

Proof of Lemma 7.1 Since F is continuous, it reaches its minimum on the compact set defined by $X \geq \mathbf{0}$ and $\sum X_i = x$. Define

$$\theta_{ij}(\chi) = F([X_{-i-j}, \chi, X_i + X_j - \chi]).$$

Then, for any pair (i, j), the allocation χ of spending $X_i + X_j$ between i and j should be optimal. It follows that

$$X_i \in \arg\max_{0 \leq \chi \leq X_i + X_j} \theta_{ij}(\chi) \tag{A.7.3}$$

is a necessary condition for X to maximize F. More in detail:

$$\theta_{ij}(\chi) = \sum_{k \neq i, k \neq j} \psi_k(\xi_k^*) + \psi_i(\max\{x - \chi, \xi_i\}) + \psi_j(\max\{x - X_i - X_j + \chi, \xi_j\})$$

The function θ is convex over $\chi \geq 0$. It is the sum of a constant and two functions. Recall that ψ_i is convex. Then the first function is

$$\psi_i(\max\{x - \chi, \xi_i\}) = \begin{cases} \psi_i(x - \chi) & \text{if } 0 \leq \chi < x - \xi_i, \\ \psi_i^* & \text{if } x - \xi_i \leq \chi. \end{cases}$$

It is convex since $\psi_i(x - \chi)$ decreases toward the minimum ψ_i^* as χ is increased. The second function is

$$\psi_j(\max\{x - X_i - X_j + \chi, \xi_j\})$$

$$= \begin{cases} \psi_j^* & \text{if } 0 \leq \chi < \xi_j - x + X_i + X_j. \\ \psi_j(x - X_i - X_j + \chi) & \text{if } X_i + X_j - (x - \xi_j) \leq \chi. \end{cases}$$

It is convex since $\psi_j(x - X_i - X_j + \chi)$ increases from the minimum ψ_j^* as χ is increased. The sum of convex functions is convex so that θ_{ij} is convex.

It follows that, in our search for an X that maximizes F, we may assume that, for any pair (i, j), either X_i or X_j equals zero. For this to be true for every possible pair it must be that $X = X^i(x)$. Then, when $x > \xi_N > \xi_i$,

$$F(X^i(x)) = \psi_i(\xi_i) + \sum_{j \neq i} \psi_j(x),$$

$$= \sum_j \psi_j(x) - [\psi_i(x) - \psi_i(\xi_i)],$$

That last expression is maximized when the bracketed term is minimized. Suppose that $i < N$; then

$$\psi_i(x) - \psi_i(\xi_i) = \int_{\xi_i}^x (P'(\chi)V_i + 1)d\chi.$$

To the right of ξ_i, $P'(\chi)V_i + 1$ is a positive function; hence,

$$> \int_{\xi_N}^x (P'(\chi)V_i + 1)d\chi,$$

$$> \int_{\xi_N}^x (P'(\chi)V_N + 1)d\chi,$$

$$= \psi_N(x) - \psi_N(\xi_N).$$

The inequalities are strict because $x > \xi_N > \xi_i$ and $V_N > V_i$. Hence, if we restrict X to $X^i(x)$, F is uniquely maximized in $X^N(x)$.

Finally, consider the possibility that X maximizes F and that $X_i > 0$ for a group M of more than one player. With $x > \xi_N$ and for (A.7.3) to hold, it must be that

$$M \equiv \{i : X_i > 0\} = \{i : x - X_i < \xi_i\}.$$

To understand this step, go back to the definition of θ_{ij} for i and j in M. If $x - X_i \geq \xi_i$, then $\psi_i(\max\{x - \chi, \xi_i\})$ has a strictly convex (decreasing) portion on the left when $0 \leq \chi < x - \xi_i$. If $x - X_j \geq \xi_j$, then $\psi_j(\max\{x - X_i - X_j + \chi, \xi_j\})$ has a strictly convex (increasing) portion on the right when $X_i + X_j - (x - \xi_j) \leq \chi \leq X_i + X_j$. If any of these two portions is present, then θ_{ij} is maximized either in $\chi = 0$ so that $X_i = 0$, or in $\chi = X_1 + X_2$ so that $X_2 = 0$. In both cases we get a contradiction.

Then

$$F(X) = \sum_{i \in M} \psi_i^* + \sum_{k \notin M} \psi_k(x). \qquad (A.7.4)$$

Now, for i and j in M, consider lowering X_j to zero and raising X_i to $X_i + X_j$. We would still have $i \in M$ but $j \notin M$. Then F would be raised by $\psi_j(x) - \psi_j^* > 0$. Hence (A.7.4) is not a maximum. $\blacksquare$

Proof of Proposition 7.4 We have already shown that $X^N(x^*)$ and L^* implements x^*. To get to x^*, we had to maximize F. Since $x^* > \xi_N$, Lemma 7.1 implies that $X^N(x^*)$ uniquely implements x^*. Proposition 7.2 establishes that L^* uniquely implements $X^N(x^*)$ with a negligence rule. $\blacksquare$

References

Bergstrom, T., L. Blum, and Varian H., 'On the private provision of public goods', *Journal of Public Economics*, 29(1) (1986): 25-49.

Bergstrom, T., L. Blum, and Varian H., 'Uniqueness of Nash equilibrium in private provision of public goods', *Journal of Public Economics*, 49(3) (1992): 391-392.

Brown, J. P., 'Toward economic theory of liability', *Journal of Legal Studies*, 2 (1973): 323-350.

Emons, W. and J. Sobel, 'On the effectiveness of liability rules when agents are not identical', *Review of Economic Studies*, 58(2) (1991): 375-390.

Kornhauser, L. A. and Revesz R. L., 'Apportioning damages among potentially insolvent actors', *Working Paper 89-22* (1989), C.V. Starr Center for Applied Economics.

Kornhauser, L. A. and Revesz R. L., 'Apportioning damages among potentially insolvent actors', *Journal of Legal Studies*, XIX (1990): 617-651.

Kornhauser, L. A. and Revesz R. L., 'Multidefendant settlements under joint and several liability: the problem of insolvency', *Journal of Legal Studies*, XXIII (1994): 517-542.

Shavell, S., 'The Judgment Proof Problem', *International Review of Law and Economics*, 6(1) (1986): 45-58.

Shavell, S., *Economic Analysis of Accident Law*, (Harvard: Harvard University Press, 1987).

Warr, P. G., 'The private provision of a public good is independent of the distribution of income', *Economic Letters*, 13 (1983): 207-211.

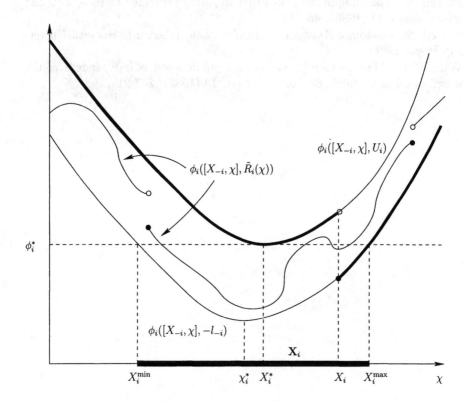

Figure 7.1: The optimality of the negligence rule

Chapter 8

A Tort for Risk and Endogenous Bankruptcy[*]

Thomas J. Miceli
University of Connecticut

Kathleen Segerson
University of Connecticut

8.1 Introduction

Environmental accidents often involve mass exposure to a toxic substance that creates (or increases) the risk of future illness. Examples include accidental chemical releases such as the one that occurred in Bhopal, India in 1984 (Fischer, 1996), nuclear accidents such as Three Mile Island and Chernobyl, and prolonged exposure to asbestos. Under traditional tort law, victims of such exposure cannot file damage claims until they actually develop symptoms of illness.[1] However, the long latency period of many illnesses, the difficulty of proving causation, and the possibility of injurer bankruptcy, all act as effective bars against recovery. Thus, some scholars have argued that victims should have the option to sue for expected damages at the time of exposure, in effect treating the exposure itself as a tort–what we will call a "tort for risk" (TFR) (Love, 1996).

Several previous authors have examined the tradeoffs involved in allowing a tort for risk.[2] A key concern is the impact on litigation costs. While Robinson (1985) has argued that allowing a tort for risk would drastically increase the number of lawsuits and hence total litigation costs, advocates suggest an offsetting deterrence benefit. In many toxic tort contexts, there is

---

[*]We acknowledge the very useful comments of David Martimort and an anonymous referee.

[1]See Keeton et al. (1984), paragraph 30, p. 165. In some jurisdictions, victims can sue for emotional distress and/or the costs of medical monitoring (Valk, 1995; Miceli and Segerson, forthcoming).

[2]See, for example, Landes and Posner (1984), Robinson (1985), and Miceli and Segerson (2003b). Also, see the related analysis by Rose-Ackerman (1989).

a long latency period between the time of exposure and the time a resulting disease is manifested. As a result, requiring victims to wait until the time of illness to file suit could actually bar victims from receiving any compensation if the injurer is insolvent or judgment-proof at the time the disease is contracted, or if the victim cannot establish a causal connection between her illness and the injurer's actions (Shavell, 1985). This affects both the extent of victim compensation and the injurer's incentive to take care to reduce the magnitude or likelihood of exposure. While victim compensation is primarily a distributional issue,[3] increased injurer care affects the expected damages from exposure and hence social welfare.

The concern about insolvency that underlies the above argument is based on an assumption that the risk that the injurer will be insolvent or judgment-proof at the time of illness is exogenous, i.e., driven by factors unrelated to liability. However, in large toxic tort cases where total liability can be large relative to a firm's assets, it is also possible for the liability itself to trigger bankruptcy if the firm has insufficient assets to cover the liability-related claims against it. A well-known example is the bankruptcy of the Johns-Manville Sales Corporation, which was triggered by the costs stemming from asbestos litigation (Note, 1983). This possibility raises an additional concern about allowing a tort for risk, namely, that it might drive a firm into bankruptcy when a traditional rule of only allowing suits at the time of illness would not have. Bankruptcy can entail a loss of social welfare (i.e., a reduction in economic efficiency), if it implies that an activity that on balance is socially beneficial (i.e., that makes an expected net positive contribution to social welfare) cannot be undertaken. This effect must be combined with the effect on litigation costs and deterrence to determine the overall impact on economic efficiency of allowing a tort for risk.

In this paper, we explore the relationship between a tort for risk and bankruptcy. Specifically, we ask whether the introduction of a tort for risk ever induces bankruptcy when it would not have occurred under the traditional rule. The reason that a tort for risk might increase the likelihood of bankruptcy is the possibility that victims will "race to file" at exposure for fear that the injurer will have insufficient assets later to pay actual damage claims.[4]

Our analysis of a race to file is related to other contexts in which a race to claim a firm's limited assets triggers bankruptcy. The classic example is a bank run, in which depositors rush to withdraw their assets from a bank because they expect other depositors to do the same and want to beat them

---

[3]Victim compensation is a reallocation of wealth from injurers to victims. Ex post, this reallocation affects the distribution of welfare within society but not society's total welfare. Of course, to the extent that a requirement of victim compensation causes injurers to invest more in prevention or care, it will indirectly affect social welfare.

[4]In this sense, bankruptcy is endogenous, whereas in most previous tort models it is treated as exogenous (Shavell, 1986; Miceli and Segerson, 2003a).

to ensure a greater claim on the bank's limited assets (Diamond and Dybvig, 1983). In a litigation context, Spier (2002) examines settlement negotiations between a defendant and multiple plaintiffs whose collective damages (if they win at trial) exceed the defendant's assets. Both of these papers are related to our analysis, but focus on different issues. Diamond and Dybvig (1983) focus on the liquidity service provided by banks and the impact of runs on risk sharing among risk averse depositors, while Spier (2002) focuses on the settlement-trial decisions of the plaintiffs rather than on the timing of their filing decisions.

The paper is organized as follows. In Section 8.2 we discuss in more detail the impact of allowing a tort for risk on litigation costs and deterrence when bankruptcy is endogenous. This discussion draws heavily on the TFR model developed in Miceli and Segerson (2003b). In the context considered there, litigation costs play a key role in inducing a possible race to file. In addition, the results depend on the assumption that there are a large number of victims. This implies that each victim thinks her individual filing decision will not have a sufficiently large impact to trigger bankruptcy. In addition, under this assumption, once an exposure has occurred the injurer can reasonably assume that a given share of victims will actually contract the illness. In Section 8.3 we consider an alternative scenario. We abstract from litigation costs by assuming that they are zero and focus instead on the "small" numbers case. In this setting, an individual victim's filing decision can trigger bankruptcy. In addition, the injurer must consider the possibility that all exposure victims will contract the illness (the "worst case" scenario).[5] We develop a simple model to examine the filing equilibria under this scenario when a tort for risk is allowed. We again ask whether allowing a TFR will trigger a race to file and bankruptcy, thereby preventing the firm from continuing to produce a product or service that could potentially be socially beneficial. We first describe the basic model and the equilibrium under the traditional tort rule. We then derive the equilibrium under the TFR rule and compare it to the traditional rule in terms of efficiency and compensation of victims. Finally, Section 8.4 concludes.

8.2 The Large Numbers Case with Litigation Costs

8.2.1 The Basic Model without Bankruptcy

Consider a scenario in which a (potential) injurer engages in an activity that creates the possibility of an accidental exposure of a large population of vic-

---

[5]While the worst case scenario is technically possible under the large number case as well, the probability that this will occur is sufficiently low when the number of exposure victims is high that the injurer can reasonably ignore it.

tims to a harmful substance.[6] The injurer can undertake activities that reduce the likelihood of an accident, but these activities are costly. If an exposure occurs, a given victim may or may not develop a related illness in the future. Let q be the probability that a victim ultimately develops the illness, which we assume results in losses equal to D dollars.[7] Assume that all victims suffer the same loss in the event of illness, but that the probability of developing the disease (q) varies across victims. In general, victims can be expected to differ in terms of the intensity of their exposure (e.g., the duration of exposure or the proximity to the point of release) and hence in their probabilities of contracting the illness. We assume that each victim's q is observable to the court so that it can calculate that victim's expected damages, qD, if necessary.

In terms of assessing liability on the injurer, we will consider two rules: the *traditional rule*, under which a victim can only seek damages in period two if she sustains an actual loss, and a *tort for risk* (TFR), under which a victim can sue in period one (immediately following exposure) for expected damages of qD. We assume throughout that liability is strict.[8] In this case, under the traditional rule, and in the absence of any bankruptcy consideration, exposure victims would be required to wait until they sustain actual damages before filing suit. If they ultimately contract the disease and file suit, they will receive a net return equal to their losses, D, less their litigation costs, denoted c_v. We assume that $D > c_v$, so that filing suit if and when the illness occurs would be profitable for all illness victims who expect to recover their full damages, yielding a net return of $D - c_v$. In contrast, if a tort for risk is allowed, some victims could choose to file at exposure instead. Victims who file at exposure would be able to recover only their expected losses, qD. We assume (for simplicity) that the victim's litigation costs are the same under the traditional rule and under the tort for risk. Then, the net return from filing at exposure would be $qD - c_v$.

Clearly, in the absence of bankruptcy, if there were no litigation costs ($c_v = 0$), then exposure victims would be indifferent between filing at exposure and at illness, since both would yield the same expected return (qD).[9] However, with positive litigation costs ($c_v > 0$), in the absence of bankruptcy all accident victims would prefer to wait rather than file at exposure since waiting yields a higher net return. Both options yield the same expected award (qD). However, if an exposure victim files suit at the time of exposure,

---

[6]See Miceli and Segerson (2003b) for a detailed description and analysis of the issues discussed in this section.

[7]If the losses take the form of lost wages, then losses are measured directly in dollars. If, on the other hand, losses stem from non-monetary impairment, then D represents the monetary equivalent of those losses.

[8]Thus, the injurer is liable for damages resulting from his activities, regardless of the level of care taken to prevent the accidental release. We do not consider negligence-based rules, under which an injurer would be liable only if he failed to meet the due standard of care.

[9]This assumes that victims are risk neutral. We also ignore discounting for simplicity.

she will incur litigation costs with certainty, while if she waits, she will incur these costs only if she contracts the disease. Thus, if there is no possibility of bankruptcy, allowing a tort for risk will have no impact on filing decision (i.e., the decision will be the same as under the traditional rule), and hence no impact on the number of suits (which determines total litigation costs) or deterrence. This result does not hold, however, if liability can bankrupt the injurer.

8.2.2   The Filing Decision with Potential Bankruptcy

When liability can bankrupt the injurer, then under a tort for risk the victim faces the following tradeoff in deciding whether to file at exposure or wait until the time of illness. If she waits, she will save on litigation costs (in expected terms), as in the case without bankruptcy. This is the benefit of waiting. The cost of waiting is the potential reduction in the damage award if the injurer's assets have been reduced or depleted by other, earlier suits. Of course, this cost depends on the number of victims who choose to file early (at exposure). Note that, since the expected savings in litigation costs decreases with the probability that they will be incurred (i.e., the probability that the illness will occur), the advantage of waiting will be lower for victims who are more likely to contract the disease. Thus, the potential for bankruptcy creates a "threshold" result under which high-exposure victims (i.e., those with a higher value of q) choose to file at exposure while low-exposure victims choose to wait. The "cut-off" occurs at the exposure level that makes a victim indifferent between filing at exposure and waiting.

Does this threshold result imply that a race to file will ensue if victims are allowed to sue at exposure? The answer depends on the injurer's asset level. If the injurer would not have had sufficient assets to cover his expected liability-related costs even in the absence of a tort for risk, then allowing a tort for risk will trigger a race to file. At least some exposure victims (those with high exposure levels) will choose to file at exposure in an attempt to secure a larger share of the injurer's limited assets. However, the race will be partial, meaning that some exposure victims (those with low exposure levels) will choose to wait in the hope that they will not contract the illness and hence never have to incur the litigation costs associated with a suit. This implies that a tort for risk effectively creates a priority rule that gives high exposure victims first claim on the injurer's limited assets.

In contrast, if the injurer would have sufficient assets to cover the expected cost of all illness suits (and hence would not be bankrupted under a traditional rule), then a race to file would not necessarily result if a tort for risk were allowed. If a victim does not expect other victims to file early, then she will have no incentive to file early either. However, if she expects others to file early, then she has an incentive to file early as well, and a race to file that results in bankruptcy can ensue. Thus, allowing a tort for risk can induce filing behavior that leads to bankruptcy when a traditional rule would not, although it is also possible that it will not affect filing behavior at all.

8.2.3   The Impact on Litigation Costs and Deterrence

What, then, are the implications of the effect on filing behavior for litigation costs and deterrence? Although a tort for risk induces some victims to file at exposure, it does not necessarily increase the total number of lawsuits (and hence total litigation costs). If exposure suits sufficiently reduce or exhaust the injurer's assets, then some illness victims will find suits at the time of illness unprofitable. Thus, while a tort for risk would allow exposure suits, if the injurer has a sufficiently low asset level it could decrease the number of illness suits and thus decrease the total number of suits as well.

The impact of filing on the injurer's choice of care depends on its impact on his expected liability-related costs. If allowing a tort for risk is expected to induce bankruptcy while the traditional rule would not, then the tort for risk will increase total expected costs for the injurer and hence provide the injurer with an incentive to increase his care level in an effort to reduce those costs. However, if the injurer would have faced bankruptcy under the traditional rule as well, then bankruptcy induced by allowing a tort for risk (albeit at an earlier date) would generate the same expected total costs for the injurer and hence the same incentives for care. Thus, while it is possible that allowing a tort for risk will increase care incentives, this outcome is not guaranteed.

The results summarized above regarding the impact of allowing a tort for risk were based on several assumptions: (1) that litigation costs are positive; (2) that there are a large number of victims so that any individual victim's filing decision cannot trigger bankruptcy, and injurers can reasonably predict the share of exposure victims who will actually contract the illness; and (3) that capital markets are perfect so that injurers can borrow against future earnings to pay current liability-related costs. This last assumption implies that whether or not bankruptcy occurs depends not on current assets but on the stream of assets over time. In practice, capital market imperfections may prevent inter-temporal borrowing against future assets, implying that bankruptcy occurs when current claims exceed current assets.

In the following section, we consider the implications of allowing a tort for risk under an alternative scenario, namely, where (i) the number of victims is small (hence injurer's face the real possibility that all victims would contract the disease, and a single victim's filing decision can trigger bankruptcy), and (ii) capital markets are imperfect so that bankruptcy occurs when current liability-related costs exceed current assets. The inability to borrow against future earnings implies that bankruptcy in an earlier period could reduce efficiency by preventing the injurer from engaging in a future activity that might be socially valuable. In order to focus on these two issues, we ignore litigation costs, which played a crucial role in the results derived above. In addition, to simplify the model, we assume that all victims face the same probability of contracting the illness if exposed, i.e., all have the same exposure intensity (same value of q). Thus, the threshold result discussed above (which hinged on differences in exposure intensity) no longer plays a role in the equilibrium. As before, we examine the equilibrium filing strategy for the victims, and its implications for bankruptcy and the existence of a race to file. Because litigation costs are assumed to be zero, we do not consider the impact of a tort for risk on total litigation costs. Likewise, we do not consider the implications for injurer care, since they would follow closely the principles driving the deterrence effects under the previous scenario.

8.3 The Small Numbers Case with Imperfect Capital Markets

8.3.1 The Basic Model

For simplicity, we consider a model with two periods and two victims. At the start of the first period the two victims are exposed to a toxic substance. As a result, both face a probability q of becoming ill in the second period $(0 < q < 1)$, in which case they will sustain damages of D dollars. Thus, their expected damages as of period one are qD. The injurer has A_1 assets in period one with which to pay damages, and if it remains in business in period two, it will generate an additional A_2 in assets (net profit).[10] However, if the injurer goes bankrupt in period one, it will not realize the A_2 assets. Thus, we can think of the foregone period two assets as the efficiency cost of bankruptcy. (Since the model is limited to two periods, bankruptcy in period two has no efficiency effects.)

As before, we consider two possible liability rules, the traditional rule and a tort for risk. Under both rules, we assume that filing suit is costless and continue to assume that liability is strict. If the injurer is bankrupted in either period as a result of lawsuits, its remaining assets are fully distributed (in equal shares if there are multiple plaintiffs). As noted, we assume that capital market imperfections prevent the injurer from borrowing in period one against its expected period two assets.

Finally, to make the model interesting, we assume that a single victim filing a TFR suit in period one cannot bankrupt the injurer. However, if both victims file in period one, the injurer will be bankrupted. Thus,

$$qD < A_1 < 2qD. \tag{8.1}$$

---

[10]We assume that only net profits are available to pay damages. Thus, for example, input suppliers have a prior claim on the injurer's revenues compared to victims.

This assumption isolates the effect of the joint decisions of the victims on the injurer's solvency. (Specifically, only a "race to file" will bankrupt the injurer in period one.) We do not, however, place any a priori restrictions on the magnitude of A_2. Thus, we characterize the equilibrium behavior of victims for different values of A_2, or equivalently, for different values of total assets, $A \equiv A_1 + A_2$, given (8.1).

8.3.2 Equilibrium under the Traditional Rule

We first consider the outcome under the traditional rule. Here, the only decision of victims is whether or not to file in period two if they become ill. Since filing is costless and liability is strict, they will always do so. Thus, the injurer faces the following possible period-two outcomes: (1) neither victim becomes ill, which occurs with probability $(1 - q)2$; (2) one victim becomes ill, which occurs with probability $2q(1 - q)$; and (3) both victims become ill, which occurs with probability $q2$. Summing the damages in each case weighted by the probabilities yields expected damages as of period one equal to $2qD$. Thus, if the injurer's total assets over the two periods are such that $A > 2qD$, then it is solvent in an expected sense since its total assets exceed its expected liability. The injurer may nevertheless go bankrupt in actual terms, depending on its total assets and which of the above outcomes actually occurs. Although bankruptcy in this case has no efficiency effects, it may limit the injurer's ability to compensate victims who become ill, a purely distributional concern.[11]

The possible cases are as follows:

$$A > 2D, \text{ the firm is never bankrupt;} \tag{8.2a}$$

$$D < A < 2D, \text{ the firm is bankrupt only if both victims become ill;} \tag{8.2b}$$

$$A < D, \text{ the firm is bankrupt if one or both victims become ill.} \tag{8.2c}$$

In the next section, we compare these outcomes to those that can occur under the TFR rule.

8.3.3 Equilibria under the Tort for Risk Rule

Under a TFR, an exposure victim can file in period one for damages of qD, or wait until she becomes ill in period two and file for D. The two victims make their filing decisions simultaneously and either choose to file "now" (at exposure) or "wait" to file until they actually become ill. Thus, there are four possible outcomes: (wait, wait), (now, wait), (wait, now), and (now, now). The outcome where both wait corresponds to the traditional rule, while the

---

[11]Obviously, if deterrence of the original exposure were an issue, an inability to fully compensate victims in either period would have efficiency effects due to the judgment proof problem.

other three involve a TFR suit by at least one victim. It is easy to show that the injurer's expected liability under each outcome is $2qD$, which is the same as under the traditional rule. Thus, absent the threat of bankruptcy, the two rules are equivalent in expected terms. As before, however, the actual outcomes in each case involve different amounts of liability, and may lead to bankruptcy, depending on the injurer's assets. This can lead to different outcomes under the two rules.

If both victims wait, they behave identically under the TFR and the traditional rule, and the possible outcomes are those described in (8.2a) to (8.2c). If one victim files at exposure and the other waits, the following two outcomes are possible:

$$A > (q+1)D, \quad \text{the firm is never bankrupt;} \tag{8.3a}$$
$$A < (q+1)D, \quad \text{the firm is bankrupt in period two} \tag{8.3b}$$
$$\text{if the victim who waits becomes ill.}$$

As under the traditional rule, only period-two bankruptcy is possible in this case, which, as noted, has purely distributional implications. Finally, if both victims file at exposure, the injurer is bankrupted in period one, given $A_1 < 2qD$. This last case (the race to file) is especially interesting because it is the only one in which bankruptcy in period one occurs, producing an efficiency loss (failure to realize A_2), as well as possible distributional effects.

In order to derive the equilibrium outcomes under a TFR, consider the normal form of the victims' filing game as shown in Figure 8.1. (The first payoff in each cell is for victim one, and the second is for victim two.) Given that $A_1 < 2qD$, if both victims choose to file now, the injurer goes bankrupt in period one and each victim receives half of his first period assets, or $A_1/2$. If one victim files now and the other waits, the one filing now gets qD (since $A_1 > qD$), while the other has an expected payoff of $qD - x$, where $x = 0$ if the injurer has enough assets in period two to pay the victim's damages if she becomes ill, and $x > 0$ if it does not. Finally, if both victims wait to file, each has an expected payoff of $qD - y$, where $y = 0$ if the injurer's assets are expected to cover its total liability and $y > 0$ if not. Obviously, the equilibrium of the filing game depends on the specific magnitudes of x and y, which in turn depend on A. We consider several cases.

Case 1: $A \geq 2D$. In this case, the injurer is never bankrupted by liability, even in the worst case ("catastrophic") scenario where both victims wait to file suit and both become ill. Thus, $x = y = 0$, and there are three pure strategy equilibria of the filing game: (wait, wait), (now, wait), (wait, now).[12] In addition, there are an infinite number of mixed strategy equilibria where one

---

[12]In the two pure strategy equilibria where one victim files now and one waits, there is nothing in the model to determine which victim adopts which strategy. In the case where victims differ in their risk of developing illness (i.e., they differ in their values of q) and

player plays the pure strategy "wait" and the other randomizes between the two strategies with an arbitrary probability.[13] Under all of these equilibria, both victims are fully compensated (either in actual terms if they wait, or in expected terms if they filé a TFR suit), and the injurer remains in business for both periods. Thus, the equilibria are also efficient in the sense that the injurer realizes its period-two assets. In this case, there is no effective difference between the traditional and TFR rules (except that some victims are compensated in expected terms and others in actual terms).[14]

Case 2: $(q + 1)D \leq A < 2D$. In this case, the injurer is only bankrupted if both victims wait to file and both become ill, in which case each receives half of the injurer's total assets, or $A/2$. Thus, if both wait, each has an expected payoff of

$$q(1 - q)D + q^2(A/2) = qD - q^2[D - (A/2)],$$

which implies

$$y = q^2[D - (A/2)] > 0. \tag{8.4}$$

However, if only one victim waits and becomes ill the injurer can cover the liability. Thus, $x = 0$.

In this case, there are two pure strategy equilibria of the filing game: {(now, wait), (wait, now)}, and one mixed strategy equilibrium where each player files now with probability

$$p^* = \frac{y}{y + (qD - (A_1/2))}, \tag{8.5}$$

and waits with the complementary probability.[15] Note that all of these equilibria are efficient in the sense that the injurer is never bankrupted in period one. However, there is an important distributional difference between the pure strategy equilibria and the mixed strategy equilibrium. In the former, victims are fully compensated because the injurer is not bankrupted in either period (as in Case 1). In contrast, under the mixed strategy equilibrium both victims may end up waiting to file in period two, and if both become ill, the injurer is bankrupted. Thus, although the equilibria are all efficient (because only period two bankruptcy is possible), the pure strategy equilibria are preferable to the mixed strategy equilibrium because they guarantee full compensation of both victims.

---

litigation is costly, we showed above that victims with lower risk wait to file while those with higher risk file at exposure.

[13] If we let $p_i \in [0, 1]$ be the probability that player i files now, then the reaction functions for the two players coincide with the axes of the unit square in (p_1, p_2) space.

[14] If a victim who receives expected damages uses it to buy market insurance against a future loss, then she will receive compensation for her actual loss in the event of illness.

[15] Note that p^* is strictly between zero and one given $y > 0$ in this case and (8.1). In this case, the reaction functions intersect at three points in (p_1, p_2) space: $(1, 0)$, $(0, 1)$, and (p^*, p^*).

Intuitively, the pure strategy equilibria avoid bankruptcy by allowing the injurer to pay expected damages of qD to one victim in period one with certainty, thereby leaving it with enough assets in period two to pay actual damages of D should the other victim become ill. Interestingly, to the extent that the pure strategy equilibria are expected to emerge in this case, the TFR rule actually lowers the risk of bankruptcy compared to the traditional rule under which period two bankruptcy is always possible when $A < 2D$.

Case 3: $(A_1/2q) + qD < A < (q+1)D$. In this case, the injurer goes bankrupt in period two if even one victim waits to file and becomes ill. Thus, the expected payoff from waiting to file, given that the other victim files now, is

$$q(A - qD) = qD - q[(q+1)D - A].$$

If follows that

$$x = q[(q+1)D - A] > 0, \tag{8.6}$$

while y continues to be given by (8.4). Although $x > 0$, the types of equilibria in this case are the same as in Case 2 if $qD - x > (A_1/2)$, or, substituting from (8.6), if

$$(A_1/2q) + qD < A, \tag{8.7}$$

which defines the lower bound on A in this case. Thus, there are again two pure strategy equilibria: {(now, wait), (wait, now)}, and a single mixed strategy equilibrium where each player files now with probability

$$p^* = \frac{y}{y + [(qD - x) - (A_1/2)]}. \tag{8.8}$$

As in Case 2, the equilibria are all efficient in the sense that the firm is never bankrupted in period one, but in contrast to Case 2, even the pure strategy equilibria in this case result in period two bankruptcy if the victim who waits becomes ill. Thus, the victim who files suit at exposure is assured full compensation (in expected terms), while the one who waits is undercompensated if he becomes ill. In comparison, if period two bankruptcy occurs under the traditional rule (cases (8.2b) and (8.2c)), any victims who become ill are undercompensated.

Case 4: $A = (A_1/2q) + qD$. In this case, if one victim files now, the other victim is indifferent between waiting and filing now (i.e., $qD - x = (A_1/2)$). As a result, there are three pure strategy equilibria: {(now, wait), (wait, now), (now, now)}, and an infinite number of mixed strategy equilibria where one player plays the pure strategy "now" and the other randomizes between the two strategies with an arbitrary probability. (This case thus mirrors Case 1.)[16]

---

[16]That is, the reaction functions coincide with the outer edges of the unit square in (p_1, p_2) space.

The injurer is bankrupted in period two if one party waits and becomes ill (as in Case 3), and is bankrupted in period one if both file now. This therefore represents the first case in which period one bankruptcy can occur as a result of a "race to file." If it does, it not only leaves victims undercompensated (even in expected terms), it is also inefficient in that the injurer does not realize his period two assets. In this case, the TFR rule is welfare-reducing and is also inferior to the traditional rule in terms of compensation.

Case 5: $A < (A_1/2q) + qD$. In this case, victims strictly prefer to file now, regardless of the other victim's choice. Thus, (now, now) is a dominant strategy, and a race to file is the only equilibrium. Again, the TFR is inferior to the traditional rule, both in terms of efficiency and compensation of victims.

To summarize, the preceding cases have shown that the equilibrium impact of a TFR rule varies depending on the level of the injurer's inter-temporal assets. This dependence is depicted graphically in Figure 8.2, which shows the regions where each case is relevant in (A_2, A_1) space. When total assets are sufficiently large (case 1), the TFR has no real effect either in terms of efficiency or compensation of victims. For intermediate asset levels (cases 2 and 3), the TFR has no efficiency effect, and it may actually increase the ability of injurers to compensate victims by allowing them to pay expected damages up front to some victims and possibly to avoid period two bankruptcy. However, for sufficiently low asset levels (cases 4 and 5), a TFR potentially has detrimental effects on both efficiency and compensation by possibly triggering a "race to file" among victims that bankrupts the firm prematurely.

8.4   Conclusion

Conventional tort law bars victims of exposure to a toxic substance from filing suit for damages until they actually become ill. This rule often has the practical effect, however, of denying victims compensation because, by the time the illness arises, the injurer may have gone bankrupt for reasons unrelated to liability. A possible solution to this problem is to allow victims to file at exposure –that is, to create a tort for risk. The tradeoff is that this rule may trigger a race to file among exposure victims (who fear future bankruptcy), thereby itself inducing bankruptcy.

Our comparison of a tort for risk with the traditional rule under both scenarios we considered showed that a race to file can indeed arise in equilibrium under certain conditions, particularly for firms that have relatively low inter-temporal asset streams. If the injurer's asset level is sufficiently high that there is no threat of bankruptcy, then allowing a tort for risk will have no effect on filing behavior since there will be no incentive for victims to file early. Thus, the impact of allowing a tort for risk stems from the possibility of bankruptcy.

The filing behavior induced by allowing a tort for risk can have several implications. First, when a firm's total assets are sufficiently low, it can actually trigger bankruptcy (perhaps prematurely), implying that bankruptcy is endogenously determined by the liability rule. Such an outcome is undesirable, both because it can be inefficient (if it prevents a future activity that is socially beneficial), and because it may leave victims under-compensated. However, for healthier firms, the rule may actually have the desirable effect of staving off bankruptcy from future tort suits by allowing the firm to pay some of its liability in expected terms, thereby leaving it enough assets to pay any future illness claims in full. In this way, the rule functions like liability insurance for firms.

Second, allowing a tort for risk can affect the total number of suits brought by victims. However, the results of Section 8.2 imply that the impact on litigation costs is unclear. In some cases, total litigation costs could increase, while in others they might decrease. Ceteris paribus, litigation costs are more likely to increase under a tort for risk when the injurer's asset level is high (but the threat of bankruptcy still exists).

Finally, if allowing a tort for risk increases total expected liability-related costs, it can also increase the injurer's incentives to take care. This can occur if the tort for risk triggers bankruptcy when it would not have occurred under the traditional rule. However, it is also possible that allowing a tort for risk will actually lower the injurer's risk of bankruptcy compared to the traditional rule.

These conclusions suggest that, in addition to affecting both the amount and the nature of victim compensation, a tort for risk can have several welfare effects, which can work in opposite directions. Thus, taken together, the results from Sections 8.2 and 8.3 imply that the welfare impacts of allowing a tort for risk are ambiguous, and likely to depend on the injurer's asset level.

References

Douglas, D. and Dybvig, P., 'Bank Runs, Deposit Insurance, and Liquidity', *Journal of Political Economy*, 91 (1983): 401-419.

Fischer, M., 'Union Carbide's Bhopal Incident: A Retrospective', *Journal of Risk and Uncertainty*, 12 (1996): 257-269.

Keeton, W. P., D. Dobbs, R. Keeton, and Owen D., *Prosser and Keeton on Torts*, 5th Edition, (St. Paul, Minn.: West Publishing Co, 1984).

Landes, W. and Posner, R., 'Tort Law as a Regulatory Regime for Catastrophic Personal Injuries', *Journal of Legal Studies*, 13 (1984): 417-434.

Love, T., 'SPECIAL PROJECT: Environmental Reform in an Era of Pollitical Discontent: Deterring Irresponsible Use and Disposal of Toxic Substances: The Case for Legislative Recognition of Increased Risk for Causes of Action,' *Vanderbilt Law Review*, 49 (1996), 789-823.

Miceli, T. and Segerson, K., 'A Note on Optimal Care by Wealth-Constrained Injurers', *International Review of Law and Economics*, 23 (2003a): 273-284.

Miceli, T. and Segerson, K., 'Do Exposure Suits Produce a 'Race to File'? An Economic Analysis of a Tort for Risk', Department of Economics *Working Paper*, (2003b), Univ. of Connecticut.

Miceli, T. and Segerson, K., (forthcoming), 'Should Victims of Exposure to a Toxic Substance Have an Independent Claim for Medical Monitoring?', *Research in Law and Economics*.

Note (1983), 'The Manville Bankruptcy: Treating Mass Tort Claims in Chapter 11 Proceedings', *Harvard Law Review*, 96, 1121- .

Robinson, G., 'Probabilistic Causation and Compensation for Tortious Risk', *Journal of Legal Studies*, 14 (1985): 779-798.

Rose-Ackerman, S., 'Dikes, Dams, and Vicious Hogs: Entitlement and Efficiency in Tort Law', *Journal of Legal Studies*, 18 (1989): 25-50.

Shavell, S., 'Uncertainty Over Causation and the Determination of Civil Liability', *Journal of Law and Economics*, 28 (1985): 587-609.

Shavell, S., 'The Judgment Proof Problem', *International Review of Law and Economics*, 6 (1986): 45-58.

Spier, K., 'Settlement with Multiple Plaintiffs: The Role of Insolvency', *Journal of Law, Economics & Organization*, 18 (2002): 295-323.

Valk, M., 'Emotional Distress: How I Learned to Stop Fearing Toxic Torts and Sue for the Fear', *Journal of Products and Toxics Liability*, 17 (1995): 67-79.

Victim 2

	Now	Wait
Now	$A_1/2,\ A_1/2$	$qD,\ qD-x$
Wait	$qD-x,\ qD$	$qD-y,\ qD-y$

(left side labeled **Victim 1** for rows Now and Wait)

Figure 8.1: Victims' filing game, normal form

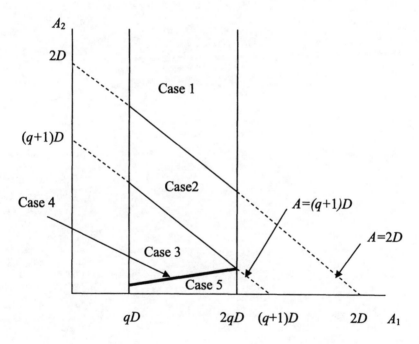

Figure 8.2: Regions where various equilibria exist

Part 3

Liability Regimes and their Consequences on Private Contracting

Chapter 9

Environmental Risk Regulation and Liability under Adverse Selection and Moral Hazard[*]

Yolande Hiriart
Université de Toulouse

David Martimort
Université de Toulouse

9.1 Introduction

The importance of an efficient design of liability and regulation for environmentally risky ventures is now well recognized and has been highlighted by
the lively debate which took place in the U.S. around the 1980 Comprehensive Environmental Response Compensation and Liability Act (CERCLA).
Among other things, this act establishes the allocation of liability between a
firm which has caused an environmental damage and its various contractual
partners in settings where the venture's assets and profits fall short of covering
the full harms caused on third-parties.[1] Conventional wisdom suggests that
the contracts signed by such risky ventures with various stakeholders take into
account this allocation of liability. To assess the full impact of environmental
risk regulation and liability rules on social welfare, public policies should thus

---

[*]We thank the *French Ministry of Ecology and Sustainable Development* for its financial support. We also thank Marcel Boyer and Pierre Dubois for comments on an earlier
version of this paper and the participants in Université de Toulouse *Lunch Seminar*, April
2002, *INRA Nantes Seminar*, April 2002, *Journées de Microéconomie Appliquée*, Rennes-St
Malo, June 2002, 5th Conference on Industrial Organization and Food Processing Industry,
Toulouse, June 2002, the *European Economic Association Annual Congress*, Venice, August
2002, *Journées de Microéconomie Appliquée*, Lille, May 2004, and the *European Association of Environmental and Resource Economics Annual Conference*, Budapest, June 2004.
The usual disclaimer applies.

[1]On this topic, see Strasser and Rodosevitch (1993) and Boyer and Laffont (1996).
CERCLA has inspired the Canadian legal framework for contaminated sites. The European
Community is also considering to develop environmental liability with the Community
Directive still under discussion.

be designed with an eye on the contracting possibilities available to the firms involved in environmentally risky activities. It seems important to delineate circumstances under which contracts are modified by liability rules and risk regulation and to understand the directions of those distortions if any.

Of course, in a world without transaction costs in private contracting, the corrective policies aimed at reducing the likelihood of an environmental damage, be they regulation or liability rules, would not have any impact on contracting. The risky venture and its contracting partners would always reach an efficient agreement. Conditionally on the level of prevention induced by public policies, contracts would achieve an efficient allocation of resources within the private sector. The only interesting issue is thus to assess the impact of ex ante regulation and ex post liability rules when private transactions are plagued by informational problems. This paper analyzes the full impact of risk regulation and extended liability on care and output levels in a framework where private transactions are perturbed by the private information that the risky venture retains at the time of contracting with stakeholders. Private information takes the form of both adverse selection on production cost and hidden choice of the level of care. It is shown that contract distortions significantly depend on whether risk regulation or liability rules are used.

To exemplify some of the issues involved, let us consider a buyer (the principal) contracting with an independent seller (the agent). The seller can be viewed as a production unit or subsidiary providing an essential input to the buyer. Production creates an environmental hazard on third-parties. For instance, it can generate a long-lasting contamination of the production site. However, the seller can take non-observable costly actions to reduce the likelihood of such harm. The environmental policy can take either the form of an ex ante regulation or an ex post liability rule, depending on the institutional context.

If there is no limit on liability, fines can be made large enough to align the agent's private incentives to exert care with the socially optimal ones. In this case, the design of the transaction between the buyer and the seller can be disentangled from the prevention incentive problem. Depending on the informational context surrounding private contracting, those contracts may not reach an efficient outcome but, at least, they are *constrained efficient* taking into account the informational constraints faced by the buyer. In any case, distortions are the same as in the absence of any environmental risk. In particular, when the seller is privately informed on his production cost, it is well-known that contracts trade off the extraction of the seller's information rent against productive efficiency to reach an interim efficient outcome.[2] Efficient sellers get an informational rent from retaining private information. Reducing this rent requires some output distortions. Despite these distortions on output away from the first-best imposed by informational

---

[2]See Laffont and Martimort (2002, Chapter 2).

constraints, contracting remains unaffected by the provision of incentives for care. As a result, outputs for an environmentally risky venture are the same as for a non-risky firm.

With limited liability, the picture is quite different. Fines cannot exceed the agent's total profits from his relationship with his principal. Of course, those profits depend on the contract signed with the buyer. Because of adverse selection, part of those profits are available to the seller under the form of adverse selection rent that can never be seized. This possibility to hide rent away from the eyes of the public authority reduces the incentives for care of the most efficient sellers. To compensate this effect, an efficient seller must, at the optimal regulation, receive an extra reward when no damage occurs. With limited liability, an agent is also rewarded for truthtelling by means of moral hazard rent. Adverse selection incentive constraints are thus relaxed by increasing effort for the efficient sellers. This creates an endogenous positive correlation between care and output even though care provision does not conflict with the minimization of short-run cost.

To sum up, at the optimal regulation with limited liability, outputs distortions are weaker than when there exists no such protection for the firms. Intuitively, relaxing the adverse selection incentive constraint requires to distort effort upwards for the most efficient firms. This has a positive social value since it reduces the likelihood of an accident. The shadow cost of the adverse selection incentive constraint has a lower value than with unlimited liability and output distortions are less attractive.

In many practical circumstances, an ex ante regulation is not feasible or not even conceivable. It must then be replaced by ex post liability rules enforced by Courts. What are the consequences of those rough rules on private contracting? Of course, the complementarity between safety care and output found under ex ante regulation remains an attractive qualitative property which should be looked for under alternative legal regimes.

In this respect, we first delineate circumstances under which the optimal regulation can still be implemented under adverse selection, moral hazard and limited liability, by simply imposing a liability payment equal to the full damage to the uninformed stakeholders of the risky venture.[3] Under this extended liability regime, both the buyer and the seller are found liable for the harm caused by the seller. When the buyer is also protected by limited liability, fines cannot exceed the whole gains from trade achieved by the private transaction. Due to his threat of losing the benefits from transacting with the seller, the buyer somewhat internalizes the environmental externality. His own incentives to promote care are aligned with the regulator's ones, even though imperfectly. The seller's contract has indeed to fill a new objective: avoiding any accident to secure gains from trade. This creates a new channel

---

[3]By implementation, we mean that the allocative consequences of regulation and extended liability are the same although they may differ with respect to the distribution of surplus they involve.

by which output distortions are affected by the liability regime. These gains depend of course on the seller's cost, i.e., on the adverse selection variable. When uncertainty on costs is sufficiently small, the gains from trade with nearby types of the seller are also close enough. Different types choose almost the same levels of care, making impossible to achieve truthtelling without further distortions on effort levels. The complementarity between care and output is maintained. Extending liability towards the buyer preserves then the most important property of the optimal regulation. However, rent extraction, output and care distortions are now excessive compared with the socially optimal ones. This points at an obvious weakness of this ex post liability regime in comparison with the optimal ex ante regulation.

By endogenizing the gains from trade in vertical relationships subject to risk regulation, this paper fills a gap in the literature. First, some authors, following Pitchford (1995), have analyzed how incentive problems between a principal and an agent are affected by the liability environment.[4] These authors focus mostly on the case of financial relationships between a polluting borrower and his lender. They analyze the impact of bargaining power at the contracting stage on the financial transaction (Balkenborg (2001)), the impact of the initial resources of the lender (Lewis and Sappington (2001)), and the degree of control that the lender exerts on the borrower (Boyer and Laffont (1997)). This literature has focused on pure moral hazard environments where the level of safety care is non-verifiable. In such contexts, there exists a conflict of interests between the lender and the regulator in the level of safety care they would like to induce. The choice of a liability rule might reduce, at least partially, this conflict. A caveat of this approach is that it takes as given the value of the private transaction and assumes away the possible effects of this moral hazard problem on the size of the financial returns. In other words, this approach implicitly assumes that there is no adverse selection between the principal and his agent, so that liability rules and risk regulation have no impact on these financial returns. Our paper focuses instead on the endogenity of these returns. We show that the lessons of this earlier literature should be taken with caution when there is adverse selection between the risky venture and its stakeholders. This endogenity should be recognized at the time of assessing the performances of various regulatory and legal frameworks. As soon as there is adverse selection, moral hazard and limited liability, output distortions are affected by public policies towards care provision sometimes in a rather complex manner.

The impact of liability rules on the whole set of transactions a risky venture is part of has been first investigated by Boyd and Ingberman (1997). In a complete information environment where a first-best regulation could be feasible if policy instruments were unrestricted, they analyze how different liability

---

[4]See Shavell (1986) for an early discussion of the so-called *judgment-proof problem* and also Segerson and Tietenberg (1992).

regimes affect cost minimization and buyer-seller transactions. Laffont (1995) analyzes instead the regulation of a public utility which exerts safety care to avoid an environmental accident in a model involving, like ours, adverse selection and moral hazard. An important technological assumption he makes is that a positive effort level increases production costs and reduces thus output, even in a first-best world. The analysis becomes extremely complex under asymmetric information because of the substitutability between safety care and cost minimizing effort. Dionne and Spaeter (2003) propose also a pure moral hazard model in which there is such a multitask externality. The agent (a borrower) can allocate his investment between directly improving the distribution of the returns of his project and reducing, also in a stochastic sense, the distribution of damages. In fact, our analysis shows that limited liability creates instead some complementarity between output and safety care, even when cost minimization is not directly affected by safety care. By simplifying the technological side, we are able to go further towards characterizing optimal contracting under adverse selection, moral hazard and limited liability. Finally, in a companion paper (Hiriart and Martimort (2003)), we investigate the impact of liability rules on contracting with safety care incurred before information on cost parameter is learned. Output distortions due to the liability regime arise then only when the buyer (principal) has no bargaining power and must recover the extra liability cost through price distortions.

Section 9.2 presents the model. Section 9.3 characterizes the optimal regulation of a buyer-seller relationship when the polluting seller has unlimited liability. In this benchmark, we show that outputs are set at the same level as if the seller's activity created no risk for the environment. There is a dichotomy between regulating output and inducing safety care. Section 9.4 focuses on the optimal regulation when the seller is protected by limited liability. Now, output and care distortions are endogenously linked and some kind of complementarity appears. We show how the optimal regulation can sometimes be implemented with extended liability if the buyer is wealthy enough. Section 9.5 analyzes the benefits of extending liability when, instead, the buyer has limited wealth. We compare the qualitative results obtained there with those of the optimal regulation and show that the complementarity between care and output still prevails. Section 9.6 concludes. Proofs are relegated to an Appendix.

9.2 The Model

Following the analysis of Boyd and Ingberman (1997 and 2001), we consider a buyer-seller relationship. However, the lessons of our work are more general and apply to other vertical relationships between an agent exerting an environmentally risky activity and some stakeholder with whom he is linked through contract. One may think for instance of shareholders-workers relationships, regulators-public utilities hierarchies or lender-borrower transactions. The buyer derives a monetary benefit $S(q)$ from using q units of the good, with $S' > 0$, $S'' < 0$, and $S(0) = 0$. To always ensure positive and interior outputs, we assume that the Inada conditions $S'(0) = +\infty$ and $S'(+\infty) = 0$ both hold. The buyer's utility fonction is

$$V = S(q) - t,$$

where t is the payment made to the seller. The buyer is risk-neutral and has a reservation payoff exogenously normalized at zero.

The risk-neutral seller has a constant marginal production cost θ that he privately knows. This random variable is distributed on $\Theta = \{\underline{\theta}, \bar{\theta}\}$ with respective probabilities ν and $1 - \nu$. We denote by $\Delta\theta = \bar{\theta} - \underline{\theta} > 0$ the size of cost uncertainty.

The production process generates an environmental hazard. The seller can nevertheless exert a level of safety care that reduces the probability of an accident. We assume that the damage size h is greater than the first-best surplus in both states of nature. We focus thus on accidents which have a substantial size. If a comprehensive ex ante regulation is not feasible, this may justify using extended liability towards deep-pocket stakeholders if needed in order to compensate (even if it is partially) harmed third-parties.

Production is exchanged even if an accident occurs; this is not output per se which is risky but the actual production process. For instance, there may be pollution leakages during or after the production process which affect a nearby river or contamine the production site without undermining the ability of the seller to produce.

The probability of an environmental damage is $1 - e$ where e is the agent's effort level which costs him a non-monetary disutility $\psi(e)$. We assume that $\psi' > 0$, $\psi'' > 0$, $\psi''' > 0$, with the Inada conditions $\psi'(0) = 0$ and $\psi'(1) = +\infty$ to ensure that effort is always interior and avoid uninteresting corner solutions. The seller's expected utility can thus be written as

$$U = t - \theta q - e z_n - (1 - e) z_a - \psi(e),$$

where z_a (resp. z_n) is the (regulatory or the liability) payment made if (resp. no) harm occurs. We should stress at this point that the level of safety care has no direct technological impact. Contrary to Laffont (1995) and Dionne and Spaeter (2003), exerting an effort to prevent an accident neither increases production cost nor decreases the damage size. To motivate this assumption, note that in many circumstances, technological choices which put a risk on the environment are related to sunk costs (choice of a production site, of a technological process, etc.) and not to short-run variable costs.

When an ex ante incentive regulation is used, a risk-neutral regulator maximizes a social welfare function that takes into account both the net cost of the environmental damage and the buyer's and the seller's utilities, namely

$$W = -(1 - e)h + ez_n + (1 - e)z_a + \alpha(U + V),$$

where $\alpha < 1$ represents the weight given to the private sector by the regulator.[5]

9.3 Regulation with Unlimited Liability

Let us first consider the normative case in which a regulatory authority offers a comprehensive grand-contract to both the buyer and the seller before any harm occurs. Of course, this complete contractual setting is highly hypothetical. It supposes that the regulator has a strong commitment power to design ex ante the rewards and fines offered to the private sector. It also assumes that private transactions can be regulated and thus, implicitly, that economic and environmental regulations are jointly designed. Nevertheless, this normative setting gives us an important benchmark before analyzing extended liability in a similar environment (see Section 9.5).

9.3.1 Full Information

To start with, we suppose that the level of safety care e and the seller's marginal cost θ are both observable by the regulator who can recommend the level of output. The regulator's problem can then be written as:

$$(P^*): \quad \max_{\{e,q,t,z_n,z_a\}} \quad -(1 - e)h + ez_n + (1 - e)z_a + \alpha(U(\theta) + V(\theta))$$

subject to

$$V(\theta) \geq 0, \tag{9.1}$$
$$U(\theta) \geq 0, \tag{9.2}$$

where (9.1) and (9.2) are the respective participation constraints of both the buyer and the seller. Replacing transfers t, z_n and z_a by their values as a function of U and V, the regulator's problem can be rewritten as:

$$\max_{\{e,q,U,V\}} \quad S(q) - \theta q - (1 - e)h - \psi(e) - (1 - \alpha)(U(\theta) + V(\theta))$$

subject to (9.1) and (9.2).

---

[5]We follow Baron and Myerson (1982) in specifying such a social welfare function with redistributive concerns towards the private sector. Among other things, those concerns can be justified when the regulator is somehow captured by the industry. In this respect, a case of particular interest is when $\alpha = 0$; the regulator can then be interpreted as an uncorruptible judge. See Boyer and Porrini (2004) for a model which distinguishes between the judge and the regulator along similar lines.

Since the rents left to the private sector are viewed as socially costly, both participation constraints above must be binding at the optimum. The first-best outputs $q^*(\theta)$ and levels of safety care e^* (independent of the seller's cost) are thus respectively given by:

$$S'(q^*(\theta)) = \theta, \qquad (9.3)$$
$$\psi'(e^*) = h. \qquad (9.4)$$

Under full information, the marginal surplus of the buyer is equal to the marginal cost of production, and the marginal disutility of safety care covers exactly the damage. To implement this outcome, the regulator can simply recommend the first-best allocation $(q^*(\theta), e^*)$ and punish harshly the agent if the recommended output, safety care or transfers are not observed.

When (9.1) is binding, the payment t from the buyer to the seller is equal to the gross surplus from trade, namely $S(q) = t$. Everything happens thus as if the optimal regulation shifted all bargaining power in favor of the informed party in private contracting.[6] Given this value of the price paid by the buyer for the good, (9.2), when it is binding, only defines the expected regulatory payment $ez_n + (1 - e)z_a$ paid by the seller to the regulator. Many such payments are thus feasible as long as the seller breaks even in expectation.

9.3.2 Asymmetric Information with Unlimited Liability

Let us now assume that neither the level of care e nor the marginal cost θ are observable by the regulator and the buyer. If instead the buyer could observe those variables, the regulator could use a "revelation scheme" à la Maskin (1999) to have both the buyer and the seller revealing these pieces of shared information at no cost. Then, clearly, the first-best optimal outcome would be implemented, and contracting between the buyer and the seller would be efficient. Of course, such a complete contracting environment is highly hypothetical. However, as long as exogenous constraints on contracting are not imposed, complete contracts cannot be ruled out a priori. This extreme efficiency result shows that, within the realm of complete contracting, the most interesting case to study is when the buyer is uninformed on the seller's cost, so that private contracting is plagued by an adverse selection problem.

Under complete contracting and asymmetric information, a regulatory contract specifies ex ante the transfers made to the seller in any event, i.e. a system of fines and rewards depending on whether an accident occurs or not. By the Revelation Principle, there is no loss of generality in assuming that an incentive regulation is a mechanism $\{t(\hat{\theta}), z_a(\hat{\theta}), z_n(\hat{\theta}), q(\hat{\theta})\}_{\hat{\theta} \in \Theta}$ stipulating a price paid by the buyer to the seller, regulatory transfers and an output as a function of the seller's report $\hat{\theta}$ on his cost.

---

[6]See Hiriart and Martimort (2003) for a similar result.

The regulatory contract must first satisfy the uninformed buyer's participation constraint

$$\underset{\theta}{E}(V(\theta)) \geq 0, \qquad (9.5)$$

where $\underset{\theta}{E}(\cdot)$ is the expectation operator with respect to θ, and

$$V(\theta) = S(q(\theta)) - t(\theta)$$

is the buyer's net profit in state θ.

Second, the seller's ex post participation constraints must hold since the seller is privately informed on his cost at the time of accepting the regulatory contract. To write down these conditions, let us first notice that his expected profit in state θ is

$$U(\theta) = t(\theta) - \theta q(\theta) - \min_{e} \{ ez_n(\theta) + (1-e)z_a(\theta) + \psi(e) \} .$$

Clearly, the optimal effort level induced by an incentive compatible mechanism solves

$$\psi'(e(\theta)) = z_a(\theta) - z_n(\theta). \qquad (9.6)$$

This effort level trades off, from the seller's viewpoint, the cost of marginally increasing effort with the benefit of reducing the expected payment made to the regulator.

To get a more compact expression of $U(\cdot)$, it is useful to define the seller's moral hazard rent as

$$R(e) = e\psi'(e) - \psi(e).$$

Note that $R(\cdot)$ is increasing and convex with the assumptions made on $\psi(\cdot)$.

Then, the seller's total rent in state θ can be written as

$$U(\theta) = t(\theta) - \theta q(\theta) - z_a(\theta) + R(e(\theta)) \geq 0, \text{ for all } \theta \text{ in } \Theta. \qquad (9.7)$$

This is the sum of his adverse selection rent coming from private information on the technology and his moral hazard rent coming from his non-observable effort level.

The more stringent participation constraint is, of course, the least efficient seller's one

$$U(\bar{\theta}) \geq 0. \qquad (9.8)$$

Finally, the regulatory contract must be incentive compatible to induce the seller to truthfully reveal his marginal cost. This yields

$$U(\theta) \geq t(\hat{\theta}) - \theta q(\hat{\theta}) - z_a(\hat{\theta}) + R\left(\varphi(z_a(\hat{\theta}) - z_n(\hat{\theta}))\right), \text{ for all } (\theta, \hat{\theta}) \text{ in } \Theta^2,$$

where $\varphi = \psi'^{-1}$. Putting it differently, we get

$$U(\theta) \geq U(\hat{\theta}) + (\hat{\theta} - \theta)q(\hat{\theta}) \text{ for all } (\theta, \hat{\theta}) \text{ in } \Theta^2.$$

As usual in two-type adverse selection problem,[7] the relevant incentive compatibility constraint corresponds to an upward deviation where an efficient seller wants to mimic an inefficient one, namely

$$U(\underline{\theta}) \geq U(\bar{\theta}) + \Delta\theta q(\bar{\theta}). \tag{9.9}$$

Indeed, by pretending to be a less efficient seller, the efficient one can produce at a lower cost the same amount and save some extra rent.

Therefore, under asymmetric information, the optimal incentive regulation must solve:

$$(R): \quad \max_{\{U(\theta), V(\theta), q(\theta), z_a(\theta), e(\theta)\}} \mathop{E}_{\theta} \Big[-(1 - e(\theta))h - \psi(e) + S(q(\theta))$$
$$-\theta q(\theta) - (1 - \alpha)(U(\theta) + V(\theta)) \Big],$$

subject to (9.5), (9.8) and (9.9).

We can summarize this optimization in the next proposition.

Proposition 9.1 *With unlimited liability, the optimal regulation with adverse selection and moral hazard entails:*

- *An efficient level of care for both types, $e^{SB}(\theta) = e^*$, for all θ in Θ.*

- *The first-best output for an efficient seller and a downward distortion below the first-best for an inefficient one:*

$$q^{SB}(\underline{\theta}) = q^*(\underline{\theta})$$

and

$$S'(q^{SB}(\bar{\theta})) = \bar{\theta} + \frac{\nu}{1 - \nu}(1 - \alpha)\Delta\theta. \tag{9.10}$$

- *The buyer's expected profit is zero, $\mathop{E}_{\theta}(V(\tilde{\theta})) = 0$. Only the efficient seller gets a positive rent, $U^{SB}(\underline{\theta}) = \Delta\theta q^{SB}(\bar{\theta})$ and $U^{SB}(\bar{\theta}) = 0$.*

Since the weight of the private sector in the social welfare is less than one, transferring wealth from the rest of society towards the private sector is socially costly. The regulated prices of the transaction in both states of nature can be fixed so that the buyer's participation constraint is binding. Instead, the optimal regulatory policy under asymmetric information must leave some rent to the efficient seller to induce revelation on his cost parameter. This

---

[7]See Laffont and Martimort (2002, Chapter 2).

rent is increasing in the inefficient seller's level of output. Hence, to reduce the socially costly adverse selection rent, the regulator distorts downwards the production of the inefficient seller. As shown in (9.10), the marginal benefit of consumption for the buyer is now equal to the *virtual cost* of the inefficient seller. As it is standard in the literature, this virtual cost captures the existing extra cost of informational rents. Intuitively, starting from the first-best output $q^*(\bar{\theta})$ and reducing the inefficient agent's production by a small amount dq is beneficial since it reduces the efficient agent's information rent to the first-order and it has only a second-order impact on efficiency in state $\bar{\theta}$. Hence, the virtual costs depends on the ratio between the probabilities of having an efficient or an inefficient seller, $\frac{\nu}{1-\nu}$. Note that the virtual cost decreases with α, the weight of the seller's utility in the social welfare function. Indeed, as α increases, the private sector receives more weight and giving up information rent to the seller is viewed as being less costly by the regulator. Output distortions are thus less needed.

Given these output distortions, the regulator can structure the regulatory payments $z_n(\theta)$ and $z_a(\theta)$ to induce the first-best level of safety care. Typically, a differential $z_a(\theta) - z_n(\theta)$ just equal to the harm level h suffices to achieve an efficient level of care. Moreover, structuring rewards and punishments so that this condition holds is costless for the risk-neutral regulator since only the expected regulatory payment he receives matters from his own point of view.

As a matter of fact, all the randomness in the seller's payments needed to induce effort can be included into the regulatory payments $z_n(\theta)$ and $z_a(\theta)$. The price $t(\theta)$ paid by the buyer for the good can be made independent on whether an accident occurs or not.[8] As a result, when the seller has unlimited wealth, there is a complete dichotomy between output distortions and incentives for safety care. These distortions are the same as those arising in the optimal economic regulation of a firm generating no risk.

It is worth stressing that the prices $t^{SB}(\bar{\theta})$ and $t^{SB}(\underline{\theta})$ are not uniquely pinned down at the optimal regulation above. Indeed, as long as the buyer breaks even in expectation, many such pairs are possible. One possibility is that the buyer gets zero profit in each state of nature, i.e., $V^{SB}(\theta) = 0$ for all θ in Θ. The prices paid by the buyer to the seller are thus defined by $t^{SB}(\theta) = S(q^{SB}(\theta))$. Two simple justifications can be given for this choice. First, the buyer may be competing à la Bertrand with other similar buyers so that the buyer's profit in each state is driven to zero. Second, the buyer may have a tiny degree or risk-aversion and full insurance requires that his returns in front of different types of sellers are the same and thus identically equal to zero. This simple choice shows that, at the optimal regulation, the liability constraints of the principal are never relevant even when he has no asset on

---

[8]The fact that prices are non-conditional on the shock θ is particularly attractive when trade between the buyer and the seller takes place long before any pollution leakage takes place.

his own. This feature will of course remain even at the optimal regulation with limited liability on the seller's side.

For further references, we will sometimes mention the optimal regulation in the absence of liability constraint as an *interim efficient outcome*, since it maximizes a weighted sum of all the utilities subject to the regulator's informational constraints.

9.4   Regulation with Limited Liability

It is well known that, in pure moral hazard environments, inducing the first-best level of care may not always be feasible when the seller is protected by limited liability. To see that point in our context, notice that the participation constraint of the inefficient seller is binding at the optimum of Proposition 9.1. This yields

$$t^{SB}(\bar{\theta}) - \bar{\theta}q^{SB}(\bar{\theta}) - z_a^{SB}(\bar{\theta}) = -R(e^*) < 0,$$

and thus the inefficient seller, if an accident occurs, must pay a fine $z_a^{SB}(\bar{\theta})$ so large that he gets bankrupt (assuming he has no asset to start with).

Instead, for an efficient seller, the existence of an adverse selection rent $\Delta\theta q^{SB}(\bar{\theta})$ creates a buffer of liabilities which reduces the risk of bankruptcy. More precisely, we have

$$t^{SB}(\underline{\theta}) - \underline{\theta}q^{SB}(\underline{\theta}) - z_a^{SB}(\underline{\theta}) = \Delta\theta q^{SB}(\bar{\theta}) - R(e^*) < 0$$

only if the damage h and thus the first-best level of care e^* are large enough or, alternatively, if the uncertainty on cost $\Delta\theta$ is small enough.

In the sequel, we will assume that h is large enough so that bankruptcy in the event of an accident is a concern whatever the type of the seller. This assumption simplifies the analysis by getting rid of mixed cases.[9]

To avoid bankruptcy, the following seller's limited liability constraints must thus be satisfied:

$$u_a(\theta) = t(\theta) - \theta q(\theta) - z_a(\theta) \geq 0, \text{ for all } \theta \text{ in } \Theta. \tag{9.11}$$

9.4.1   Pure Moral Hazard

Let us start by considering the case where the marginal cost θ is common knowledge. Since h is large enough, (9.11) will be binding in both states of nature. We can thus rewrite

$$U(\theta) = \max_e \{e(t(\theta) - \theta q(\theta) - z_n(\theta)) - \psi(e)\}$$

---

[9]Laffont (1995) makes a similar assumption.

or

$$U(\theta) = R(e(\theta)) \tag{9.12}$$

where $\psi'(e(\theta)) = u_n(\theta) = t(\theta) - \theta q(\theta) - z_n(\theta)$ is positive to induce a positive level of care.

With limited liability and complete information on θ, the regulator's problem rewrites as

$$(R): \quad \max_{\{e(\theta),U(\theta),V(\theta)\}} \quad -(1-e(\theta))h - \psi(e(\theta)) + S(q) - \theta q - (1-\alpha)(U(\theta) + V(\theta)),$$

subject to (9.1) and (9.12).

Proposition 9.2 *With limited liability and moral hazard only, the optimal regulation entails:*

- *The first-best production levels $q^{MH}(\theta) = q^*(\theta)$ for all θ in Θ.*

- *A downward distortion in the level of care $e^{MH}(\theta) = e^{MH} < e^*$ which is the same for both seller types:*

$$h = \psi'(e^{MH}) + (1-\alpha)e^{MH}\psi''(e^{MH}). \tag{9.13}$$

- *Both types of the seller receive the same limited liability rent*

$$U^{MH}(\theta) = U^{MH} = R(e^{MH}), \tag{9.14}$$

and there is no rent left to the buyer.

Under pure moral hazard, the second-best effort trades off the social benefit against the cost of diminishing the accident probability. This cost has two components: first, as under complete information, the disutility of effort incurred by the seller; second, the cost of leaving a moral hazard rent to the seller to induce his effort when it is non-observable. Indeed, because of moral hazard and limited liability, the regulator can no longer threaten the seller with large fines in case of an accident to provide him costless incentives towards safety care. Only rewards are available and a moral hazard rent $U = R(e)$ must be left to the seller to induce effort e. This rent is again socially costly (with a negative weight $-(1-\alpha)$). To reduce the social cost of this rent, the second-best effort e^{MH} must be downward distorted below the first-best level. This distortion is greater when the seller's utility has little weight in the social welfare function (e^{MH} increases with α).

9.4.2 Moral Hazard and Adverse Selection

Let us now suppose that the regulator does not observe the firm's marginal cost θ. As before, adverse selection has an impact on the quantity that should be traded. However, this imperfect knowledge of the seller's profit also affects the amount that can be seized by the regulator when an accident occurs. Indeed, when he considers overstating his marginal cost, the efficient seller takes into account the fact that, if a damage occurs, a lower profit can be seized. In fact, upon such an event, the efficient seller can still save the adverse selection rent $\Delta\theta q(\bar{\theta})$ that he may grasp from mimicking an inefficient seller. This leaves only the inefficient seller's profit as possible liability payments. The possibility of saving this adverse selection rent when an accident occurs undermines much of the efficient seller's incentives to exert care.

With both adverse selection and moral hazard, incentive compatibility for an efficient seller can be written as:[10]

$$U(\underline{\theta}) = \max_e \{e(t(\underline{\theta}) - \underline{\theta}q(\underline{\theta}) - z_n(\underline{\theta})) - \psi(e)\}$$
$$\geq \max_e \{e(t(\bar{\theta}) - \underline{\theta}q(\bar{\theta}) - z_n(\bar{\theta})) + (1-e)\Delta\theta q(\bar{\theta}) - \psi(e)\},$$

or putting it differently,

$$U(\underline{\theta}) = R(e(\underline{\theta})) \geq \Delta\theta q(\bar{\theta}) + R(e(\bar{\theta})). \tag{9.15}$$

This incentive compatibility constraint is important and drives much intuition behind the forthcoming results. Compared with the case with unlimited liability, the price received by the seller for the good is a less effective tool to induce revelation since, with some probability, the sales revenue will be seized by the regulator. The seller has to be rewarded for truthtelling by means of moral hazard rents. These rents are less efficient means of transferring wealth to the private sector to relax incentive constraints since they have also an allocative impact on the levels of safety care. The incentive constraint (9.15) shows that the adverse selection and moral hazard parts of the incentive problem cannot be disentangled under limited liability.

The relevant participation constraint for the inefficient seller is

$$U(\bar{\theta}) \geq R(e(\bar{\theta})). \tag{9.16}$$

This participation constraint is also hardened with respect to the case with unlimited liability. There must be a positive rent left even to the least efficient seller if one wants any effort to be exerted.

The regulator's problem can now be rewritten as:

$$(R): \quad \max_{\{e(\theta),q(\theta),U(\theta)\}} E_\theta \Big(-(1-e(\theta))h - \psi(e(\theta)) + S(q(\theta))$$
$$-\theta q(\theta) - (1-\alpha)(U(\theta) + V(\theta)) \Big),$$

---

[10]It can be checked ex post that this is the only relevant constraint.

subject to (9.5)-(9.15) and (9.16).

Proposition 9.3 *Assume that h is large enough so that the seller's limited liability constraints are binding whatever his type. Then, the optimal regulation entails:*

- *All constraints (9.5)-(9.15) and (9.16) are binding.*

- *There exists $\lambda > 0$, the multiplier of the incentive constraint (9.15), such that the optimal effort levels $e^R(\underline{\theta})$ and $e^R(\bar{\theta})$ verify $e^R(\underline{\theta}) > e^R(\bar{\theta})$ and satisfy*

$$h = \psi'(e^R(\underline{\theta})) + \left(1 - \alpha - \frac{\lambda}{\nu}\right) e^R(\underline{\theta})\psi''(e^R(\underline{\theta})), \qquad (9.17)$$

$$h = \psi'(e^R(\bar{\theta})) + \left(1 - \alpha + \frac{\lambda}{1-\nu}\right) e^R(\bar{\theta})\psi''(e^R(\bar{\theta})), \qquad (9.18)$$

and

$$R(e^R(\underline{\theta})) = \Delta\theta q^R(\bar{\theta}) + R(e^R(\bar{\theta})). \qquad (9.19)$$

- *The efficient seller produces the first-best output $q^R(\underline{\theta}) = q^*(\underline{\theta})$ whereas the inefficient one's output is downwards distorted, $q^R(\bar{\theta}) < q^R(\underline{\theta})$ with*

$$S'(q^R(\bar{\theta})) = \bar{\theta} + \frac{\lambda}{1-\nu}\Delta\theta. \qquad (9.20)$$

- *The buyer obtains zero rent $\underset{\theta}{E}(V^R(\theta)) = 0$. The seller gets a positive rent whatever his type*

$$U^R(\underline{\theta}) = \Delta\theta\bar{q}^R(\bar{\theta}) + R(e^R(\bar{\theta})) > 0, \qquad (9.21)$$

$$U^R(\bar{\theta}) = R(e^R(\bar{\theta})) > 0. \qquad (9.22)$$

Note first that in a pure moral hazard environment, the same moral hazard rents are left to both seller types. When costs are instead non-observable, this is no longer possible. Doing so would indeed always make attractive for an efficient seller to underestimate his profit in order to systematically "save" the adverse selection rent $\Delta\theta q(\bar{\theta})$. This forces the regulator to give an extra reward to the efficient agent on top of the amount given under pure moral hazard. This extra reward corresponds to the non-verifiable informational rent $\Delta\theta q(\bar{\theta})$ that can never be seized by the regulator.

With limited liability, these extra rewards increase in fact the level of care exerted by an efficient seller. At the same time, the efficient seller is less tempted to mimic an inefficient one if the latter's moral hazard rent is downwards distorted. The level of care exerted by this agent is thus reduced to facilitate truthtelling.

Finally, as with unlimited liability, reducing the production of the inefficient seller also helps relaxing the incentive constraint (9.15). However, limited liability impacts on this output distortion as it can be seen by comparing the r.h.s. of (9.10) and (9.15). At a rough level, it is still true that the regulator trades off the efficiency gain from raising $q(\bar{\theta})$ against its incentive cost. The marginal cost of raising $q(\bar{\theta})$ is now given by the shadow cost λ of the incentive constraint (9.15). With unlimited liability, this shadow cost is simply the social cost of the efficient firm's information rent, namely $\nu(1-\alpha)$. Instead, with limited liability, efforts and outputs are linked altogether. Raising the output $q(\bar{\theta})$ has also an extra social value which is to increase the effort $e(\underline{\theta})$ performed by the most efficient firm and to reduce the likelihood of an accident by this type.

Through the limited liability constraints, everything happens thus as if the buyer's value of trade was made explicitly dependent on the probability of accident. To get further intuition on the nature of the output distortion, it is useful to rewrite (9.20) taking into account (9.19). We get

$$(1-\nu)(S'(q^R(\bar{\theta})) - \bar{\theta}) + \nu\left(\frac{h - \psi'(e^R(\underline{\theta}))}{e^R(\underline{\theta})\psi''(e^R(\underline{\theta}))}\right)\Delta\theta = \nu(1-\alpha)\Delta\theta. \quad (9.23)$$

The first term on the l.h.s. of (9.23) is the marginal surplus an inefficient seller times the probability that marginal cost is high. If output $q^R(\bar{\theta})$ is increased by dq, the expected surplus increases thus by this term multiplied by dq. At the same time, such an increase raises the socially costly rent of an efficient seller by an amount $\nu(1-\alpha)\Delta\theta dq$ which explains the r.h.s. of (9.23). However, with limited liability, the efficient seller can only be rewarded for truthtelling in terms of moral hazard rent. Thus, the effort of an efficient seller is also increased, reducing thereby the likelihood of an accident for that type. This second effect appears as the second term on the l.h.s. of (9.23).

Increasing production of the inefficient seller has not only an impact on productive surplus but it has also an environmental value. Everything happens as if output had an environmental impact directly incorporated into the consumer's utility function. It is striking that even though production and care do not interact directly in the production function, incentive compatibility creates such an endogenous link through the liability constraints. One cannot design an environmental policy without keeping an eye on its impact on production. With risk regulation, this impact is positive and output distortions are reduced.

At the optimal regulation, the shadow cost of the incentive constraint *with* limited liability is lower than *without* limited liability, at least for small cost uncertainty. Far from exacerbating output distortions, limited liability reduces them. The marginal price paid for production by the inefficient firm is thus closer to its value under complete information. However, this seemingly efficiency gain is somewhat of a mirage. Under asymmetric information, the right notion of efficiency is *interim efficiency*, which should account for the existing informational constraints. Compared to the interim efficient outcome obtained in Proposition 1, the binding liability constraints move us away from the optimal outcome.

Comparative Statics: The difficulty in computing explicitly the value of the shadow cost λ of the incentive constraint (9.15) makes it hard to get general comparative statics. However, we have:

Proposition 9.4 *With adverse selection, moral hazard and limited liability, more efficient sellers produce more and exert more care than less efficient ones. Moreover*

- $e^R(\underline{\theta}) > e^{MH} > e^R(\bar{\theta})$,

- $q^*(\bar{\theta}) > q^R(\bar{\theta}) > q^{SB}(\bar{\theta})$ *when $\Delta\theta$ is small enough.*

Our model predicts therefore that efficient sellers are less likely to create an environmental harm than inefficient ones under risk regulation. Production and safety care are positively correlated under limited liability. Note that there does not exist such a correlation without limited liability. Indeed, both types of the seller exert then the same first-best level of safety care even though they produce different outputs.

To understand the lesser magnitude of the output distortion under limited liability, it is useful to come back to (9.24) to explain better the social value of raising output. As long as $e^R(\underline{\theta})$ remains below the first-best level of effort e^* (and this is the case for instance when $\Delta\theta$ is small enough since then $e^R(\underline{\theta})$ is close to e^{MH}), the environmental benefit of raising $q(\bar{\theta})$ and $e(\underline{\theta})$ by the same token, is positive. This extra value of production justifies less output distortions than without liability constraint.

Implementation: The optimal regulation found above is quite demanding. Indeed, it requires communication between the regulator and the seller, observability of the private transaction between the buyer and the seller (and most noticeably control of output), and also commitment to a regulatory scheme. Nevertheless, this regulation can sometimes be implemented by using only ex post liability even when such comprehensive grand-contract is no longer feasible (maybe because economic and environmental regulations are split or because output is hardly verifiable by the regulator).

Suppose that, ex post, a judge imposes a fine equal to the harm, $z_a = h$, on the seller. Assume also that the buyer has all bargaining power in designing the private transaction with the seller, and that he has unlimited wealth so that he may end up paying whatever harm done. Then, it is easy to see that everything happens as if the fine is paid by the buyer himself.[11] The design

---

[11]See Segerstrom and Tietenberg and (1992) and Bontemps, Dubois and Vukina (2003) for this "equivalence principle".

of the private transaction solves a problem very similar to (R) except that the buyer does not take into account the social value of the seller's rent in his own objective function. The optimal regulation can be implemented with an *ex post* liability rule only if $\alpha = 0$, by asking the seller (or the buyer) to pay for the full damage. When the regulator has no redistributive concerns, the buyer shares with the latter the same desire for extracting the seller's rent and will thus implement the same output. Of course, contrary to the regulation case, the buyer's expected payoff is non-zero in the liability regime.

The case $\alpha = 0$ corresponds actually to a regulator who does not give any weight to the private sector in his objective function. He is thus only interested in collecting damages and counting the expected harm caused to third-parties. This leads exactly to the same objective as if he was acting as a judge forced to balance the cost of this harm with the payments requested from the private sector.

For $\alpha > 0$, the buyer definitively extracts too much rent from a social welfare viewpoint. Equation (9.23) is still useful to understand how the liability rule might be modified in this environment. Indeed, to find out the optimal output and effort chosen by the buyer for the efficient seller when the former must pay a damage D for the harm done, it suffices to replace respectively h by D and α by zero in that formula. Diminishing D below the full harm reduces efforts on both types and, due to the convexity of $R(\cdot)$, hardens the efficient seller's truthtelling incentive constraint, making output distortions even more valuable. This suggests that reducing the liability of stakeholders is of little help to reduce output distortions.

9.5 Extended Liability with Shallow Pocket

Even when $\alpha = 0$, the simple ex post liability rule proposed above may not be feasible when the buyer-seller coalition has not enough assets to cover harm h. This is typically the case when the buyer has himself few assets available or can easily hide them and the level of harm is much larger than the first-best surplus, $S(q^*(\theta)) - \theta q^*(\theta)$. Then, the ex post intervention of the judge can at most seize from the private sector the total value of the gains from trade.

In such an environment, the expected payoff of the buyer (still assuming he has all bargaining power in designing private transaction) becomes

$$\underset{\theta}{E}\left(e(\theta)(S(q(\theta)) - \theta q(\theta) - u_n(\theta))\right),$$

where $u_n(\theta) = t_n(\theta) - \theta q(\theta) = \psi'(e(\theta))$ is the seller's payoff when no accident occurs.

Now, the buyer may want to distort production to protect the benefits of his transaction with the seller from the threat of being seized. This adds a new distortion. To understand this new distortion, it is useful to start with the case of pure moral hazard.

9.5.1 Pure Moral Hazard

Suppose that the marginal cost θ is common knowledge within the buyer-seller coalition. The benefit of a transaction effectively accrues to the buyer only if an accident does not occur so that the optimal contract solves:

$$(P): \quad \max_{\{e_n(\theta),q(\theta)\}} \quad e(\theta)(S(q(\theta)) - \theta q(\theta) - \psi'(e(\theta))).$$

Proposition 9.5 *Assume that the buyer-seller coalition is subject to ex post environmental liability but that both the buyer and the seller are protected by limited liability. Then the optimal private transaction entails:*

- *A level of care $e^L(\theta)$ such that $e^L(\bar{\theta}) < e^L(\underline{\theta})$ with*

$$S(q^*(\theta)) - \theta q^*(\theta) = \psi'(e^L(\theta)) + e^L(\theta)\psi''(e^L(\theta)). \tag{9.24}$$

- *The first-best outputs $q^L(\theta) = q^*(\theta)$.*

Conditionally on the fact that no accident takes place, the buyer finds no reason to distort output under complete information. Trade remains always efficient. Imposing liability on both the buyer and the seller has no impact on the traded volume under complete information. The often heard criticism that extending liability towards principals modifies contracting and output should be qualified. This is not the case when the stakeholder has complete information on the agent's adverse selection parameter. Complete information between the buyer and the seller gives thus some foundations to the assumption made in the earlier literature that modifying the level of care exerted by the seller has no impact on the value of the transaction.

However, under the extended liability regime, the levels of care are far too low with respect to their levels at the optimal regulation (even in the most extreme case where $\alpha = 0$). The private value of the gains from trade is, by assumption, less than the harm level. Protecting those gains does not give enough stake to incentivize the seller to exert effort. The levels of care are even far below the second-best levels found in Section 9.4.

Note also that different seller types choose different effort levels because the first-best surpluses associated with those types are different. With extended liability and pure moral hazard, we recover the positive correlation between care and output even though its origins are quite different from what we found in Section 9.4.2. This is now the fact that a more efficient seller creates more surplus that increases his incentives for care within a buyer-seller coalition protected by limited liability. Private contracts have an impact on care but not the reverse under complete information on costs.

9.5.2 Moral Hazard and Adverse Selection

Let us now turn to the case where θ is not known by the buyer. His problem becomes:

$$(P): \quad \max_{\{e(\theta),q(\theta)\}} \; \mathop{E}_{\theta} \left(e(\theta)(S(q(\theta)) - \theta q(\theta)) - \psi'(e(\theta)) \right),$$

subject to (9.15).

Since the first-best surpluses for $\underline{\theta}$ and $\bar{\theta}$ may be far away from each other, it is not immediately clear whether the incentive constraint is binding or not at the optimum of (P). When the uncertainty on cost is not significant,[12] however, the incentive constraint (9.15) is in fact always binding at the optimum of (P). Indeed as uncertainty decreases, $e^L(\underline{\theta})$ and $e^L(\bar{\theta})$ come close to each other, and even though the adverse selection informational rent $\Delta\theta q^*(\bar{\theta})$ becomes small, one can show that the first effect dominates so that (9.15) is violated by the solution proposed when neglecting this constraint.

Proposition 9.6 *Assume that h is large enough. Then, for $\Delta\theta$ small enough, the optimal contract between the buyer and the seller is such that there exists $\mu > 0$, the multiplier of the adverse selection incentive constraint (9.15), such that:*

- *Only the efficient seller produces the first-best output, $q^A(\underline{\theta}) = q^*(\underline{\theta})$. For the inefficient seller, production is downward distorted with*

$$S'(q^A(\bar{\theta})) = \bar{\theta} + \frac{\mu}{(1-\nu)e^A(\bar{\theta})}\Delta\theta. \tag{9.25}$$

- *The levels of care $e^A(\underline{\theta})$ and $e^A(\bar{\theta})$ are respectively above and below their values in the pure moral hazard case; $e^A(\underline{\theta}) > e^L(\underline{\theta}) > e^L(\bar{\theta}) > e^A(\bar{\theta})$.*

$$S(q^*(\underline{\theta})) - \underline{\theta}q^*(\underline{\theta}) = \psi'(e^A(\underline{\theta})) + (1-\mu)e^A(\underline{\theta})\psi''(e^A(\underline{\theta})), \tag{9.26}$$
$$S(q^A(\bar{\theta})) - \bar{\theta}q^A(\bar{\theta}) = \psi'(e^A(\bar{\theta})) + (1+\mu)e^A(\bar{\theta})\psi''(e^A(\bar{\theta})). \tag{9.27}$$

- *Finally, (9.15) is binding so that*

$$R(e^A(\underline{\theta})) = R(e^A(\bar{\theta})) + \Delta\theta q^A(\bar{\theta}). \tag{9.28}$$

The qualitative features of the solution are quite similar to those of the optimal regulation. There still exists a positive correlation between effort and output which is reinforced by the fact that the adverse selection incentive compatibility constraint is now binding. The most efficient seller also exerts more care.

---

[12]Which was an implicit assumption made when we looked at the conditions under which both types may get bankrupt at the optimal regulation.

Again, as in the optimal regulation, the buyer solves the adverse selection problem by rewarding an efficient seller through an extra moral hazard rent, whereas an inefficient seller sees that rent being reduced to facilitate truthtelling.

Simultaneously, the buyer reduces the inefficient seller's output to relax (9.15). However, since the benefits from trade only go to the buyer when there is no environmental damage, the efficiency cost of distorting the inefficient seller's output downwards is not viewed as so important by the buyer. Indeed, with some probability, trade with this inefficient seller will not be beneficial to the buyer. This forces him to reduce output more than what a regulator would do (even in the extreme case where $\alpha = 0$). The marginal price paid for the output of an inefficient seller is quite low because the buyer has to account for a premium paid for the risk of losing all the benefits of the transaction. Adverse selection introduces a feed-back effect of care on output distortions which are now exacerbated.

With liability being extended to the buyer, and under the conditions of small cost uncertainty, strong allocative distortions appear and contracting forms that look quite inefficient from an interim efficiency viewpoint emerge.

9.6   Conclusion

In this paper, we have first explored the optimal risk regulation of a buyer-seller hierarchy in a framework with limited liability, moral hazard and adverse selection on some technological parameter which is a priori unrelated to care. At the optimal regulation, one cannot solve separately the moral hazard and the adverse selection sides of the incentive problem. Even in the absence of any technological interaction, the second-best optimal policy endogenously creates such a positive relationship between care and output. Efficient sellers exert more care.

Starting from this characterization of the optimal regulation, we then asked under which conditions it can be implemented through a simple liability rule. Such an implementation requires that the regulator has no redistributive concerns at all towards the private sector of the economy. In that case, the optimal regulation can be implemented with a liability rule imposing to either trading partner (the principal or the agent) a fine equal to the harm caused to third-parties. Whenever the principal is wealthy enough, a fine just equal to the full harm induces the second-best optimal level of care even when the agent is protected by limited liability.

However, when the harm size is large with respect to the gains from trade and the principal has a limited amount of assets that can be seized, such a liability rule cannot be used to the same extent. We investigated the impact of having both the buyer and the seller being subject to limited liability on the design of a private transaction subject to ex post legal intervention. In such a context, the principal and the agent may lose all their gains from trade if an accident occurs. The private transaction is designed with an eye on that threat. Extended liability still distorts contracting. Even though they are qualitatively similar and exhibit again a complementarity between care and output, distortions are more severe than at the optimal regulation.

The directions in which output is distorted by risk regulation and liability are not as intuitive as it could seem at first glance. Risk regulation tends to reduce output distortions compared with the interim efficient outcome obtained in the absence of liability constraints. Instead, extended liability tends to increase those distortions quite significantly. This points at the different impacts that risk regulation and liability rules have on production.

Appendix

• **Proof of Proposition 9.1** As standard in two-type adverse selection model (see Laffont and Martimort (2002), Chapter 2, for instance), (9.8) and (9.9) are both binding at the optimum. Moreover, (9.5) is also obviously binding. From those binding constraints, we derive $U(\underline{\theta}) = \Delta\theta q(\bar{\theta}), U(\bar{\theta}) = 0$ and $E_\theta(V(\theta)) = 0$. Inserting into the principal's objective function and optimizing with respect to efforts and outputs yields Proposition 9.1.

Note that $S(q(\underline{\theta})) - t(\underline{\theta}) = 0$ and $U(\underline{\theta}) = \Delta\theta q^{SB}(\bar{\theta})$ define only the expected payment of the efficient seller:

$$e^* z_n(\underline{\theta}) + (1 - e^*) z_a(\underline{\theta}) = S(q^*(\underline{\theta})) - \underline{\theta} q^*(\underline{\theta}) - \psi(e^*) - \Delta\theta q^{SB}(\bar{\theta}).$$

Given this expected value, we can find the values of $z_n(\underline{\theta})$ and $z_a(\underline{\theta})$ also satisfying $z_n(\underline{\theta}) - z_a(\underline{\theta}) = h = \psi'(e^*(\underline{\theta}))$, as it is needed to implement the first-best effort.

Similarly, $V(\bar{\theta}) = 0 = S(q(\bar{\theta})) - t(\bar{\theta})$ and $U(\bar{\theta}) = 0$ define only the expected payment to the inefficient seller

$$e^* z_n(\bar{\theta}) + (1 - e^*) z_a(\bar{\theta}) = S(q^{SB}(\bar{\theta})) - \bar{\theta} q^{SB}(\bar{\theta}) - \psi(e^*).$$

Again, we can easily find the values of $z_n(\bar{\theta})$ and $z_a(\bar{\theta})$ satisfying also $z_n(\bar{\theta}) - z_a(\bar{\theta}) = \psi'(e^*) = h$.

• **Proof of Proposition 9.2** (9.1) is obviously binding. Moreover, inserting (9.12) into the objective function and optimizing yields first-best outputs and the distorted effort given by (9.13).

• **Proof of Proposition 9.3** First observe that (9.5) must necessarily be binding. Even if (9.15) was slack, optimization would lead to $e(\bar{\theta}) = e(\underline{\theta}) = e^{MH}$ and we would get a contradiction when $q(\bar{\theta}) = 0$. Hence, (9.15) is also necessarily binding.

Denote by λ the corresponding positive multiplier. The Lagrangean writes as:

$$\underset{\theta}{E}\left(-h(1 - e(\theta)) - \psi(e(\theta)) + S(q(\theta)) - \theta q(\theta) - (1 - \alpha)U(\theta)\right)$$
$$+\lambda\left(R(e(\underline{\theta})) - R(e(\bar{\theta})) - \Delta\theta q(\bar{\theta})\right).$$

Optimizing and using the slackness condition yields (9.17) to (9.20).

• **Proof of Proposition 9.4** Because $\lambda > 0$, we have

$$h - \psi'(e^R(\underline{\theta})) - (1 - \alpha)e^R(\underline{\theta})\psi''(e^R(\underline{\theta})) < \quad 0 \quad < h - \psi'(e^R(\bar{\theta}))$$
$$-(1 - \alpha)e^R(\bar{\theta})\psi''(e^R(\bar{\theta})).$$

Using the fact that $(\psi'(e) + (1 - \alpha)e\psi''(e))' > 0$, we immediately get $e^R(\underline{\theta}) > e^{MH} > e^R(\bar{\theta})$.

Let us show also that $\lambda < \nu(1 - \alpha)$ when $\Delta\theta$ is small enough.

First, let us make explicit for $e^R(\bar{\theta}, \lambda), e^R(\underline{\theta}, \lambda)$ and $q^R(\bar{\theta}, \lambda)$ the dependence on λ obtained through equations (9.17), (9.18) and (9.20).

The value of λ is then obtained from solving

$$H(\lambda) = R(e^R(\underline{\theta}, \lambda)) - R(e^R(\bar{\theta}, \lambda)) - \Delta\theta q^R(\bar{\theta}, \lambda) = 0. \qquad \text{(A.9.1)}$$

Note of course that $H'(\lambda) > 0$ and thus that the solution to (A.9.1) is unique.

By definition, we have $H(0) = -\Delta\theta q^*(\bar{\theta}) < 0$. Moreover, for $\lambda = \nu(1 - \alpha)$, we have $e^R(\underline{\theta}, \lambda) = e^*$, $q^R(\bar{\theta}, \lambda) = q^{SB}(\bar{\theta})$ and

$$h = \psi'(e^R(\bar{\theta}, \lambda)) + \frac{1 - \alpha}{1 - \nu}e^R(\bar{\theta}, \lambda)\psi''(e^R(\bar{\theta}, \lambda)),$$

thus $e^R(\underline{\theta}, \lambda) > e^R(\bar{\theta}, \lambda)$. Finally, $H(\nu(1 - \alpha)) > 0$ when $\Delta\theta$ is small enough.

• **Proof of Proposition 9.5** It is immediate and follows from direct optimization.

• **Proof of Proposition 9.6** Suppose that the solution is given as in Proposition 6. Denote $F(e) = \psi'(e) + e\psi''(e)$ and $G = F^{-1}$, we want to prove that, for $\Delta\theta$ small enough,

$$R(G(W^*(\underline{\theta}))) - R(G(W^*(\bar{\theta}))) < \Delta\theta q^*(\bar{\theta}), \qquad \text{(A.9.2)}$$

where $W^*(\theta) = S(q^*(\theta)) - \theta q^*(\theta)$ so that we will have a contradiction with the fact that (9.15) cannot be slack.

By the Theorem of Intermediate Values, there exists $\tilde{W} \in (W^*(\underline{\theta}), W^*(\bar{\theta}))$ such that

$$R(G(W^*(\underline{\theta}))) - R(G(W^*(\bar{\theta}))) = (R \circ G)'(\tilde{W})(W^*(\underline{\theta}) - W^*(\bar{\theta})).$$

Moreover, we have $R'(e) = e\psi''(e)$

$$G'(\tilde{W}) = \frac{1}{2\psi''(\tilde{e}) + \tilde{e}\psi'''(\tilde{e})}$$

for some $\tilde{e}$ in $(e^*(\bar{\theta}), e^*(\underline{\theta}))$. Hence, we get

$$(R \circ G)'(\tilde{W}) = \frac{\tilde{e}\psi''(\tilde{e})}{2\psi''(\tilde{e}) + \tilde{e}\psi'''(\tilde{e})} \leq \frac{1}{2}$$

because $e \in [0, 1]$.

Finally, (A.9.2) holds when

$$\frac{1}{2}(W^*(\underline{\theta}) - W^*(\bar{\theta})) < \Delta\theta q^*(\bar{\theta}). \tag{A.9.3}$$

but for $\Delta\theta$ small, we have

$$W^*(\underline{\theta}) - W^*(\bar{\theta}) \approx \frac{|S^D(q^*(\underline{\theta}))|}{2}(q^*(\underline{\theta}) - q^*(\bar{\theta}))^2 + \Delta\theta q^*(\bar{\theta})$$

and (A.9.3) is clearly satisfied because $(q^*(\underline{\theta}) - q^*(\bar{\theta}))^2$ is $0(\Delta\theta^2)$.

References

Baron, D. and Myerson , R., 'Regulating a Monopolist with Unknown Costs', *Econometrica*, 50 (1982): 911-930.

Balkenborg, D., 'How Liable Should the Lender be? The Case of Judgement-Proof Firms and Environmental Risks: Comment', *American Economic Review*, 91 (2001): 731-738.

Bontemps, P., Dubois, P. and Vukina, T., 'Optimal Regulation of Private Production Contracts with Environmental Externalities', *Journal of Regulatory Economics*, 26 (2004): 284-298.

Boyd, J. and Ingberman, D., 'The Search for Deep Pockets: Is 'Extended Liaiblity' Expensive Liability?', *Journal of Law, Economics and Organization*, 13 (1997): 232-258.

Boyd, J. and Ingberman, D., 'The Vertical Extension of Environmental Liability through Chains of Ownership, Contract and Supply', in A. Heyes ed. *The Law and Economics of the Environment*, (2001), 44-70.

Boyer, M. and Laffont, J.J., 'Environmental Protection, Producer Insolvency and Lender Liability', in A. Xepapadeas ed. *Economic Policy for the Environment and Natural Resources*, (1996), 1-29, Edward Elgar.

Boyer, M. and Laffont, J.J., 'Environmental Risk and Bank Liability', *European Economic Review*, 41 (1997): 1427-1459.

Boyer, M. and Porrini, D., 'Modelling the Choice Between Regulation and Liability in Terms of Social Welfare,' *Canadian Journal of Economics*, 37 (2004): 590-612.

Dionne, G. and Spaeter, S., 'Environmental Risks and Extended Liability: The Case of Green Technologies', *Journal of Public Economics*, 87 (2003): 1025-1060.

Hiriart, Y. and Martimort, D., 'The Benefits of Extended Liability', *Rand Journal of Economics*, forthcoming.

Laffont, J.J., 'Regulation, Moral Hazard and Insurance of Environmental Risks', *Journal of Public Economics*, 58 (1995): 319-336.

Laffont, J.J. and Martimort, D., *The Theory of Incentives: The Principal-Agent Model*, (Princeton and Oxford: Princeton University Press, 2002).

Lewis, T. and Sappington, D., 'How Liable Should the Lender be? The Case of Judgement-Proof Firms and Environmental Risks: Comment', *American Economic Review*, 91 (2001): 724-730.

Pitchford, R., 'How Liable Should the Lender be? The Case of Judgement-Proof Firms and Environmental Risks', *American Economic Review*, 85 (1995): 1171-1186.

Maskin, E., 'Nash Equilibrium and Welfare Optimality', *The Review of Economic Studies*, 66(1) (1999): 23-38.

Segerson, K. and Tietenberg, T., 'The Structure of Penalties in Environmental Enforcement: An Economic Analysis', *Journal of Environmental Economics and Management*, 23 (1992): 179-200.

Shavell, S., 'The Judgement-Proof Problem', *International Review of Law and Economics*, 6 (1986): 45-58.

Strasser, K. and Rodosevich, D., 'Seeing the Forest for the Trees in CERCLA Liability', *Yale Journal on Regulation*, 10 (1993): 493-560.

Chapter 10

Judgment-Proofness and Extended Liability in the Presence of Adverse Selection[*]

Dieter Balkenborg
University of Exeter

10.1 Introduction

Liability rules are an important element in many systems of law. Liability laws are appealing to practitioners because they are easy and seemingly cost-less to implement. However, their impact on the incentives of economic agents remains much a matter of debate, in particular when one of the parties concerned can be judgement-proof. A firm becomes judgment proof if the damage costs of an environmental accident caused by the firm exceeds her own capital base. For such cases many countries considered the introduction of extended liability where lenders to such firms can be made liable for the residual damage costs which the firm cannot pay (see Boyer and Laffont (1997)). Starting with Pitchford (1995), Heyes (1996) and Boyer and Laffont (1997) a growing literature has studied the incentive effects of extended liability. Pitchford (1995) and Boyer and Laffont (1997) study the impact on the incentives of firms to prevent accidents. Boyer and Laffont (1997) also consider the role of private information with respect to the profitability of projects. Heyes (1996) studies a model with adverse selection concerning the environmental riskiness of projects and with moral hazard in regard to the efforts firms invest to prevent accidents. Boyer and Laffont as well as Hayes assume that the lender has all the bargaining power when selecting the lending contract. However, this is at odds with much of the finance literature where lenders are typically

---

[*]I would like thank for helpful comments and suggestions: the editor of this volume, David Martimort; an anonymous referee; David DeMeza; Miltos Makris; Roger Myerson; Jean Tirole; the seminar participants at Exeter, York and Royal Holloway; the participants of the 1st CIRANO-IDEI-LEERNA conference on Regulation; Liability and the Management of Major Industrial / Environmental Risks, Toulouse 2003 and the participants of the EAERE annual conference, Budapest 2004.

assumed to make zero profits. As shown in Balkenborg (2001) assumptions about the bargaining power a lender has vis-à-vis a firm when bargaining on a loan contract can be crucial when evaluating the impact of extended liability.

This paper determines the optimal extended liability rule for the adverse selection problem that arises when the firm has private information on the accident probability of the project she seeks to get financed. We will use the same base model as used in Balkenborg (2001) to study moral hazard, except that the probability of an environmental accident does not depend on the effort of the firm, but is exogenously given and known only to the firm.

The comparative statics result for the moral hazard model and for our adverse selection model could not be more different. A joint analysis of the moral hazard and the adverse selection problem in a single model promises hence to be rather subtle and is not attempted here.

The first difference concerns the question whether social welfare is higher when the lender or when the firm has all the bargaining power. In the moral hazard model the accident probability is decreasing in the bargaining power of the lender. A monopoly lender is the worst-case scenario because a monopoly lender extracting all the surplus from the project is bad for the incentives of the firm to take care (see Balkenborg (2001), Shavell (1997)).

In contrast, only a monopoly lender guarantees a first-best outcome in the adverse selection case. In our model, a monopoly lender is able to extract all the surplus, regardless of the private knowledge of the firm. Hence a firm who knows that the accident probability of her project is high cannot gain by pretending to have a low accident probability. Therefore high-risk projects do not jeopardize low-risk projects, the potential distortions due to adverse selection do not arise and first best can be achieved. As soon as the firm can capture some of the surplus, adverse selection causes distortions away form first best.

The second difference concerns the optimal joint liability. If the firm has all the bargaining power it is in the moral hazard model never socially optimal to use the deep pockets of the lender. The lender should contribute zero to the damage costs (see Pitchford (1995)). In the adverse selection model, however, not only should the lender contribute to the damage costs, but the overall joint liability imposed on firm and lender should typically exceed the actual damage costs (the joint liability should be "punitive"). Conversely, when the lender has all the bargaining power, the optimal liability is punitive in the moral hazard model but full joint liability (i.e., a joint liability equal to the actual damage costs) is optimal in the adverse selection case.

The distortion in our adverse selection model is, in essence, due to signalling. Bargaining should settle on a contract where it is not possible for low-risk types of the firm to propose an alternative contract that harms high-risk types and makes low-risk types of the firm and the lender better off. Such alternative contracts, if offered, would not lead the lender to conclude that he is facing a high-risk type and would hence be accepted. This would destabilize the initial agreement. Considerations of this type lead in our model to the selection of the contract that maximizes the expected payoff of the type of firm with the lowest accident probability subject to the constraint that the lender gets a fixed share of the maximal joint surplus.[1] This contract is shown to be typically an option contract with three different options.[2]

1. The first option is to run the project and give a high share of the surplus to the firm when no accident occurs and nothing when an accident occurs. This option favors low-risk types of the firm. In equilibrium it is taken up by those firms for whom the project promises (after taking account of the joint liability) a surplus in expectation.

2. The second option is a compensation payment to prevent firms with a high accident probability from running the project. This is necessary here because firms are assumed to have no own wealth and can hence only gain from running the project.[3] In equilibrium, this option is taken by the high-risk types of the firm for whom the project would yield a loss in expectation.

3. There is a third option where the project is only run with a small probability. The third option is for the medium-risk types of the firm for whom the project yields a loss in expectation but where it is cheaper to have them run the project than to pay them a compensation for staying out of business. They would have to be given a higher compensation than the high-risk types. However, this higher compensation, if offered, would be taken by all high-risk types as well. The total expected payment for bribes could become so high that the surplus available for the low-risks or the lender would have to be reduced.

It is the third option and the medium-risk types which create the distortion in our model.[4] Suppose the joint liability to the firm and the lender is set equal to the actual damage costs. Then the low-risk types are those producing a social surplus in expectation and they run the project. For the high-risk types the project would yield a social loss, but they take the compensation. All this is first-best. However, for the medium types the project also yields a social loss and for them the project is run with positive probability. This would not happen if the accident probability were public information.

---

[1] I skip here the case of a joint liability that the firm could pay on its own.

[2] In the formal model we work with "direct mechanisms" where the firm first announces the accident probability of her project and a neutral mediator then selects the option.

[3] To have such "bribes" in a loan contract may seem odd at first, but all it means is that the lender finances a nice new office and a rather decent salary to the owner/manager that does not have to be paid back if the firm later (after some further costly "research") decides to withdraw from the project.

[4] The precise classification into low-risk, medium-risk and high risk types is part of the definition of the optimal contract and hence dependent on the joint liability.

In Balkenborg (2001) we used a weighted version of the Nash bargaining solution (Nash (1950), Myerson (1991)) to solve the bargaining stage of the model. In this paper we have to analyze a bargaining problem with incomplete information to which the Nash bargaining solution does not apply. In particular, when the firm has all the bargaining power we have a bargaining problem with an informed principal (see Myerson (1983), Maskin and Tirole (1990) and Maskin and Tirole (1992)). In this paper we use a weighted version of the neutral bargaining solution Myerson (1983), Myerson (1984)) to solve the bargaining stage, primarily because it is the most generally applicable solution concepts for such problems. Technically, the determination of this solution is the main contribution of the paper. The relation to other approaches is considered in Subsection 10.1.

Section 11.3 introduces the model and the notations. Section 10.3 determines the weighted neutral bargaining solutions and compares it with other approaches. Comparative statics results are given in Section 10.4. In the conclusions in Section 10.5 we discuss some limitations of our approach. The appendix contains most of the proofs.

10.2 The Model

A wealth-constrained risk-neutral firm with no own wealth would like to run an environmentally risky project. This project requires an initial investment of size K and would yield a gross profit $v + K$. The net value of the project is hence v. Since the firm has no own wealth she needs a loan to run the project. A risk neutral lender with large but finite wealth $X > 0$ would be able to finance the project.

As stated, the project is environmentally risky. With probability p the project might cause an environmental accident. An accident would be a pure externality causing damage costs $h > v$ on potential victims. We assume that the wealth of the lender plus the surplus generated by the project would be more than enough to pay for the total damage costs, so $v + X > h > v$.

The accident will not directly affect the firm or the lender unless some form of liability is imposed. We study here the effects of a *joint and strict liability* $c \geq 0$. By this we mean that, if an accident occurs, the firm and the lender are liable with all their joint wealth for the amount c. First the firm is liable with all her wealth up to the amount c. If the firm cannot cover the full amount c, the lender has to pay for the remainder. Effectively the joint liability cannot exceed $v + X$, the total amount of cash available. Hence we assume $c \leq v + X$ in the following.

In this paper we study the case of adverse selection with respect to the accident probability $0 \leq p \leq 1$. Thus the accident probability p is the private knowledge of the firm. Firms with different accident probabilities correspond to different *types* of the firm. We consider here the case of finitely many types

$$0 \le p_0 < p_1 < \cdots < p_t < \cdots < p_T \le 1$$

and denote by $0 < q_t \le 1$ the ex-ante probability for the firm to be of type p_t. Sometimes we refer to the index t rather than the probability p_t as "the type" of the firm.

We assume either that there are only three types or that the following *monotone hazard rate* condition is satisfied for $t < T$:[5]

$$\frac{(p_{t+1} - p_t) \sum_{\tau=t+1}^{T} q_\tau}{q_t} \text{ is decreasing in } t. \tag{10.1}$$

We consider here the case of a bilateral monopoly with a single firm and a single lender. Admittedly, this analysis does not immediately carry over to a scenario where several firms compete for a loan and / or where several lenders compete to finance the project. The timing of the model is as follows:

1. The social planner announces the liability $0 \le c \le v + X$.

2. Nature selects with probability q_t the level of safety $p = p_t$ of the project which is then revealed to the firm.

3. The lender and the firm bargain over a mechanism. Following the revelation principle (Myerson (1979)), we assume that direct mechanisms are used.

4. If a mechanism is agreed upon, the firm announces a type $\hat{p}$ which, of course, does not have to be her true type.

5. The direct mechanism determines the outcome conditional on the type $\hat{p}$ announced by the firm. If the project is run, an accident occurs with probability p. If an accident occurs, the joint and strict liability c must be paid.

In our case a direct mechanism can be described by a 4-tuple of functions[6]

$$\mu = (\mu(\hat{p}))_{0 \le \hat{p} \le 1} = \left(Q(\hat{p}), w^+(\hat{p}), w^-(\hat{p}), w^0(\hat{p}) \right)_{0 \le \hat{p} \le 1}$$

---

[5]Notice that if the discrete distribution approximates a differentiable cdf $F(p)$ with density $f(p)$ then $\frac{1-F(p)}{f(p)} = \frac{(p_{t+1}-p_t)\sum_{\tau=t+1}^{T} q_\tau}{q_t}$. Thus our monotone hazard rate condition is the familiar one from the literature (see, e.g., Laffont and Martimort (2002)). It is made to ensure that it suffices to check the "local" incentive constraints (type t does not want to imitate type $t+1$ or $t-1$) in order to prove overall incentive compatibility. It turns out that we need this assumption only when there are four types or more.

[6]Without loss of generality we can assume, whenever convenient, that the mechanism is defined for all accident probabilities $0 \le p \le 1$. See Footnote 8 below.

If the firm announces to be of type $\hat{p}$, i.e. if she claims that her project has accident probability $\hat{p}$, then $Q(\hat{p})$ is the probability with which the project is undertaken.

If, with probability $1 - Q(\hat{p})$, the project is *not* undertaken, the expected final wealth of the firm is $w^0(\hat{p})$. The lender loses this amount.

If, with probability $Q(\hat{p})$, the project is undertaken, then the firm's expected final wealth is $w^+(\hat{p})$ if no accident occurs and $w^-(\hat{p})$ if an accident occurs. The lender gains $v - w^+(\hat{p})$ or $v - c - w^-(\hat{p})$, respectively.

The firm can only end up with a non-negative wealth, i.e. $0 \le w^+(\hat{p})$, $w^-(\hat{p})$, $w^0(\hat{p})$ must hold. Because the project generates at most the amount of capital v and since the lender's own wealth is X, we must overall impose the restrictions $0 \le Q(\hat{p}) \le 1$, $0 \le w^+(\hat{p}) \le v + X$, $0 \le w^-(\hat{p}) \le v + X - c$ and $0 \le w^0(\hat{p}) \le X$ on the mechanism.

Notice that the mechanism can be, but does not have to be, interpreted as menu of wage-contracts.

Let

$$U_f^*(\mu, \hat{p}|p) = (1 - Q(\hat{p})) w^0(\hat{q}) + Q(\hat{p}) \left[(1 - p) w^+(\hat{p}) + p w^-(\hat{p}) \right]$$

denote the expected profit of the firm from the direct mechanism μ if she is of type p and announces to be of type $\hat{p}$. Let

$$U_f(\mu|p) = U_f^*(\mu, p|p)$$

denote her expected utility from truthfully announcing her type. If all types of the firm announce her type truthfully the expected profit to the lender is, conditional on the firm being of type p,

$$U_l(\mu|p) = Q(p)(v - pc) - U_f(\mu|p)$$

and overall

$$U_l(\mu) = \sum_{t=0}^{T} U_l(\mu|p_t) q_t$$

A mechanism is called *incentive compatible* if it satisfies for all $p, \hat{p} \in [0, 1]$ the *incentive constraint*

$$U_f(\mu|p) \ge U_f(\mu, \hat{p}|p)$$

We assume that the reservation utility of the lender and of each type of the firm is zero. To be *individually rational* the mechanism must hence satisfy the *participation constraint* $U_l(\mu) \ge 0$ for the lender. Because the firm has no own wealth the participation constraint $U_f(\mu|p) \ge 0$ is automatically satisfied when the non-negativity constraints $0 \le w^+(\hat{p})$, $w^-(\hat{p})$, $w^0(\hat{p})$ hold. We call a mechanism *feasible* if it is individually rational and incentive compatible.

It will be important to distinguish three types of interim efficiency. A mechanism μ is *interim efficient* (among all types of all players) if it is feasible and if there exists no feasible mechanism ν such that $U_l(\nu) \geq U_l(\mu)$ and such that $U_f(\nu|p_t) \geq U_f(\mu|p_t)$ holds for all $0 \leq t \leq T$ whereby at least one of these inequalities is strict. μ is *interim efficient for the firm* if it is feasible and if there exists no feasible mechanism ν such that $U_f(\nu|p_t) \geq U_f(\mu|p_t)$ holds for all $0 \leq t \leq T$ whereby at least one of these inequalities is strict. In our model $U_l(\mu) = 0$ must hold for all mechanisms μ that are interim efficient for the firm. Finally, a mechanism is *interim efficient for the lender* if it maximizes the profit of the lender among all feasible mechanisms.[7] In our model, mechanisms that are interim efficient for the firm or the lender are also overall interim efficient.

When setting the joint liability c the social planner wants to maximize the utilitarian social welfare

$$\sum_{t=0}^{T} q_t Q(p_t)(v - p_t h)$$

where $Q(p_t)$ is the probability that the project is run given the firm's type and given the direct mechanism chosen by the lender and the firm conditional on the joint liability being c. In our model, the first-best level of social welfare that can be achieved is

$$\sum_{t=0}^{T} q_t (v - p_t h)^{+}$$

where we make use of the notation

$$(a)^{+} = \begin{cases} a & \text{for} \quad a > 0, \\ 0 & \text{for} \quad a \leq 0. \end{cases}$$

10.3 The Weighted Neutral Bargaining Solution

We start with a basic observation on incentive compatible (but not necessarily individually rational) mechanisms.[8]

Lemma 10.1 *The following holds for any incentive compatible mechanism* μ

---

[7]There is only one type of lender and hence no need to compromise between different types.

[8]Strictly speaking, the mechanism has only to be defined and to be incentive compatible with respect to the values of p in the support of the distribution of types. However, one can always extend an incentive compatible mechanism μ defined only for the types $p_0, \cdots, p_T$ to an incentive compatible mechanism μ' defined for all types $0 \leq p \leq 1$ by choosing for any $0 \leq p \leq 1$ $\mu'(p) = \mu(\hat{p}_t)$ such that $U_f^*(\mu, \hat{p}_t|p) = \max_{0 \leq t \leq T} U_f^*(\mu, p_t|p)$.

a) The inequality

$$-Q\left(p\right)\left(w^+\left(p\right)-w^-\left(p\right)\right) \geq \frac{U_f\left(\mu|p\right)-U_f\left(\mu|\hat{p}\right)}{p-\hat{p}} \geq -Q\left(\hat{p}\right)\left(w^+\left(\hat{p}\right)-w^-\left(\hat{p}\right)\right)$$

is satisfied for any $0 \leq \hat{p} < p \leq 1$.

b) The firm's type contingent payoff $U_f\left(\mu|p\right)$ *is convex in* p, *i.e.*

$$\frac{U_f\left(\mu|p'\right)-U_f\left(\mu|p\right)}{p'-p} \leq \frac{U_f\left(\mu|p''\right)-U_f\left(\mu|p'\right)}{p''-p'}$$

is satisfied for all $0 \leq p < p' < p'' \leq 1$.

c) The inequality

$$U_f\left(\mu|p\right) \geq \frac{1-p}{1-\hat{p}}U_f\left(\mu|\hat{p}\right)$$

is satisfied for $0 \leq \hat{p} < p \leq 1$ *while the inequality*

$$U_f\left(\mu|p\right) \geq \frac{p}{\hat{p}}U_f\left(\mu|\hat{p}\right)$$

is satisfied for $0 \leq p < \hat{p} \leq 1$.

Proof: (a and b) From the definition of U_f^* and U_f we obtain for any $p \neq \hat{p}$

$$\begin{aligned}
&U_f^*\left(\mu,\hat{p}|p\right) - U_f\left(\mu|\hat{p}\right)\\
=\ &\left(1-Q\left(\hat{p}\right)\right)w^0\left(\hat{q}\right) + Q\left(\hat{p}\right)\left[\left(1-p\right)w^+\left(\hat{p}\right)+pw^-\left(\hat{p}\right)\right]\\
&-\left(\left(1-Q\left(\hat{p}\right)\right)w^0\left(\hat{q}\right)+Q\left(\hat{p}\right)\left[\left(1-\hat{p}\right)w^+\left(\hat{p}\right)+\hat{p}w^-\left(\hat{p}\right)\right]\right)\\
=\ &-Q\left(\hat{p}\right)\left(p-\hat{p}\right)\left(w^+\left(\hat{p}\right)-w^-\left(\hat{p}\right)\right)
\end{aligned}$$

and hence, since μ is incentive compatible,

$$U_f\left(\mu|p\right) - U_f\left(\mu|\hat{p}\right) \geq -Q\left(\hat{p}\right)\left(p-\hat{p}\right)\left(w^+\left(\hat{p}\right)-w^-\left(\hat{p}\right)\right) \tag{10.2}$$

Interchanging the roles of p and $\hat{p}$ we get

$$U_f\left(\mu|\hat{p}\right) - U_f\left(\mu|p\right) \geq -Q\left(p\right)\left(\hat{p}-p\right)\left(w^+\left(p\right)-w^-\left(p\right)\right)$$

or

$$U_f\left(\mu|p\right) - U_f\left(\mu|\hat{p}\right) \leq -Q\left(p\right)\left(p-\hat{p}\right)\left(w^+\left(p\right)-w^-\left(p\right)\right) \tag{10.3}$$

Hence a) follows for $\hat{p} < p$ and then immediately b) follows for $p < p' < p''$.

(c) The definition of U_f and the nonnegativity constraints on the mechanism yield $Q\left(\hat{p}\right)\left(1-\hat{p}\right)w^+\left(\hat{p}\right) \leq U_f\left(\mu|\hat{p}\right)$. From Inequality (10.2) we obtain for $p > \hat{p}$

$$U_f\left(\mu|p\right) - U_f\left(\mu|\hat{p}\right) \geq -Q\left(\hat{p}\right)\left(p - \hat{p}\right)w^+\left(\hat{p}\right) \geq -\frac{p - \hat{p}}{1 - \hat{p}}U_f\left(\mu|\hat{p}\right)$$

and $U_f\left(\mu|p\right) \geq \frac{1-p}{1-\hat{p}}U_f\left(\mu|\hat{p}\right)$. Symmetrically $Q\left(\hat{p}\right)\hat{p}w^-\left(\hat{p}\right) \leq U_f\left(\mu|\hat{p}\right)$ and Inequality (10.3) imply $U_1\left(\mu|p\right) \geq \frac{p}{\hat{p}}U_1\left(\mu|\hat{p}\right)$ for $p < \hat{p}$. ∎

Given any function $f\left(p\right) \geq 0$ that has all the properties described for the function $U_f\left(\mu|p\right)$ in the lemma, it is not difficult to construct an incentive compatible mechanism μ with $U_f\left(\mu|p\right) = f\left(p\right)$ for all $0 \leq p \leq 1$. Of course, the mechanism may violate the participation constraint of the lender. Still, one can construct a plethora of interim efficient mechanism where $U_f\left(\mu|p\right)$ is U-shaped or increasing or constant in the accident probability p. In particular, let $S\left(c\right) = \sum_{t=0}^{T} q_t\left(v - p_t c\right)^+$ denote the ex-ante maximally available surplus for the firm and the lender. Then the mechanism $\mu = \left(Q, w^+, w^-, w^0\right)$ defined by $Q\left(\hat{p}\right) = 1$ for $\hat{p} < v/c$, $Q\left(\hat{p}\right) = 0$ for $\hat{p} \geq v/c$, $w^+\left(\hat{p}\right) = w^-\left(\hat{p}\right) = w^0\left(\hat{p}\right) = S\left(c\right)$ for all $0 \leq \hat{p} \leq 1$ is interim efficient and distributes the gain $S\left(c\right)$ equally among all types. Mechanism like these are implausible because they do not reward productive types more than unproductive types.

To rule out such mechanisms and to select among the many interim efficient mechanisms we use the concept of the neutral bargaining solution, due to Myerson (1983) and Myerson (1984). Myerson's concept is an extension of the Nash bargaining solution to bargaining problems with incomplete information. It is based on an axiomatic approach and uses elements of both cooperative and non-cooperative game theory. It is the solution concept with the smallest solution sets satisfying the axioms described below. We use here a weighted version depending on a parameter $0 \leq \alpha \leq 1$ which we interpret as the *bargaining power* of the lender. The firm has bargaining power $1 - \alpha$. The case $\alpha = 0$ corresponds to the case where the firm has all the bargaining power and can essentially make a take-it-or-leave-it offer to the lender. The neutral bargaining solution for this case is developed in Myerson (1983). $\alpha = 0.5$ is the case of equal bargaining power discussed in Myerson (1984).

After we have described in this section the neutral bargaining solution and the mechanisms it selects in our model, we will compare it to other refinement approaches. I will take quite some freedom in describing the neutral bargaining solution and its axiom, partly to give an alternative exposition of the ideas and partly because some rephrasing is needed to fit with the model here. Sometimes I have to be a bit vague because I do not want to develop Myerson's general framework. Readers familiar with Myerson's papers will not find it difficult to verify the equivalence with his original formulations.

10.3.1 Efficiency

Just as Nash required his bargaining solution to be Pareto-efficient, Myerson requires that *the neutral bargaining solution is interim efficient* as an axiom. If we expect the bargaining to end efficiently and if we assume that the firm knows her types when bargaining, this is clearly the right notion of efficiency. However, even if bargaining occurs ex-ante, before the firms learns her type, interim efficiency seems appropriate if there is the possibility to renegotiate.

10.3.2 Strong Solutions and Random Dictatorship

a) Strong solutions for the firm are of interest if the firm has all the bargaining power. A *strong solution* for the firm is a mechanism μ with the following three properties:

1. The payoff to the lender from the mechanism neither depends on the type announced by the firm nor on her true type: $U_l(\mu) = U_l^*(\mu, \hat{p}|p)$ for all $0 \leq p, \hat{p} \leq 1$. At a strong solution, asymmetric information is not a problem.

2. The lender's expected payoff is zero, as we would expect it if the firm has all the bargaining power: $U_l(\mu) = 0$.

3. The mechanism is interim efficient for the firm.

There is general agreement that the strong solution, if it exists, is the appropriate solution concept if the lender has all the bargaining power. It is an interim efficient Rothschild-Stiglitz-Wilson allocation relative to the initial endowment and yields hence the unique perfect Bayesian equilibrium outcome in the three-stage game where the informed principal can make a take-it-or-leave-it offer to select among direct mechanisms (see Maskin and Tirole (1992) in particular Section 9). It is consistent with many equilibrium refinements for this game (see Myerson (1983), Theorem 1 and Maskin and Tirole (1992), Proposition 7).

The next axiom imposed by Myerson is hence: *A strong solution for the firm is the neutral bargaining solution if the firm has all the bargaining power.*

If the joint liability c does not exceed the net value of the project, our model has a strong solution. Namely, the firm always pays for the liability herself and pays exactly the investment costs K back to the lender. The lender makes zero profit and his deep pockets are not employed for liability payments. Formally,

Proposition 10.1 *If $c \leq v$, the following mechanism $\mu_0 = \left(Q_0, w_0^+, w_0^-, w_0^0\right)$ is a strong solution for the firm and hence the neutral bargaining solution for the firm: $Q_0(\hat{p}) = 1$, $w_0^+(\hat{p}) = v$, $w_0^-(\hat{p}) = v - c$ for all $0 \leq \hat{p} \leq 1$.*

b) Because there is only one type of lender, the strong solution for the lender (and hence the neutral bargaining solution if the lender has all the bargaining power) is the mechanism which maximizes his expected payoff among all feasible mechanisms. In our model it is the following simple mechanism where the lender appropriates all surplus.

Proposition 10.2 *The strong solution for the lender (and hence the neutral bargaining solution if the lender has all the bargaining power) is the mechanism* $\mu_1 = \left(Q_1, w_1^+, w_1^-, w_1^0\right)$ *defined by* $Q_1\left(\hat{p}\right) = 1$ *if* $v - \hat{p}c > 0$, $Q_1\left(\hat{p}\right) = 0$ *if* $v - \hat{p}c < 0$, $w_1^+\left(\hat{p}\right) = w_1^-\left(\hat{p}\right) = w_1^0\left(\hat{p}\right) = 0$ *for* $0 \leq \hat{p} \leq 1$.

c) For intermediate bargaining power it is not clear what a "strong solution" should be. Myerson (1984) avoids the problem with his *"random dictatorship" axiom*, which extends to arbitrary bargaining power $0 \leq \alpha \leq 1$ as follows:

If μ_0 *is a strong solution for the firm and* μ_1 *a strong solution for the lender, then the randomized mechanism* $(1 - \alpha)\mu_0 + \alpha\mu_1$, *where* μ_0 *is played with probability* $1 - \alpha$ *and* μ_1 *with probability* α, *is, if interim efficient, the weighted neutral bargaining solution for the bargaining power* α *of the lender.*

It follows immediately that:

Proposition 10.3 *If* $c \leq v$ *the following mechanism* $\mu_\alpha = \left(Q_\alpha, w_\alpha^+, w_\alpha^-, w_\alpha^0\right)$ *is a neutral bargaining solution when the lender has bargaining power* $0 \leq \alpha \leq 1$: $Q_\alpha\left(\hat{p}\right) = 1$, $w_\alpha^+\left(\hat{p}\right) = (1 - \alpha)v$ *and* $w_\alpha^-\left(\hat{p}\right) = (1 - \alpha)(v - c)$.

Thus the project is always run and the proceeds, whatever they are, are divided according to the bargaining power.

10.3.3 Extended Models and Limits

When the joint liability c exceeds the net value v of the project, a strong solution for the firm does not exist for our model. When strong solutions do not exist, Myerson's idea is to look at extensions of the model for which a strong solution exists. He shows that there exists always an interim efficient mechanism μ of the original model for which one can find a sequence of extended models with strong solutions that approach in the limit the given mechanism "from below". A neutral bargaining solution for $\alpha = 0$ is any mechanism that can be approached in this way.

It suffices to consider here only extensions of our model where a "safe project" D is added that always gives the lender zero regardless of the type of firm, $u_l\left(D|p\right) = 0$ for all $0 \leq p \leq 1$, and where the payoff to the firm $u_f\left(D|p\right)$ can be dependent on the type, but not on any announcements of types. D is thus trivially a feasible mechanism. Provided it is interim efficient, it is a strong solution. A general mechanism of the extended model is in our case given by five functions $\left(R\left(\hat{p}\right), Q\left(\hat{p}\right), w^+\left(\hat{p}\right), w^-\left(\hat{p}\right), w^0\left(\hat{p}\right)\right)$ where $1 - R\left(\hat{p}\right)$ is the probability with which the safe project D is chosen if the firm announces to be of type $\hat{p}$ and where $\left(Q\left(\hat{p}\right), w^+\left(\hat{p}\right), w^-\left(\hat{p}\right), w^0\left(\hat{p}\right)\right)$ determines, with the same interpretations as for the original model, what happens if D is not chosen. To connect the original model with the extended model, we need the following *Independence of Irrelevant Alternatives* axiom: *If the mechanism μ of the original model is not a neutral bargaining solution of the extended model, then it is also not a neutral bargaining solution of the original model.*

Next we have to consider a sequence of extensions for each integer $k = 1, 2, 3, \ldots$ The extensions in the sequence differ only with respect to the type-dependent payoffs $u_f^k\left(D|p\right)$ which the firm gets from the safe project D. We can assume that these payoffs converge to a limit $\lim_{k \to \infty} u_f^k\left(D|p\right) = u_f\left(D|p\right)$. To connect the solutions in the sequence of models with the solution for the limit model we impose the following *continuity axiom*: *Suppose that μ_α is an interim efficient mechanism of the limit model and that the k-th model in the sequence has the neutral bargaining solution μ_α^k for the bargaining power α. Then, if*

$$\lim_{k \to \infty} u_f\left(\mu_\alpha^k | p\right) \leq u_f\left(\mu_\alpha | p\right) \text{ holds for all } 0 \leq p \leq 1,$$

μ_α *is a weighted neutral bargaining solution for the bargaining power α of the limit model.*[9]

The natural way to find the neutral bargaining solution for $\alpha = 0$ is to approximate a solution candidate "from below" by a sequence of strong solutions in extended models. Only very few mechanisms can be approximated in this way, which is why Myerson's approach selects very sharply among the mechanisms.

It may be instructive to see in the simplest setting why the mechanism μ which distributes the surplus equally among all types cannot be approximated from below. Suppose there are only two types. For the low-risk type $t = 0$ the accident probability is p_0 is zero, for the high-risk type $t = 1$ it is strictly positive. Assume that the joint liability exceeds the net value of the project $(c > v)$ but that it is still worthwhile for the high-risk type to produce $(v - p_1 c > 0)$. Suppose the mechanism μ described above where both types get $S\left(c\right) = q_0 v + q_1\left(v - p_1 c\right)$ could be approached by strong solutions from below. Suppose first that $\lim_{k \to \infty} u_f^k\left(D|p_1\right) > \left(1 - p_1\right) S\left(c\right)$. Then the following mechanism would Pareto-dominate D in all extended models for large k. If the firm announces to be of the low-risk type, the original project is run. The firm receives $S\left(c\right) + \varepsilon$ if no accident occurs and zero otherwise. If she announces to be of the high risk type, the safe project D is selected. Hereby $\varepsilon > 0$ is chosen such that $u_f^k\left(D|p_1\right) > \left(1 - p_1\right)\left(S\left(c\right) + \varepsilon\right)$ for large k. This mechanism is feasible in the extended models and gives the low-risk type

---

[9] Myerson (1984) imposes an additional probability invariance axiom which we do not need here and which I therefore skip.

$S(c) + \varepsilon$ for arbitrarily large k. This contradicts the assumption that D is interim efficient and $\lim_{k \to \infty} u_f^k(D|p_0) \leq S(c)$.

So we must have $\lim_{k \to \infty} u_f^k(D|p_1) \leq (1 - p_1) S(c)$. However, then we can consider a mechanism where the original project is always run and where both types get $S(c) + \varepsilon$, with $\varepsilon > 0$ sufficiently small, if no accident occurs and zero otherwise. We have $\lim_{k \to \infty} u_f^k(D|p_0) < S(c) + \varepsilon$ and $\lim_{k \to \infty} u_f^k(D|p_1) < (1 - p_1)(S(c) + \varepsilon)$ for large k. Moreover, $(q_0 + q_1(1 - p_1))(S(c) + \varepsilon) < S(c)$ for small $\varepsilon > 0$. Thus we obtain the contradiction that for k large D is not interim efficient for the firm.

10.3.4   Characterization of the Neutral Bargaining Solution for $c > v$

To find, first for $\alpha = 0$, the neutral bargaining solution of the original model we want to choose the payoffs $u_f^k(D|p)$ such that D is a strong solution and such that

$$\omega(p) = \lim_{k \to \infty} u_f^k(D|p) \leq u_f(\mu|p)$$

holds for a mechanism μ of the original model. μ is then the neutral bargaining solution if the firm has all the bargaining power. The numbers $\omega(p)$ are called *warranted claims*.

Since types are ordered by the accident probability in our model it is not surprising that we can determine the warranted claims jointly with the neutral bargaining solution in an inductive procedure starting with the highest risk type.

To do this, recall that we have only finitely many types $p_0, \ldots, p_T$ in our model. For any type $0 \leq t \leq T$ we call the following model the t-bargaining problem. The model differs from the given one only in the prior over the types. The new prior (q_τ^t) is the posterior to which the lender would update if he learned that the type of the firm is t or higher. Thus

$$q_\tau^t = \begin{cases} 0 & \text{for } 0 \leq \tau < t \\ \frac{q_\tau}{q_t + \ldots + q_T} & \text{for } t \leq \tau \leq T \end{cases}$$

Theorem 10.1 *The neutral bargaining solutions $\mu^{0,t}$ in the t-bargaining problems when the firm has all the bargaining power and the warranted claims $\omega(p_t)$ are inductively defined for $t = T, T - 1, \ldots, 0$ as follows:*

$\mu^{0,t}$ is the feasible mechanism in the t-bargaining problem that maximizes type p_t's expected payoff among all feasible mechanisms that give all types $\tau > t$ at least their warranted claims $\omega(p_\tau)$ as expected payoff. Type t's warranted claim is his expected payoff in the mechanism $\mu^{0,t}$.

In particular, $\omega(p_T) = (v - p_T c)^+$.

Notice that the theorem implies Proposition 10.1 for the case $v > c$. For $c > v$ we obtain the following result. Recall that a type t' with $t' > t$ can guarantee himself in any incentive compatible mechanism a fraction $\frac{1-p_{t'}}{1-p_t}$ of the expected payoff of type t.

Proposition 10.4 *Suppose $v < c$. Then the warranted claims satisfy $w(p_t) = 0$ for all t with $v - p_t c \leq 0$. For all $t < t'$ with $v - p_t c > 0$ and $p_{t'} < 1$ we have.*

$$w(p_t) > \frac{1 - p_t}{1 - p_{t'}} w(p_{t'})$$

Therefore the neutral bargaining solution when the firm has all the bargaining power is the mechanism that maximizes the expected payoff of type $t = 0$ among all feasible mechanisms.

We give next a more explicit description of this mechanism. This mechanism is basically an option contract with three options. The first option is chosen by low-risk types $t < t_1$, the second by medium-risk types $t_1 \leq t < t_2$ and the third by high-risk types $t_2 \leq t$. Hereby t_1 and t_2 are chosen as follows. If no t with $v - p_t c \leq 0$ exists (so the project yields a gain for all types), set $t_1 = t_2 = T + 1$. Otherwise, let t_1 be the smallest index with $v - p_t c \leq 0$ and define t_2 as the smallest integer for which

$$q_t (v - p_t c) + (p_{t+1} - p_t)(v + X) \sum_{\tau = t+1}^{T} q_\tau \leq 0 \qquad (10.4)$$

Clearly, $t_2 \geq t_1$. Next, we determine a "wage" w^+ to be paid if the project is run without accident. Let

$$p_t^* = \begin{cases} p_t & \text{for} \quad t_2 \leq t_2 \\ p_{t_2} & \text{else} \end{cases}$$

and

$$r_t = \begin{cases} q_t \frac{p_t(c-v)+(1-p_t)X}{(1-p_t)(v+X)} & \text{for} \quad t_1 \leq t < t_2 \\ q_t & \text{else} \end{cases} .$$

Let

$$w^+ = \frac{\sum_{t=0}^{T} q_t (v - p_t c)^+}{\sum_{t=0}^{T} r_t (1 - p_t^*)}$$

and let $w^0 = (1 - p_{t_2}) w^+$

Theorem 10.2 *The neutral bargaining solution μ^0 when the firm has all the bargaining power is for $c > v$ the following mechanism $(Q(p), w^0(p), w^+(p), w^-(p))$:*

1. $Q(p_t) = 1$, $w^+(p_t) = w^+$, $w^-(p_t) = 0$ for $0 \leq t < t_1$.

2. $Q(p_t) = \frac{w^+}{v+X}$, $w^+(p_t) = v + X$, $w^-(p_t) = w^0(p_t) = 0$ for $t_1 \leq t < t_2$.

3. $Q(p_t) = 0$, $w^0(p_t) = w^0$ for $t_2 \leq t \leq T$.

Figure 10.1 shows the expected gain $U_f(\mu^0, p)$ of each type of the firm in this mechanism. The graph of this function is piecewise linear with a kink at $p = p_{t_2}$, it is downward sloping to the left and constant to the right of the kink.

The mechanism is an option contract with the following three options:

- One option is to run the project with certainty and to receive the net payment $w^+ > 0$ when no accident occurs and zero otherwise. This is the option chosen by all low-risk types $t < t_1$ for whom the project yields a gain in expectation $(v - pc \geq 0)$. Because this option is always available to the firm, each type p of the firm must gain at least $w^+ (1 - p)$.

- Another option is not to run a project and to receive a compensation. This is the option chosen by the high-risk types with accident probability $p \geq p_{t_2}$. For these types running the project would yield a loss $(v - pc < 0)$. If the compensation w^0 would not be offered in exchange for not running the project, these types would choose to run the project negligently since they have no own money at stake and can hence only win. The compensation is chosen such that type t_2 is indifferent between running the project and taking the compensation. Since the size of the compensation cannot be conditioned on the accident probability, the graph of $U_f(\mu^0, p)$ is flat for $p \geq p_{t_2}$.

- The types $t_1 < t < t_2$ are of "medium risk". For them the project is run with a small probability $Q = \frac{w^+}{v+X}$, in which case they receive all available wealth $v + X$. Running the project for this type brings a loss $(v - pc < 0)$ and therefore the project is run only with a small probability. If it is run and no accident occurs, the firm wins the maximal available prize $v + X$. This lottery is chosen such that the medium-risk types are indifferent between accepting the contract with the lottery and accepting the contract for the low-risk types. With both contracts they would gain $(1 - p) w^+$ in expectation. In equilibrium they select the one with the lower accident probability.

  The medium types are not "bribed out of business" because they require a higher compensation than the high-risk types. Since any compensation given to them can also be taken up by the high-risk types, it is, from the perspective of the low-risk types, cheaper to keep them working than to bribe them out of business. Even better is the described lottery.

When determining the solution, the interesting part is to determine t_2, i.e. to find out which types who run a loss $(v - pc < 0)$ should play the "lottery" and which ones should be bribed out of business. Thanks to the monotone hazard rate condition (10.1), this critical number t_2 is determined by the inequality (10.4).

Example 10.1 *Although our analysis is restricted to the case of finitely many types, we can, for instance, approximate the uniform distribution of types by setting $p_t = \frac{t}{T}$ and $q_t = \frac{1}{T+1}$ $(t = 0, \cdots, T)$ and then take the limit $T \to \infty$. In the limit p_{t_1} converges to v/c, p_{t_2} to $\frac{2v+X}{v+c+X}$ and w^+ to*

$$\frac{v^2 (v + c + X)}{v^2 + c^2 + c(c - v) + cX}$$

Since $\frac{2v+X}{v+c+X} > \frac{v}{c}$ there is in the limit for $c > v$ always an interval of medium-risk types for whom the project is run with positive probability.

We have now described the neutral bargaining solutions μ_0 when the firm has all the bargaining power and μ_1 when the lender has all the bargaining power. It is not difficult to show that the mechanism $\mu_\alpha := (1 - \alpha)\mu_0 + \alpha\mu_1$, where nature chooses with probabilities $1 - \alpha$ and α between the two mechanisms μ_0 and μ_1, is interim efficient. The random dictatorship axiom hence implies:

Proposition 10.5 *The neutral bargaining solution when the lender has bargaining power $0 \le \alpha \le 1$ is the mechanism μ_α just described.*

10.3.5   Comparison with Other Approaches

Bargaining before the private information is received

Does the firm learn the accident probability before or after she bargains with the lender? Both scenarios are meaningful in our model, although the neutral bargaining solution seems to be tailored for bargaining problems where the firm already has her private information.

Let us again consider the case where the firm has all the bargaining power and let us suppose that she can make a take-it-or-leave-it offer before she learns her type. She is restricted, however to propose interim efficient contracts. Ex ante all the firm cares about is her expected payoff averaged over all her possible types. The maximal payoff she can gain is $S(c) = \sum_{t=0}^{T} q_t (v - p_t c)^+$ and this maximum is, for instance, achieved with the mechanism μ where $S(c)$ is distributed equally among all types. It is hence an equilibrium if the firm proposes this mechanism and the lender accepts. Any other ex-ante efficient feasible mechanism gives an equilibrium as well. The neutral bargaining solution is, however, not ex-ante efficient whenever there exist "medium-risk" types who are not "bribed our of business" and for whom the project yields a loss $(v - pc < 0)$. Thus the results from bargaining ex-ante seem to conflict with the neutral bargaining solution.

However, this argument does not take account of the possibility of renegotiation. Suppose ex-ante bargaining has led to the mechanism where the surplus is shared equally between all types. Now the firm gets her private information and learns that her project has a very low accident probability. She could then address the lender and propose the neutral bargaining solution μ_0. Since middle- and high-risk types can only lose if this proposal gets accepted, the lender should infer that he is facing a low-risk type firm and accept the offer. (After having excluded the possibility that high-risk types would make such an offer, he would actually expect to win.)

Thus the simple bargaining model where a party makes an offer ex-ante before learning her type can give interesting insights (see Laffont and Martimort (2002) for examples), but it is problematic if renegotiation could lead to a mechanism that is not ex-ante efficient. Once renegotiation is taken into account, the simplicity of the ex-ante approach is lost and it may be easier to analyze the interim bargaining problem, after the type has been revealed.

Perfect Bayesian Equilibria

Maskin and Tirole (1990) and Maskin and Tirole (1992) characterize perfect Bayesian equilibria in the three-stage game where an informed principal can make a take-it-or leave-it offer for the (direct) mechanism to be selected. Their analysis is based on the Rothschild-Stiglitz-Wilson allocation (short: RSW allocation). As remarked above, their approach also selects the strong solution as the unique perfect Bayesian equilibrium outcome for $c < v$. However, when the joint liability c exceeds the net value v of the project, results for the RSW allocation relative to the "no deal" allocation, where the lender gets zero for every type, are somewhat disappointing.[10] In particular, when there are types with accident probability less than 1 for whom the project does not yield a gain $(v - pc < 0)$, the RSW allocation gives zero to each type of the firm. This is so because in any feasible mechanism where one type of the firm makes a positive gain, the types for whom the project yields a loss must also gain a positive amount because they can imitate. In any such mechanism the lender must hence make a loss on the highest-risk type. Since the lender must get at least zero in the RSW allocation conditional on each type, each type of the firm must gain zero. The results of Maskin and Tirole (1992) imply that *every feasible* mechanism of the model yields a perfect Bayesian equilibrium in the three-stage game. The RSW allocation is interim efficient for the firm and it is the unique perfect Bayesian equilibrium of the three-stage if and only if the prior puts positive probability only on types of the firm for whom the project generates a loss. The equilibrium is then, of course, the one where trade never occurs $(Q(p) = 0$ and $w^0(p) = 0$ for all $p)$.

---

[10]Further insights, though, might be obtained from studying the renegotiation game discussed in Maskin and Tirole (1992) and RSW allocations relative to other intial allocations.

When there are only types for whom the project yields a gain, the results have a little more structure. In the RSW allocation the lender cannot make a loss on the highest type T, so the highest risk type cannot gain more than $v - p_T c$. Lemma 10.1 c) implies that the RSW allocation yields $\frac{1-p_t}{1-p_T}(v - p_T c)$ for type t. The following simple calculation shows that the RSW allocation is interim efficient (and so the three-stage game has a unique perfect Bayesian equilibrium) if and only if there is only a single type and hence no problem of incomplete information.

$$\sum_{t=0}^{T} q_t \frac{1-p_t}{1-p_T}(v - p_T c) \leq \sum_{t=0}^{T} q_t (v - p_t c) \Leftrightarrow \sum_{t=0}^{T} \frac{q_t}{1-p_t}(p_T - p_t)(c - v) \leq 0$$

whereby equality occurs if and only if $p_t = p_T$ for all t and hence $T = 0$.

If there is more than one type, every feasible mechanism which Pareto-dominates for all types of the firms the RSW allocation yields an equilibrium outcome for the three-stage game. Again, the selection is not sufficiently strong to allow for an insightful comparative statics analysis.

Refinements for Signalling Games

The three-stage game just discussed is a signalling game. The findings in this subsection suggest that there is indeed a connection between Myerson's weighted bargaining solution and the refinement concepts for signalling games discussed in the literature. We need, however, a fairly strong refinement criterion, namely the FGP criterion based on Farrell (1985) and Grossman and Perry (1986) as discussed in Maskin and Tirole (1992), Section 5B. The relation to other refinement criteria needs further research.

The consistency of the neutral bargaining solution μ_0 with the FGB criterion follows in the case $v \leq c$ from Proposition 7 in Maskin and Tirole (1992). The explicit description of the neutral bargaining solution for $c > v$ implies:

Lemma 10.2 *Let μ be any interim efficient mechanism for the firm different from μ_0, the neutral bargaining solution if the lender has no bargaining power. Then there exists a type $0 < t \leq T$ with $v - p_t c > 0$ such that $U_f(\mu|p_\tau) \leq U_f(\mu_0|p_\tau)$ for all $\tau < t$ and $U_f(\mu|p_\tau) > U_f(\mu_0|p_\tau)$ for all $\tau \geq t$.*

As a straightforward consequence of the definitions and the lemma we obtain.

Proposition 10.6 *Consider the three stage game, where first the firm can propose a mechanism which is interim efficient for the firm, the lender can then accept or reject and thereafter the proposed mechanism, if accepted, is implemented. Then the neutral bargaining solution μ_0 when the firm has all the bargaining power is the only pure strategy perfect Bayesian equilibrium outcome of this game consistent with the FGP criterion.*

Proof: We show first that μ_0 defines such an equilibrium. This equilibrium is as follows. All types of the firm propose μ_0. The lender accepts this mechanism. If the firm would deviate and propose a different interim efficient mechanism μ, the lender would update his believes by assuming that all types who would strictly gain by having μ rather that μ_0 accepted have an equal likelihood to have deviated. Let t be the type determined for μ and μ_0 in the previous lemma. The lender's posterior is then given by $\{q_\tau^t\}$ from the t-bargaining problem. This belief is consistent with the FGP criterion. Since μ is interim efficient for the firm, the lender expects to break even when μ is played given his initial prior. He expects to make a loss from μ given his posterior. (The types $\tau < t$ are the most productive and they gain even less in μ than in μ_0. Since his posterior rules out these types, he must expect to make a loss.) Hence it is optimal for him to reject. Thus the described behavior describes a perfect Bayesian equilibrium consistent with the FGP criterion.

Consider now a perfect Bayesian equilibrium consistent with the FGP criterion where an interim efficient mechanism μ is offered. Suppose the firm deviates and offers μ_0 instead. The FGP criterion then requires the lender to update his belief by assuming that the deviation must have come with probability zero from the types who gain less in μ_0 than in μ, with equal probability from the types who gain more in μ_0 than in μ and with an equal or possibly lower probability from types who are indifferent. Since, by the lemma, he is eliminating only high risk types from his belief by updating, he expects to strictly gain in expectation by accepting μ_0. The equilibrium must, by the FGP criterion, hence be such that he accepts μ_0. However, then we cannot have an equilibrium because the lowest-risk type would always deviate and propose μ_0. This is a contradiction. ∎

One can similarly "justify" the neutral bargaining solution for the bargaining power α by assuming that the firm must offer an interim efficient mechanism that gives the lender at least the fraction α of the maximally available surplus.

10.4 The Optimal Liability

We can now present our main result concerning the optimal joint liability in the pure adverse selection case studied here.

Theorem 10.3 *Either a full or a punitive joint liability maximizes social welfare. In particular, the deep pockets of the lender have to be employed to achieve the social optimum.*

Proof: The neutral bargaining solution is described in Theorems 10.2 and 10.5. Assume that initially the full liability $c = h > v$ is imposed and then gets reduced to a liability $\tilde{c} \geq v$. (The case $\tilde{c} < v$ is obvious.) Initially, the project is run with certainty only if it is socially worthwhile, i.e., when $v - ph \geq 0$ Let us write $t_1(c)$, $t_2(c)$, and $w^+(c)$ to indicate the dependence of these numbers on c. It is immediate that $t_1(c)$ and $t_2(c)$ are weakly increasing in c. We must have $w^+(\tilde{c}) > w^+(h)$ because, by Theorem 10.4, $w^+(\tilde{c})$ is the solution to a less constrained optimization problem than $w^+(h)$ (i.e., the same objective function is maximized over a smaller set). For all types t with $v - p_t c \geq 0$ the project is run with certainty both at the liabilities h and $\tilde{c}$, for types $t_2(\tilde{c}) \leq t$ it is never run. For all other types the probability of running the project increases from 0 to $(1-\alpha)\frac{w^+(\tilde{c})}{v+X}$ or 1 or from $(1-\alpha)\frac{w^+(h)}{v+X}$ to $(1-\alpha)\frac{w^+(\tilde{c})}{v+X}$ or 1.Thus social welfare cannot increase and hence either a full or a joint liability must be optimal. ∎

Whether the optimal joint liability is strictly punitive or just equal to the damage costs is harder to say in general. If $t_1(h) = t_2(h)$ then full liability is optimal because it yields first best since $\sum_{t=0}^{T} q_t(v - p_t h)^+$ is the maximal social welfare. This is, for instance, always the case when there are only two types. Due to the discreteness of the type space we can, however, not deduce that the optimal liability is unique. The range of optimal liabilities may contain values above and below the damage costs, at least one value cannot be below.

To avoid these integer problems, let us look at the uniform distribution of types as a limit, as described in Example 10.1.[11] For this example we are going to show that a) the optimal liability is strictly punitive unless the lender has all the bargaining power and b) the optimal liability is decreasing and social welfare increasing in the bargaining power of the lender, at least as long as the optimal liability does not jump discontinuously.

We write $P_1 = v/c$, for the limit of p_{t_1} and $P_2 = \frac{2v+X}{v+c+X}$ for the limit of p_{t_2}. $P_1(c)$ and $P_2(c)$ are strictly increasing since $\frac{\partial P_1}{\partial c} = -\frac{v}{c^2} < 0$ and $\frac{\partial P_2}{\partial c} = -\frac{2v+X}{(v+c+X)} < 0$. We have $w^+ = \frac{v^2(v+c+X)}{v^2+c^2+c(c-v)+cX}$ and hence $\frac{\partial w^+}{\partial c} = -v^2\frac{2c^2+4cv-2v^2+4cX+X^2}{(v^2+2c^2-cv+cX)^2} < 0$.

When, for $c > v$, the weighted neutral bargaining solution for the bargaining power α of the lender is played, utilitarian social welfare is

$$SW = \int_0^{P_1} (v - ph)\,dp + (1-\alpha)\frac{w^+}{v+X}\int_{P_1}^{P_2} (v - ph)\,dp$$

---

[11]An older script, available from my website, gives a detailed analysis for the case of three types.

Leibniz' rule yields

$$
\frac{\partial SW}{\partial c} = \frac{\partial P_1}{\partial c}(v - P_1 h) + (1 - \alpha)\left[\frac{\partial w^+}{\partial c}\frac{1}{v + X}\int_{P_1}^{P_2}(v - ph)\,dp\right.
$$

$$
\left. + \frac{w^+}{v + X}\left(\frac{\partial P_2}{\partial c}(v - P_2 h) - \frac{\partial P_1}{\partial c}(v - P_1 h)\right)\right]
$$

and substitution shows $\frac{\partial SW}{\partial c}\big|_{c=h} > 0$ as long as $\alpha < 1$. Since the proof of the previous proposition still applies here, it follows that *the optimal liability must be strictly punitive* unless the lender has all the bargaining power. Since $h > v$, it is always optimal to employ the deep pockets of the lender.

I have not yet been able to show that the social welfare function is quasi-concave. Therefore the following argument is correct only as long as there is no discontinuous jump in the optimal joint liability c (and $\frac{\partial^2 SW}{\partial c^2} < 0$ holds at the optimum).

Let $c^* > h$ denote the optimal liability given as a solution to the first-order condition $\frac{\partial SW}{\partial c} = 0$. $\frac{\partial SW}{\partial c}$ takes the form $A + (1 - \alpha)B$ where A and B do not depend on α. At the optimum A is positive since $c^* > h$ and so $B < 0$ for $\alpha < 1$ from the first-order condition. Since $\frac{\partial^2 SW}{\partial \alpha \partial c} = -B > 0$ the envelope theorem implies that *social welfare is* (at least locally) *increasing in α*. Since $\frac{\partial^2 SW}{\partial c^2} < 0$ at the optimum we obtain then

$$
\frac{\partial c^*}{\partial \alpha} = -\frac{\partial^2 SW}{\partial \alpha \partial c}\Big/\frac{\partial^2 SW}{\partial c^2} < 0,
$$

i.e., *the optimal joint liability is decreasing in the bargaining power α of the lender*.

10.5 Conclusion

We have been able to determine the weighted neutral bargaining solution for an adverse selection model where lenders lend to wealth-constrained firms who could cause a severe environmental accident. We have seen how adverse selection can create a distortion away from first-best because in the contract selected by the lender and the firm the potentially hazardous project is undertaken at a loss for medium-risk types of the firm in order to avoid the large compensation payments needed to deter these types from running the project. Overall, the comparative statics of the model considered here is simpler, but in sharp contrast to the comparative statics in Balkenborg (2001) for the moral hazard problem.

The analysis has a number of limitations. First of all, it is still a conjecture that the weighted neutral bargaining solution is unique for this model.

Secondly, the conclusions would not be as extreme if the firm had some small amount of own capital to lose from running the project. This more realistic assumption was avoided here to keep the analysis simple. Thirdly, our analysis is made for a bilateral monopoly and it is not obvious how it extends to the case of several lenders and firms in competition.

Appendix

Proofs

For $c \geq v$ we apply Theorem 4 of Myerson (1984) to verify that we have indeed found the neutral bargaining solutions. We will not address the question of uniqueness of the neutral bargaining solution.

Because all types of each player are risk neutral, every mechanism $\mu = \left(Q\left(p \right), w^{+}\left(p \right), w^{-}\left(p \right), w^{0}\left(p \right) \right)_{0 \leq p \leq 1}$ is payoff equivalent to a lottery $\left(\mu\left(d_i | p_t \right) \right)_{\substack{1 \leq i \leq 5 \\ 1 \leq t \leq T}}$ over the following six mechanisms.

d_0: The project is not run and everyone gets zero. $u_f\left(d_0 | p_t \right) = u_l\left(d_0 | p_t \right) = 0$;

d_1: The project is not run and the firm receives all available cash. $u_f\left(d_1 | p_t \right) = X$, $u_l\left(d_1 | p_t \right) = -X$;

d_2: The project is run and the lender gets all available wealth. $u_f\left(d_2 | p_t \right) = 0$, $u_l\left(d_2 | p_t \right) = v - p_t c$;

d_3: The project is run and the firm gets all available wealth if no accident occurs but nothing otherwise. $u_f\left(d_3 | p_t \right) = \left(1 - p_t \right)\left(v + X \right)$, $u_l\left(d_2 | p_t \right) = p_t\left(v - c \right) - \left(1 - p_t \right) X$;

d_4: The project is run and the firm gets all available wealth if an accident occurs but nothing otherwise. $u_f\left(d_3 | p_t \right) = p_t\left(v + X - c \right)$, $u_l\left(d_2 | p_t \right) = -p_t X + \left(1 - p_t \right) v$;

d_5: The project is run and the firm gets all available wealth. $u_f\left(d_5 | p_t \right) = v + X - p_t c$, $u_l\left(d_2 | p_t \right) = -X$.

With some abuse of notation we write $\mu = \left(\mu\left(d_i | p_t \right) \right)_{\substack{0 \leq i \leq 5 \\ 1 \leq t \leq T}}$.

For $k = 1, 2, \cdots$ let $\left(\lambda_t^k \right)_{0 \leq t \leq T}$ be a strictly positive vector of weights with $\lambda_t^k \to 0$ for $k \to \infty$. Moreover, we assume $\lambda_t^k < r_t$ with r_t as in defined subsection 10.3.4. A mechanism μ that maximizes for given k

$$\left(\lambda_0^k + \alpha_0^k \right) U_f\left(\mu | p_0 \right) + \sum_{t=1}^{T} \lambda_t^k U_f\left(\mu | p_t \right) + U_l\left(\mu \right) \tag{A.10.1}$$

with given $\alpha_0^k \geq 0$ among all individually rational and incentive compatible mechanisms is interim efficient. The Lagrangian for this linear programming problem can be written as

$$\left(\lambda_0^k + \alpha_0^k\right) U_f\left(\mu|p_0\right) + \sum_{t=1}^{T} \lambda_t^k U_f\left(\mu|p_t\right) + U_l\left(\mu\right)$$

$$+ \sum_{t=1}^{T} \alpha_t^k \left(U_f\left(\mu|p_t\right) - U_f^*\left(\mu, p_{t-1}|p_t\right)\right)$$

$$+ \sum_{t=1}^{T} \beta_t^k \left(U_f\left(\mu|p_{t-1}\right) - U_f^*\left(\mu, p_t|p_{t-1}\right)\right)$$

$$= \sum_{t=0}^{T} sv\left(\mu|p_t\right) = \sum_{t=0}^{T} \sum_{i=0}^{5} \mu\left(d_i|p_t\right) sv\left(d_i|p_t\right)$$

whereby

$$
\begin{aligned}
sv\left(\mu|p_t\right) &= \left(\lambda_t^k + \alpha_t^k + \beta_{t+1}^k\right) U_f\left(\mu|p_t\right) + q_t U_l\left(\mu|p_t\right) \\
&\quad - \alpha_{t+1}^k U_f^*\left(\mu, p_t|p_{t+1}\right) - \beta_t^k U_f^*\left(\mu, p_t|p_{t-1}\right) \\
&= \left(\lambda_t^k + \alpha_t^k + \beta_{t+1}^k - \alpha_{t+1}^k - \beta_t^k - q_t\right) U_f\left(\mu|p_t\right) + q_t Q\left(p_t\right)\left(v - p_t c\right) \\
&\quad + \left(\alpha_{t+1}^k\left(p_{t+1} - p_t\right) - \beta_t^k\left(p_t - p_{t-1}\right)\right) Q\left(p_t\right)\left(w^+\left(p_t\right) - w^-\left(p_t\right)\right)
\end{aligned}
$$

provided the only binding constraints are the incentive constraints requiring that type $t+1$ cannot gain from imitating type t and vice versa. Hereby we use the convention $\alpha_{T+1}^k = \beta_1 = 0$ and obtain from the definitions[12]

$$
\begin{aligned}
sv\left(d_0|p_t\right) &= 0 \\
sv\left(d_1|p_t\right) &= \left(\lambda_t^k + \alpha_t^k + \beta_{t+1}^k - \alpha_t^{k+1} - \beta_t^k - q_t\right) X \\
sv\left(d_2|p_t\right) &= q_t\left(v - p_t c\right) \\
sv\left(d_3|p_t\right) &= \left(\lambda_t^k + \alpha_t^k + \beta_{t+1}^k - \alpha_{t+1}^k - \beta_t^k - q_t\right)\left(1 - p_t\right)\left(v + X\right) \\
&\quad + q_t\left(v - p_t c\right) + \left(\alpha_{t+1}^k\left(p_{t+1} - p_t\right) - \beta_t^k\left(p_t - p_{t-1}\right)\right)\left(v + X\right) \\
&= \left(\left(\lambda_t^k - q_t\right)\left(1 - p_t\right) + \alpha_t^k\left(1 - p_t\right) + \beta_{t+1}^k\left(1 - p_t\right)\right. \\
&\quad \left. - \alpha_{t+1}^k\left(1 - p_{t+1}\right) - \beta_t^k\left(1 - p_{t-1}^k\right)\right)\left(v + X\right)) + q_t\left(v - p_t c\right) \\
sv\left(d_4|p_t\right) &= \left(\lambda_t^k + \alpha_t^k + \beta_{t+1}^k - \alpha_{t+1}^k - \beta_t^k - q_t\right) p_t\left(v + X - c\right) \\
&\quad + q_t\left(v - p_t c\right) + \left(\alpha_{t+1}^k\left(p_{t+1} - p_t\right) - \beta_t^k\left(p_t - p_{t-1}\right)\right)\left(v + X - c\right) \\
sv\left(d_5|p_t\right) &= \left(\lambda_t^k + \alpha_t^k + \beta_{t+1}^k - \alpha_{t+1}^k - \beta_t^k - q_t\right)\left(v + X - p_t c\right) \\
&\quad + q_t\left(v - p_t c\right) + \left(\alpha_{t+1}^k\left(p_{t+1} - p_t\right) - \beta_t^k\left(p_t - p_{t-1}\right)\right) p_t c
\end{aligned}
$$

---

[12]In the notation we suppress the dependency of $sv\left(d_i|p_t\right)$ on α_t^k and λ_t^k.

The dual problem is the problem of minimizing for given (λ_t^k) and α_0^k the sum

$$\sum_{t=0}^{T} \max_{0 \le i \le 5} sv\,(d_i|p_t)$$

with respect to $\alpha_1^k, \cdots, \alpha_T^k, \beta_1^k, \cdots, \beta_T^k \ge 0$.

We choose inductively for $t = T, T-1, \ldots$

$$\alpha_t^k = \sum_{\tau=t}^{T} \frac{1 - p_\tau^*}{1 - p_t^*} \left(r_\tau - \lambda_\tau^k\right) + \beta_t^k \frac{1 - p_{t-1}^*}{1 - p_t^*}$$

$$\beta_t^k = \begin{cases} \frac{1 - p_t^*}{p_t^* - p_{t-1}^*} \left[r_t - q_t - \frac{p_{t+1}^* - p_t^*}{1 - p_t^*} \alpha_{t+1}^k\right]^+ & \text{for} \quad t_1 \le t < t_2 \\ 0 & \text{else} \end{cases}$$

One verifies immediately (using $p_{t_2} = 0$) the recursive formula

$$\alpha_t^k + \beta_{t+1}^k = r_t - \lambda_t + \alpha_{t+1}^k \frac{1 - p_{t+1}^*}{1 - p_t^*} + \beta_t^k \frac{1 - p_{t-1}^*}{1 - p_t^*}$$

Lemma 10.3 *With these choices we have*

$$q_t\,(v - p_t c)^+ = \max_{0 \le i \le 5} sv\,(d_i|p_t) = \begin{cases} sv\,(d_2|p_t) = sv\,(d_3|p_t) & \text{for} \quad 0 \le t < t_1 \\ sv\,(d_0|p_t) = sv\,(d_3|p_t) & \text{for} \quad t_1 \le t < t_2 \\ sv\,(d_0|p_t) = sv\,(d_1|p_t) & \text{for} \quad t_2 \le t \le T \end{cases}$$

Proof: We leave it to the reader to check that $\max_{4 \le i \le 5} sv\,(d_i|p_t) \le \max_{0 \le i \le 3} sv\,(d_i|p_t)$. We calculate $sv\,(d_1|p_t)$ and $sv\,(d_3|p_t)$ using the recursive formula for α_t^k.

Suppose $0 \le t < t_1$. Then $\beta_t^k = 0$. We obtain $sv\,(d_1|p_t) < 0$ and $sv\,(d_3|p_t) = v - p_t c$ from the recursion formula, $r_t = q_t$ and the second expression for $sv\,(d_3|p_t)$ above.

Suppose $t_1 \le t < t_2$. Then $\alpha_t^k + \beta_{t+1}^k = q_t \left(1 - \frac{q_t(v - p_t c)}{(1 - p_t)(v + X)}\right) - \lambda_t^k + \frac{1 - p_{t+1}}{1 - p_t} \alpha_{t+1}^k + \beta_t^k \frac{1 - p_{t-1}^*}{1 - p_t^*}$ and therefore $sv\,(d_3|p_t) = 0$ whereas

$$sv\,(d_1|p_t) = \left(r_t - q_t + \frac{1 - p_{t+1}}{1 - p_t} \alpha_{t+1}^k - \left[r_t - q_t - \frac{p_{t+1}^* - p_t^*}{1 - p_t^*} \alpha_{t+1}^k\right]^+\right) X \le 0$$

Suppose $t_2 \le t \le T$. Then $\alpha_t^k = q_t - \lambda_t^k + \alpha_{t+1}^k$ and so $sv\,(d_1|p_t) = 0$ whereas

$$sv\,(d_3|p_t) = q_t\,(v - p_t c) + (p_{t+1} - p_t)\,\alpha_{t+1}^k\,(v + X) < 0.$$

∎

We can choose now $w_f^k(p_t)$ and w_l^k as the solution to the simultaneous system of equations

$$\left(\lambda_t^k + \alpha_t^k + \beta_{t-1}^k\right) w_f^k(p_t) - \alpha_{t+1}^k w_f^k(p_{t+1}) - \beta_t^k w_f^k(p_{t-1}) = (1-\alpha)\, q_t\, (v - p_t c)^+$$

$$w_l^k = w_l = \alpha \sum_{t=0}^{T} q_t\, (v - p_t c)^+$$

As $k \to \infty$ we obtain the limits $\alpha_t^k \to \alpha_t = \sum_{\tau=t}^{T} \frac{1-p_\tau^*}{1-p_t^*} r_\tau + \beta_t \frac{1-p_{t-1}}{1-p_t}$,

$$\beta_t^k \to \beta_t = \begin{cases} \frac{1-p_t^*}{p_t^* - p_{t-1}^*}\left[r_t - q_t - \frac{p_{t+1}^* - p_t^*}{1-p_t^*}\alpha_{t+1}\right]^+ & \text{for} \quad t_1 \le t < t_2 \\ 0 & \text{else} \end{cases}$$

and $w_f^k(p_t) \to w_f(p_t)$ whereby

$$(\alpha_t + \beta_{t+1})\, w_f(p_t) = (1-\alpha)\, q_t\, (v - p_t c)^+ + \alpha_{t+1} w_f(p_{t+1}) - \beta_t w_f(p_{t-1})$$

from which we obtain as a unique solution (since $(v - p_t c)^+ = 0$ whenever $\beta_t \ne 0$)

$$\alpha_t w_f(p_t) = (1-\alpha) \sum_{\tau=t}^{T} q_\tau\, (v - p_\tau c)^+$$

In particular we have

$$w_l = \alpha \sum_{t=0}^{T} q_t\, (v - p_t c)^+ = \alpha \sum_{t=0}^{T} \max_{0 < i \le 5} sv\,(d_i | p_t) = U_l\,(\mu_\alpha)$$

$$w_f(p_0) = (1 - p_0) \frac{(1-\alpha) \sum_{t=0}^{T} q_t\, (v - p_t c)^+}{\sum_{t=0}^{T} r_t\,(1 - p_t^*)} = U_f\,(\mu_\alpha | p_t)$$

Since μ_α is interim efficient (see the argument below), it follows from Theorem 4 in Myerson (1984) that μ_α is a neutral bargaining solution if $w_f(p_t) \le U_f\,(\mu_\alpha | p_t)$ holds for all t.

To verify this, notice that our notation corresponds to the one used in Myerson's theorem as follows. We use p_t instead of t_i to indicate types. For the lowest risk-type p_0 we write $\lambda_0^k + \alpha_0^k$ instead of $\lambda^k(p_0)$. For all other types p_t we write λ_t^k instead $\lambda^k(p_t)$. Our analysis assumes that $\alpha^k(p_t | p_{t'})$ is zero except when $t' = t+1$ or $t' = t-1$ and so we have write α_t^k instead of $\alpha^k(p_t | p_{t-1})$ and β_t instead of $\alpha^k(p_{t-1} | p_t)$. We write $sv\,(d | p_t)$ instead of $\sum_{j=f,l} V_j\,(d, p_t, \lambda^k, \alpha^k)$.

For all $t \geq t_1$ we have $\omega_f(p_t) = 0 < U_f(\mu^\alpha | p_t)$. Now suppose $t < t_1$. Then $U_f(\mu_\alpha | p_t) = \frac{1-p_t}{1-p_0} U_f(\mu^\alpha | p_0) = \frac{1-p_t}{1-p_0} \omega_f(p_0)$. Thus we must show

$$\omega_f(p_t) \leq \frac{(1-p_t)}{(1-p_0)} \omega_f(p_0)$$

which is verified as follows

$$\omega_f(p_t) \leq \frac{(1-p_t)}{(1-p_0)} \omega_f(p_0)$$

$$\Leftrightarrow \quad \frac{(1-\alpha)\sum_{\tau=t}^{T} q_\tau (v - p_\tau c)^+}{\alpha_t} \leq \frac{(1-p_t)}{(1-p_0)} \frac{(1-\alpha)\sum_{\tau=0}^{T} q_\tau (v - p_\tau c)^+}{\alpha_0}$$

$$\Leftrightarrow \quad \frac{\sum_{\tau=t}^{T} q_\tau (v - p_\tau c)^+}{\sum_{\tau=t}^{T} r_\tau (1 - p_\tau^*)} \leq \frac{\sum_{\tau=0}^{T} q_\tau (v - p_\tau c)^+}{\sum_{\tau=0}^{T} r_\tau (1 - p_\tau^*)}$$

$$\Leftrightarrow \quad \left(\sum_{\tau=0}^{T} r_\tau (1 - p_\tau^*)\right) \left(\sum_{\tau=t}^{T} q_\tau (v - p_\tau c)^+\right)$$

$$\leq \left(\sum_{\tau=t}^{T} r_\tau (1 - p_\tau^*)\right) \left(\sum_{\tau=0}^{T} q_\tau (v - p_\tau c)^+\right)$$

$$\Leftrightarrow \quad \left(\sum_{\tau=0}^{t-1} r_\tau (1 - p_\tau^*)\right) \left(\sum_{\tau=t}^{T} q_\tau (v - p_\tau c)^+\right)$$

$$\leq \left(\sum_{\tau=t}^{T} r_\tau (1 - p_\tau^*)\right) \left(\sum_{\tau=0}^{t-1} q_\tau (v - p_\tau c)^+\right)$$

$$\Leftrightarrow \quad 0 \leq \sum_{\tau=t}^{T} \sum_{\sigma=0}^{t-1} \left[r_\tau (1 - p_\tau^*) q_\sigma (v - p_\sigma c)^+ - r_\sigma (1 - p_\sigma^*) q_\tau (v - p_\tau c)^+ \right]$$

$$\Leftrightarrow \quad 0 \leq \sum_{\tau=t_1+1}^{T} \sum_{\sigma=0}^{t-1} r_\sigma r_\tau \left[(1 - p_\tau^*)(v - p_\sigma c)^+ \right]$$

$$+ \sum_{\tau=t}^{t_1-1} \sum_{\sigma=0}^{t-1} q_\sigma q_\tau \left[(1 - p_\tau)(v - p_\sigma c) - (1 - p_\sigma)(v - p_\tau c) \right]$$

$$\Leftarrow \quad \sum_{\tau=t}^{t_1-1} \sum_{\sigma=0}^{t-1} q_\sigma q_\tau (p_\tau - p_\sigma)(c - v)$$

Except for the interim efficiency of μ_α we have now shown that μ_α is the weighted neutral bargaining solution.

For the remaining arguments of the paper one has to consider variants of the Lagrangian

$$
\alpha_0 U_f \left(\mu | p_0 \right) + U_l \left(\mu \right) + \sum_{t=1}^{T} \alpha_t \left(U_f \left(\mu | p_t \right) - U_f^* \left(\mu, p_{t-1} | p_t \right) \right)
$$

$$
+ \sum_{t=1}^{T} \beta_t \left(U_f \left(\mu | p_{t-1} \right) - U_f^* \left(\mu, p_t | p_{t-1} \right) \right)
$$

and observe that the $\mu_\alpha \left(d_i | p_t \right)$ (for the relevant α) together with the Lagrangian multipliers α_t and β_t, as calculated above, form admissible solutions for a variety of maximization problems and their duals. By the duality theory for linear programming they are hence solutions to the respective optimization problems. In this way we can show, for instance, that μ_α maximizes $U_f \left(\mu | p_0 \right)$ subject to $U_l \left(\mu \right) \geq \alpha S \left(c \right)$ and the incentive constraints. This completes together with the above lengthy calculation Proposition 10.4 and we obtain in turn Theorem 10.1. By fixing the values of $U_f \left(\mu | p_\tau \right)$ and $U_l \left(\mu | p_\tau \right)$, the dual variables α_τ etc. for all $\tau < t$ with a given t one can similarly show that μ_α maximizes the lenders expected payoff in the t-bargaining problem subject to the incentive constraints and the constraint that type t gets at least $U_f \left(\mu_\alpha | p_t \right)$. From this result it follows immediately that μ_α is interim efficient. This line of arguments can also be used to derive Lemma 10.2.

References

Balkenborg, D., 'How Liable Should A Lender Be? The Case of Judgement-Proof Firms and Environmental Risks: Comment', *American Economic Review*, 91 (2001): 731-738.

Boyer, M. and Laffont, J.J., 'Environmental Risks and Bank Liability', *European Economic Review*, 41 (1997): 1427-1459.

Farrell, J., 'Credible Neologisms in Games of Communication', *mimeo* (1985).

Grossman, S. and Perry, M., 'Perfect Sequential Equilibrium', *Journal of Economic Theory*, 39 (1986): 97-119.

Heyes, A.G., 'Lender Penalty for Environmental Damage and the Equilibrium Cost of Capital', *Economica*, 63 (1996): 311-323.

Laffont, J.J. and Martimort, D., *The Theory of Incentives: The Principal-Agent Model*, (Princeton and Oxford: Princeton University Press, 2002).

Maskin, E. and Tirole, J., 'The Principal-Agent Relationship with an Informed Principal: The Case of Private Values', *Econometrica*, 58 (1990): 379-409.

Maskin, E. and Tirole, J., 'The Principal-Agent Relationship with an Informed Principal: Common Values', *Econometrica*, 60 (1992): 1-47.

Myerson, R., *Game Theory*, (Cambridge Massachussets: Harvard University Press, 1991).

Myerson, R.B., 'Incentive Compatibility and the Bargaining Problem', *Econometrica*, 47 (1979): 61-73.

Myerson, R.B., 'Mechanism Design by an Informed Principal', *Econometrica*, 51 (1983): 1767-1798.

Myerson, R.B., 'Two-Person Bargaining Problems with Incomplete Information', *Econometrica*, 52 (1984): 461-488.

Nash, J.F., 'The Bargaining Problem', *Econometrica*, 36 (1950): 155-162.

Pitchford, R., 'How Liable Should A Lender Be? The Case of Judgement-Proof Firms and Environmental Risk', *American Economic Review*, 85 (1995): 1171-1186.

Shavell, S., 'The Optimal Level of Corporate Liability given the Limited Ability of Corporations to Penalize Their Employees', *International Review of Law and Economics*, 17 (1997): 203-213.

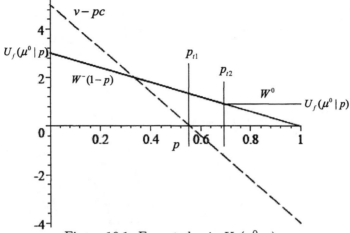

Figure 10.1: Expected gain $U_f(\mu^0, p)$

Chapter 11

Environmental Regulation of Livestock Production Contracts*

Philippe Bontems
Université de Toulouse and INRA

Pierre Dubois
Université de Toulouse and INRA

Tomislav Vukina
North Carolina State University

11.1 Introduction

During the 1980s, a substantial increase in the number of environmental clean-up cases in the U.S. has been coupled by an increase in the entry rate of small judgment proof firms into hazardous sectors (Ringleb and Wiggins, 1990). This phenomenon occurred because firms, trying to minimize the liability exposure, segregated their risky activities in small corporations. As the claimants were restricted to the assets of the small corporation typically unable to pay the associated liability damages, such segregation was found valuable. This result exposed the inefficiency of the tort liability as a primary institutional form for dealing with large-scale, long-term environmental hazards.

Given this problem, some authors have investigated the design of optimal schemes for lender's liability in the case of judgment-proof firms (e.g., Pitchford, 1995; Boyer and Laffont, 1997; and Balkenborg, 2001). However, there has been noticeably less interest in addressing these problems in a standard regulation framework. Moreover, the papers that examined the environmental regulation, similar to the above literature on vicarious liability, focused

*Financial and intellectual support from the French Ministry of Ecology and Sustainable Development is gratefully acknowledged. We thank Bob Chambers, Emma Hutchinson, David Martimort and Katleen Segerson, an anonymous referee and the participants of the 1st CIRANO-IDEI-LEERNA conference on "Regulation, Liability and the Management of Major Industrial Environmental Risks" in Toulouse, 2003, for their comments. All errors are ours.

only on cases where agents alone influence the level of pollution whereas the principal has little direct means for prevention or abatement.[1]

In this paper we address the problem of the optimal regulation of a ver-tically integrated livestock industry in which the environmentally polluting production (grow-out) of live animals is contracted with independent agents. A distinct feature of these contracts is the fact that the provision of production inputs is divided between the principal and the agents such that the resulting environmental pollution is the consequence of their joint actions. The potential impact of livestock and poultry production on environmental quality has become a major concern in areas with high density of concentrated animal feeding operations (CAFOs). It is increasingly common for environmental advocacy groups to argue that contracting is an important cause of adverse environmental quality effects in livestock production, largely because contracting increases the scale of livestock operations, simultaneously reducing opportunities for economics of scope in livestock utilization through reduced specialization.

The production and management of animal waste generates many potential external effects that need to be considered when designing regulatory policy. The most important one is the nutrient runoff and leaching from application of manure to cropland. Accidental spills and leaks from waste storage facilities and direct ambient air pollution from feedlots and storage facilities including odors and ammonia gases are also causing serious concerns. The problems associated with the design and implementation of environmental regulation of CAFOs are different than those related to regulating traditional family farms. In the later case the standard economic prescription of taxing the externality such that the polluter pays the environmental cost of his action is not feasible due the non-point source nature of the pollution problem. On the contrary, CAFOs are more similar to point source industrial polluters, hence some of the traditional regulatory instruments may prove to be adequate. However, the fact that a significant portion of CAFOs are in fact contract operations makes the design of the regulatory policy regimes substantially more complicated because the incidence of the regulatory compliance cost is not obvious.

In what follows we use the tax as a generic regulatory instrument and model the trilateral relationship between the Environmental Protection Agency (EPA), the contractor (firm) and an agent (producer), with the technology characterized by a joint production of output (live animal weight) and pollution (waste). We assume that output is observable and verifiable and hence contractible whereas pollution is observable but not verifiable and hence not contractible. From a theoretical point of view, this three-tier hierarchical model can be compared to the recent modelling of supervisory problem in a

---

[1]For example, Chambers and Quiggin (1996) modelled a non-point source pollution problem as a multi-task principal-agent problem where the agents are independent farmers producing corn and polluting the environment and the principal is the regulatory agency.

hierarchy (Faure-Grimaud, Laffont and Martimort, 2003; Faure-Grimaud and Martimort, 2001; Macho-Stadler and Pérez-Castrillo, 1998) where the principal (here the EPA) uses an intermediary agent (here the principal) to regulate a final agent (here the producer). In our regulatory model, the information structure is quite simple with either one-sided moral hazard or double-sided moral hazard, and the use of taxation of both contracting parties can be interpreted as a way to improve the regulation of producers only.

We find that in this three-tier hierarchy involving either a one-sided or a double-sided moral hazard problem, the principle of equivalence across regulatory schemes mostly obtains. In both situations, regardless of the tax legal incidence, for a given amount of tax revenue, the regulator can obtain the same outcome. The $EPA's$ only task is to determine the optimal total tax revenue in each state because any sharing of the tax burden between the principal and the agent would result in the same optimal solution. However, when the effects of regulation on the endogenous organizational choices of the industry are explicitly taken into account, the equivalence principle breaks down and the design of the optimal regulatory scheme becomes more complicated. When the regulator wants to foster contracting as a dominant mode of organizing livestock production, the optimal taxation scheme prescribes the minimal and maximal shares that the agent and the principal have to pay. In a situation where the EPA needs to simultaneously regulate independent producers and principal-agent contract organizations without being able to discriminate, a uniquely determined optimal division of the aggregate tax burden between the principal and the agent is necessary.

Our results provide an important extension of an earlier work by Segerson and Tietenberg (1992) who studied the structure of penalties in a three-tier hierarchy under the assumption of risk neutrality for all parties and moral hazard on the agent's side. Their main result is that the efficient outcome can be reached by imposing a penalty on either party, which corresponds exactly to our equivalence result. A generalization of the results obtained in this paper appear in Bontems, Dubois and Vukina (2004) where effort, pollution and production variables are all continuous and the joint distribution of production and pollution given effort can be of any form.

The rest of the paper is organized as follows. The next section is devoted to stylized facts on contracting in animal agriculture. The basic model is developed in Section 11.3. In Section 11.4 we analyze the case of a one-sided moral hazard in the relationship between the processor and the agent and in Section 11.5 the case of a double-sided moral hazard. Section 11.6 investigates the consequences of endogenous organizational choice for the equivalence results obtained earlier. Concluding remarks are given in Section 11.7.

11.2 Contracting in Animal Agriculture: Institutions and Technology

Contracts are becoming an integral part of the production and marketing of ever increasing number of agricultural commodities. The declining importance of spot markets relative to contracts can be well illustrated by observing the expansion of contracting in the U.S. swine industry. Recent data indicate that only 13.5% of pigs sold during January 2003 were sold in negotiated cash markets, compared with 16.7% a year earlier, and 35.8% in 1999 (Grimes, Plain, and Meyer, 2003). Pig production under contract in 2000 amounted to 42% of the total live production. The Southern Seaboard region, which includes North Carolina, the second largest hog production state in the U.S. after Iowa, is dominated by production contracts, with 87% of the total production volume being produced under contract (USDA-ERS, 2001).

Production contracts specify in detail the division of production inputs supplied by the two parties, the quality and quantity of a particular commodity and the type of the remuneration mechanism for the grower. In the livestock sector, a production contract is an agreement between a processing firm (also known as integrator) and a farmer (grower) that binds the farmer to specific production practices. Growers provide land, production facilities, utilities (electricity and water) and labor. Housing and waste handling units have to be constructed and equipped in strict compliance with the integrator's specifications. Growers are also fully responsible for compliance with federal, state and local environmental laws regarding disposal of dead animals and manure. An integrator company provides animals to be grown to processing weight, feed, medications and services of field men who supervise the adherence to the contract stipulations and provide production and management expertise. Typically, the company also owns and operates hatcheries, feed mills and processing plants, and provides transportation of feed and live animals. The integrator also decides on the volume of production both in terms of the rotations of batches on a given farm and the density of animals inside the house.

The most notable characteristic of modern livestock production systems based on contracts has been the shift to large-scale, intensive, specialized, confined animal operations. Opponents of such productions systems cite many negative environmental impacts of increased geographic concentration of manure stocks. Among various externalities generated by the production and management of animal waste, nutrient runoff and leaching and air quality problems (ammonia emissions) are the most pervasive ones. For both of those, nutrient management plays a critical role. The nutrients of greatest concern are nitrogen and phosphorus.

The amount of nutrients from animal waste that ends up deposited in the environment is directly related to the type of animals raised, the composition of animal feed, and the waste management technology that farmers use. Once feed composition and the waste handling and storing technology is fixed, the amount of pollution (nutrient content in manure) generated by a particular type of animal (e.g., a sow, a feeder pig, or a finished hog) is more or less deterministic.

The amount of nitrogen in manure can be reduced by substituting synthetic amino-acids for crude proteins (corn, soybeans) in animal feed. When deciding on the feed rations, the integrator's objective is to minimize costs per pound of feed subject to a series of nutritional constraints. Consequently, the cost-efficiency of feed formulations is highly dependable on relative prices of competing inputs. When the prices of corn and soybeans are high, it may be actually profitable to replace crude proteins with synthetic ones. When the prices of feed grains are low, absent any regulatory or other incentives, responding only to price signals, the feed rations will be based on crude proteins with high nitrogen in manure.

Phytic acid is the predominant form of phosphorus in corn and soybeans. However, phytic acid phosphorus cannot be digested by the monogastric animals (pigs, poultry) and is excreted in manure. Therefore, producers add inorganic phosphorus to diets to meet nutritional requirements of these animals. One way to increase the availability of phytic acid phosphorus is to add phytase to the ration with an objective to convert phytic acid into inorganic phosphorus. When phytase is added to the diets, inorganic phosphorus supplementation can be reduced and still maintain optimal performance. The rations based on phytase are more expensive than the regular inorganic phosphorus diets. While the use of phytase in animal rations benefits the society by reducing phosphorus pollution, the cost is exclusively born by the integrators.

Different types of waste management practices that growers use can produce dramatically different outcomes. Nitrogen losses like volatilization, denitrification and nitrate leaching that occur between excretion and crop uptake can vary from as little as 30% in liquid pit systems where manure is injected into the soil to greater than 90% in multistage lagoons where manure is irrigated.[2] On the other hand, phosphorus is not subject to losses between excretion and land application, so all phosphorus excreted by animals remains in the manure. Excessive phosphorus levels are known to degrade surface water quality by causing algae blooms and accelerating the eutrophication process.

An obvious solution to manure nutrient management problem is the source reduction. Conceptually, the pollution can be reduced by restricting the output (for example via a tax on output), or by reducing the amount of unusable nutrients in feed by somehow regulating the nutrient content in feed. The former regulatory scheme is easily implementable because the output is readily observable by all interested parties. The later scheme is considerably more complicated because the precise feed composition is known only to the inte-

---

[2]For details see Vukina (2003).

grator and could be discovered by the growers and the regulator only after bearing the costs of laboratory analysis. The regulatory objective can be however achieved by providing the integrator with the incentives to use environmentally friendly feed instead of the traditional environmentally unfriendly mix, even when this type of feed is less productive (more costly) in terms of feed efficiency.

In reality, to the extent that it is subject to the federal water quality regulation, the livestock and poultry production in the U.S. is regulated under the Clean Water Act of 1972. The most recent changes in regulation specific to livestock and poultry farming was introduced in February 2003 in the form of *National Pollutant Discharge Elimination System Permit Regulation and Effluent Limitation Guidelines and Standards for Concentrated Animal Feeding Operations (CAFOs); Final Rule* (EPA, 2003). Regulations under the Clean Water Act prior to the *Final Rule* had several shortcomings. First, exemptions in EPA's regulation has allowed an estimated 60% of animal feeding operations with more than 1,000 animal units to avoid regulation. Specifically, operations that discharged waste into waterways only during a 25-year, 24-hour storm event or greater were not explicitly defined as CAFOs and did not require permits. Also, chicken operations with dry manure-handling systems were not generally required to obtain permits. Third, animal waste applied to crop land and pastures were not generally regulated under the CAFO program.

EPA's Final Rule made significant modifications to the regulation of CAFOs while maintaining the basic regulatory structure. The major changes include a) the elimination of the 25-year, 24-hour storm discharge exemption, b) requirement that chicken operations that use dry manure handling systems obtain permits, and c) subjecting wastes applied to cropland and pastures under the control of the CAFO operator to permit requirements. EPA estimates the total monetized social cost of the final regulations at about $335 million annually in the pre-tax 2001 dollars. These costs include compliance costs borne by CAFOs and also administrative costs to federal and State governments. EPA estimates the total compliance costs for large CAFOs at $283 million per year, the administrative costs to federal and State governments $9 million per year, and the remainder are the compliance costs to medium and small CAFOs. Across all livestock sectors, an estimated 285 or 3% of existing large CAFOs may be vulnerable to facility closure as a result of complying with final regulation. These results are based on an analysis that does not consider the long-term effects on market adjustment and also available cost share assistance from federal and State governments (EPA, 2003).

The increase in cost of regulatory compliance for CAFOs as a result of the *Final Rule* will occur mainly through mandating the adoption of certain environmentally friendlier production practices or pollution abatement technologies and administrative costs associated with permitting. However, to keep things general and reasonably simple, in the theoretical model that follows we use the tax as a generic regulatory instrument. The obtained results will be interpreted in light of the new CAFO regulation in the concluding section.

11.3   The Basic Model

11.3.1   Assumptions and Notation

We consider an institutional structure with three players: the regulator (EPA), the principal (P) and the agent (A). This corresponds to the case where an integrator firm contracts the production of live animals with independent producers (growers). The production of output generates a negative externality that is subject to EPA's regulation.

The production process can be described as follows. An agent exerts effort $e \in \{\underline{e}, \overline{e}\}$ that is unobservable to the principal. Also the principal supplies a production input $x \in \{\underline{x}, \overline{x}\}$. In the case of livestock production this input is animal feed. The principal can choose either good feed $\underline{x}$ which is less efficient in the production of output (live weight) but environmentally friendlier or bad feed $\overline{x}$ which is highly productive but more polluting. Effort e and input x together generate output $q \in \{\underline{q}, \overline{q}\}$ with $\underline{q} < \overline{q}$ and pollution $d = h(q) \in \{\underline{d}, \overline{d}\}$ with $\underline{d} \leq \overline{d}$.

Hence, the production technology depends on input x provided by P and effort e provided by A and is described by the following stochastic process

$$P(\overline{q} \mid e, x) = 1 - P(\underline{q} \mid e, x) = \theta(x)\pi(e)$$

with the following normalization $\theta(\overline{x}) = 1$, $\theta(\underline{x}) = \theta < 1$, and notation $\pi(\underline{e}) = \underline{\pi}$ and $\pi(\overline{e}) = \overline{\pi} > \underline{\pi}$. Pollution is a production externality that is jointly determined with the production state of nature. We call the high state of nature the case where production and pollution are high and the low state of nature the case where they are low.

The cost of effort for the agent is normalized to e, the level of effort. The input cost for the principal is $C(x)$ with the normalization $C(\overline{x}) = 0$ and $C(\underline{x}) = c > 0$.

We assume that production is observable and verifiable which implies that it is contractible. In addition we assume that pollution is also observable but not verifiable implying that neither the EPA nor the principal can write contracts contingent on pollution damages. Note however that the optimal regulatory scheme would be exactly the same if pollution were verifiable because it is simply a joint outcome of the production process.

The principal and the agent contract the production of output q. Because of the moral hazard problem, the wage w received by the agent needs to be contingent on production. The contract is then simply $\{w(\underline{q}) = \underline{w}, w(\overline{q}) =$

$\overline{w}\}$. Before contracting between P and A occurs, the EPA commits to some regulatory scheme to control pollution. Given the pollution level, P and A are required to pay a fee to the EPA in the amount of $F(q) \in \{F(\underline{q}) = \underline{F}, F(\overline{q}) = \overline{F}\}$ and $T(q) \in \{T(\underline{q}) = \underline{T}, T(\overline{q}) = \overline{T}\}$. Total tax revenue is then $R(q) = F(q) + T(q) \in \{R(\underline{q}) = \underline{R}, R(\overline{q}) = \overline{R}\}$. The objective of the EPA is to maximize the expected difference between the tax revenue and the environmental damage $\mathbb{E}(R - d)$.[3]

The agent's utility function is separable in consumption and effort and equal to $U(w - T) - e$ where U is increasing concave and e is the cost of effort. The principal's utility function is $V(q - F - w - C(x))$ where V is also increasing concave. Both P and A are risk averse. The exogenous reservation utilities of the principal and the agent are respectively U_0 and V_0. Throughout the paper, we will sometimes consider a special case where P and A have constant absolute risk aversion utility functions of the form $U(Y) = -\frac{1}{\sigma_U}\exp(-\sigma_U Y)$ and $V(Y) = -\frac{1}{\sigma_V}\exp(-\sigma_V Y)$.

11.3.2 First-Best

The first-best outcome obtains in case where there is no asymmetry of information among the players, that is, both agent's effort and principal's input are observable by everybody. Assuming that it is socially optimal to always produce with low effort $\underline{e}$ and environmentally friendly feed $\underline{x}$, the regulatory agency can simply mandate the use of these inputs. In a standard procedure, the problem can be solved in two steps. First, solve the principal-agent optimal contracting problem given some regulatory scheme $\{\underline{F}, \overline{F}, \underline{T}, \overline{T}\}$ and then solve the EPA's problem taking into account the endogenous optimal contract from the first step. It is easy to see that perfect risk sharing between the principal and the agent is then achieved meaning that the payment to the agent $\overline{w}_0(\overline{F}, \overline{T})$ and $\underline{w}_0(\underline{F}, \underline{T})$ must solve the following equation

$$\frac{V'(\overline{q} - \overline{F} - \overline{w}_0(\overline{F}, \overline{T}) - c1_{x=\underline{x}})}{V'(\underline{q} - \underline{F} - \underline{w}_0(\underline{F}, \underline{T}) - c1_{x=\underline{x}})} = \frac{U'(\overline{w}_0(\overline{F}, \overline{T}) - \overline{T})}{U'(\underline{w}_0(\underline{F}, \underline{T}) - \underline{T})}.$$

Next, we study two cases, one in which input x is observable by everybody and the other where x is private information of the principal and hence unobservable by the EPA and the agent. In the first case there is a one-sided moral hazard problem associated with the agent's effort, whereas in the second case there is a double-sided moral hazard problem associated with the agent's effort and the principal's input.

---

[3] This objective is quite general and can take into account any net social surplus obtained in each state of nature simply by redefining the value of d.

11.4 Optimal Regulation under One-sided Moral Hazard

Absent any regulation, the principal would always selects the more polluting input $\overline{x}$ that results in the environmental damage $\overline{d}$. The regulatory intervention can be justified on the ground that some optimal scheme can implement the less polluting input $\underline{x}$ resulting in $\underline{d}$ at a cost lower than the expected benefit from damage reduction.[4] We start by assuming that x is observable by the EPA and therefore the principal is required to supply $\underline{x}$. The other case will be studied in section 11.5.

With respect to effort e, there are two possible cases. In the first case, the regulator wants to implement low effort $\underline{e}$ which is in conflict with the principal's preference under no regulation scenario. In the second case, the EPA wants to implement $\overline{e}$, which is the same effort level that P would implement absent any regulation, hence there is no conflict of interest. The latter is of course the case if the damage is the same in each state and, by a continuity argument, when the gap between damages $\overline{d}$ and $\underline{d}$ is sufficiently low. Then, the EPA's problem is only to extract the principal's rent knowing his reservation utility V_0. The case of no conflict of interest is less interesting and will be ignored in the following.

We start by characterizing the optimal contract between P and A given some regulatory scheme $\{\underline{F}, \overline{F}, \underline{T}, \overline{T}\}$. Then, we solve the EPA's problem taking into account the endogenous optimal contract. The characterization of optimal contracting is interesting in itself but we will see that it is not necessary to prove the equivalence principle that we expose subsequently.

11.4.1 The Optimal Contract Between P and A

With incentives to implement low effort $\underline{e}$ and the mandate from the EPA to use the good input $\underline{x}$, the principal's program is

$$\max_{\underline{w},\overline{w}} \theta \underline{\pi} V(\overline{q} - \overline{F} - \overline{w} - c) + (1 - \theta\underline{\pi})V(\underline{q} - \underline{F} - \underline{w} - c)$$

$$s.t. \quad \theta\underline{\pi}U(\overline{w} - \overline{T}) + (1 - \theta\underline{\pi})U(\underline{w} - \underline{T}) - \underline{e}$$
$$\geq U_0 \qquad\qquad (\mu)$$
$$\theta\underline{\pi}U(\overline{w} - \overline{T}) + (1 - \theta\underline{\pi})U(\underline{w} - \underline{T}) - \underline{e}$$
$$\geq \theta\overline{\pi}U(\overline{w} - \overline{T}) + (1 - \theta\overline{\pi})U(\underline{w} - \underline{T}) - \overline{e} \quad (\lambda) \text{ (11.1)}$$

where μ and λ are Lagrange multipliers associated with the participation and incentive constraints. To avoid notational clutter, let $V(\overline{q} - \overline{F} - \overline{w} - c) = \overline{V}$,

---

[4]The remaining case where using the bad input $\overline{x}$ is better for society is less interesting and will be ignored, although it can be solved in the same fashion as other problems in this paper.

$V(\underline{q}-\underline{F}-\underline{w}-c) = \underline{V}$, $U(\overline{w}-\overline{T}) = \overline{U}$, $U(\underline{w}-\underline{T}) = \underline{U}$, and $V'(\overline{q}-\overline{F}-\overline{w}-c) = \overline{V}'$, $V'(\underline{q}-\underline{F}-\underline{w}-c) = \underline{V}'$, $U'(\overline{w}-\overline{T}) = \overline{U}'$, $U'(\underline{w}-\underline{T}) = \underline{U}'$. The program is concave, so that first order conditions are sufficient and give

$$\lambda = \frac{(1-\theta\pi)\pi}{\pi-\overline{\pi}}\frac{\overline{V}'\underline{U}'-\underline{V}'\overline{U}'}{\underline{U}'\overline{U}'} \tag{11.2}$$

$$\mu = \frac{\theta\pi\overline{V}'\underline{U}'+(1-\theta\pi)\underline{V}'\overline{U}'}{\underline{U}'\overline{U}'} > 0 \tag{11.3}$$

because U and V are increasing.

The participation constraint is thus binding ($\mu > 0$). If, in addition, the incentive constraint is also binding, i.e. if $\lambda > 0$ (which is the case provided that $\overline{e}$ is not too large),[5] we obtain the following results

$$\underline{U} = U_0 + \frac{\overline{\pi}\underline{e}-\underline{\pi}\overline{e}}{\overline{\pi}-\underline{\pi}}$$

$$\overline{U} = \underline{U} + \frac{\overline{e}-\underline{e}}{\theta(\overline{\pi}-\underline{\pi})} > \underline{U}$$

since $\underline{e} < \overline{e}$ and $\overline{\pi} > \underline{\pi}$. Because the incentive constraint is binding, the net utility of the agent in the low state is lower than that in the high state (in the opposite case, $\overline{U} < \underline{U}$, the incentive constraint would be trivially strictly satisfied). Notice that $\underline{U}$ does not depend on θ whereas $\overline{U}$ is decreasing in θ indicating that the more the good input reduces the expected production, the less powerful will be the incentives given to the agent. Moreover, looking at the way the contract depends on the taxes proposed by the EPA, we get the following proposition:

Proposition 11.1 *The optimal wages $\overline{w}^*$ and $\underline{w}^*$ are such that $\overline{w}^*$ depends only on $(\overline{F},\overline{T})$, $\underline{w}^*$ depends only on $(\underline{F},\underline{T})$ and*

$$\frac{\partial\overline{w}^*}{\partial\overline{T}} = 1 + \frac{\partial\overline{w}^*}{\partial\overline{F}} = \frac{\overline{\sigma}_U}{\overline{\sigma}_U+\overline{\sigma}_V} \in (0,1) \tag{11.4}$$

$$\frac{\partial\underline{w}^*}{\partial\underline{T}} = 1 + \frac{\partial\underline{w}^*}{\partial\underline{F}} = \frac{\underline{\sigma}_U}{\underline{\sigma}_U+\underline{\sigma}_V} \in (0,1) \tag{11.5}$$

where $\overline{\sigma}_U = -\frac{\overline{U}''}{\overline{U}'}$, $\overline{\sigma}_V = -\frac{\overline{V}''}{\overline{V}'}$, $\underline{\sigma}_U = -\frac{\underline{U}''}{\underline{U}'}$, $\underline{\sigma}_V = -\frac{\underline{V}''}{\underline{V}'}$ are the rates of absolute risk aversion of P and A at consumption levels in good and bad states. Agent's net wages ($\overline{w}^* - \overline{T}$ and $\underline{w}^* - \underline{T}$) depend only on the total tax revenue

---

[5] Alternatively, the incentive constraint could be non-binding (for example, when $\overline{e}$ tends to $+\infty$), in which case the first-best would be achieved.

R, that is

$$\frac{\partial(\overline{w}^* - \overline{T})}{\partial \overline{R}} = \frac{\partial(\overline{w}^* - \overline{T})}{\partial \overline{T}} = \frac{\partial(\overline{w}^* - \overline{T})}{\partial \overline{F}} = \frac{-\overline{\sigma}_V}{\overline{\sigma}_U + \overline{\sigma}_V} \in (-1, 0)$$

$$\frac{\partial(\underline{w}^* - \underline{T})}{\partial \underline{R}} = \frac{\partial(\underline{w}^* - \underline{T})}{\partial \underline{T}} = \frac{\partial(\underline{w}^* - \underline{T})}{\partial \underline{F}} = \frac{-\underline{\sigma}_V}{\underline{\sigma}_U + \underline{\sigma}_V} \in (-1, 0)$$

Similarly, the principal's profits $(\overline{q} - \overline{F} - \overline{w}^* - c$ *and* $\underline{q} - \underline{F} - \underline{w}^* - c)$ *depend only on the total tax revenue R*

$$\frac{\partial(\overline{w}^* + \overline{F})}{\partial \overline{R}} = \frac{\partial(\overline{w}^* + \overline{F})}{\partial \overline{T}} = \frac{\partial(\overline{w}^* + \overline{F})}{\partial \overline{F}} = \frac{\overline{\sigma}_V}{\overline{\sigma}_U + \overline{\sigma}_V} \in (0, 1)$$

$$\frac{\partial(\underline{w}^* + \underline{F})}{\partial \underline{R}} = \frac{\partial(\underline{w}^* + \underline{F})}{\partial \underline{T}} = \frac{\partial(\underline{w}^* + \underline{F})}{\partial \underline{F}} = \frac{\underline{\sigma}_V}{\underline{\sigma}_U + \underline{\sigma}_V} \in (0, 1)$$

Proof 11.1 *See Appendix 11.7.*

Proposition 11.1 provides a set of important results. First, wages in each state depend only on taxes corresponding to the same state. Second, an increase in taxes T on A increases the wage but the agent is not fully compensated. Therefore, when taxes increase, net wages decrease. Similarly, a reduction in taxes decreases wages but the wage reduction is less than the tax reduction. Third, the changes in wages with respect to taxes T depend on the ratio of the absolute risk aversions of the agent and the principal. The more risk averse A relative to P, the less net wages respond to taxes on A, that is the more insurance will the principal provide to agent's net wage against changing taxes. In other words, the principal absorbs the larger part of the net wage change coming from the changes in taxes when the agent is more risk averse. Fourth, wages also depend on taxes F paid by the principal. Finally, the agent's net wage $w - T$ changes exactly in the same way with respect to taxes T or F. Actually, wages respond to the change in taxes in such a way that an increase in taxes $\overline{T}$ can be compensated exactly by a decrease in $\overline{F}$ to leave net wages $\overline{w}^*(\overline{F}, \overline{T}) - \overline{T}$ unchanged. The same is true for the other state of nature. This result follows from $\frac{\partial \overline{w}^*}{\partial \overline{T}} - \frac{\partial \overline{w}^*}{\partial \overline{F}} = \frac{\partial \underline{w}^*}{\partial \underline{T}} - \frac{\partial \underline{w}^*}{\partial \underline{F}} = 1$.

With the incentive constraint binding, i.e. $\lambda > 0$, one can show the following inequality[6]

$$\frac{\overline{V}'}{\underline{V}'} < \frac{\overline{U}'}{\underline{U}'} \tag{11.6}$$

Setting $\overline{w}_0(\overline{F}, \overline{T})$ and $\underline{w}_0(\underline{F}, \underline{T})$ equal the wages that would accomplish the perfect risk sharing between P and A for a given regulatory scheme, the

---

[6]Notice that the reverse inequality in (11.6) is obtained when the principal wants to implement $\overline{e}$. For example, this will happen if there is no regulation.

inequality (11.6) shows that wages $\overline{w}^*(\overline{F},\overline{T})$ and $\underline{w}^*(\underline{F},\underline{T})$ are such that $\overline{w}^*(\overline{F},\overline{T}) < \overline{w}_0(\overline{F},\overline{T})$ or $\underline{w}^*(\underline{F},\underline{T}) > \underline{w}_0(\underline{F},\underline{T})$.

With CARA utility functions for P and A, (11.6) implies that $\overline{w}^*(\overline{F},\overline{T}) - \underline{w}^*(\underline{F},\underline{T}) < \overline{w}_0(\overline{F},\overline{T}) - \underline{w}_0(\underline{F},\underline{T})$. Therefore, the gap between wages in good and bad states is reduced compared to the case with the perfect risk sharing. Given the fact that the risk aversion coefficients are constant, we obtain

$$\frac{\partial \overline{w}^*}{\partial \overline{T}} = \frac{\partial \underline{w}^*}{\partial \underline{T}} = 1 + \frac{\partial \overline{w}^*}{\partial \overline{F}} = 1 + \frac{\partial \underline{w}^*}{\partial \underline{F}} = \frac{\sigma_U}{\sigma_U + \sigma_V} \in (0,1)$$

and the optimal wage contracts are linear with respect to taxes, that is $\overline{w}^*(\overline{F},\overline{T}) = \frac{\sigma_U}{\sigma_U+\sigma_V}\overline{T} - \frac{\sigma_V}{\sigma_U+\sigma_V}\overline{F} + \overline{w}^*(0,0)$ and $\underline{w}^*(\underline{F},\underline{T}) = \frac{\sigma_U}{\sigma_U+\sigma_V}\underline{T} - \frac{\sigma_V}{\sigma_U+\sigma_V}\underline{F} + \underline{w}^*(0,0)$. In this case, $\overline{w}^*(0,0)$ and $\underline{w}^*(0,0)$ are the optimal wages with both taxes set to zero obtained as

$$\underline{w}^*(0,0) = U^{-1}(U_0 + \frac{\overline{\pi}\underline{e} - \underline{\pi}\overline{e}}{\overline{\pi} - \underline{\pi}})$$

$$\overline{w}^*(0,0) = U^{-1}(\underline{U} + \frac{\overline{e} - \underline{e}}{\theta(\overline{\pi} - \underline{\pi})}).$$

11.4.2 The Regulator's Problem

Given the optimal contract between the principal and the agent, the EPA's program is to choose taxes F and T to maximize the expected tax revenue net of environmental damage $\mathbb{E}(R - d)$, with $R = F + T$, subject to the participation constraint of the principal:

$$\max_{\underline{F},\overline{F},\underline{T},\overline{T}} \theta\overline{\pi}(\overline{F}+\overline{T}-\overline{d}) + (1-\theta\underline{\pi})(\underline{F}+\underline{T}-\underline{d}) \tag{11.7}$$
$$\text{s.t. } \theta\overline{\pi}\overline{V}^* + (1-\theta\underline{\pi})\underline{V}^* \geq V_0 \qquad (\eta)$$

where $\overline{V}^* = V(\overline{q} - \overline{F} - \overline{w}^*(\overline{F},\overline{T}) - c)$, $\underline{V}^* = V(\underline{q} - \underline{F} - \underline{w}^*(\underline{F},\underline{T}) - c)$ and $\overline{w}^*(\overline{F},\overline{T})$ and $\underline{w}^*(\underline{F},\underline{T})$ are the solutions to (11.1).

The program being concave, the first order conditions are sufficient

$$\theta\overline{\pi} = \eta\{\theta\overline{\pi}(1 + \frac{\partial \overline{w}^*}{\partial \overline{F}})\overline{V}^{*\prime} + (1-\theta\underline{\pi})\frac{\partial \underline{w}^*}{\partial \overline{F}}\underline{V}^{*\prime}\} \tag{11.8}$$

$$1 - \theta\underline{\pi} = \eta\{\theta\overline{\pi}\frac{\partial \overline{w}^*}{\partial \underline{F}}\overline{V}^{*\prime} + (1-\theta\underline{\pi})(1 + \frac{\partial \underline{w}^*}{\partial \underline{F}})\underline{V}^{*\prime}\} \tag{11.9}$$

$$\theta\overline{\pi} = \eta\{\theta\overline{\pi}\frac{\partial \overline{w}^*}{\partial \overline{T}}\overline{V}^{*\prime} + (1-\theta\underline{\pi})\frac{\partial \underline{w}^*}{\partial \overline{T}}\underline{V}^{*\prime}\} \tag{11.10}$$

$$1 - \theta\underline{\pi} = \eta\{\theta\overline{\pi}\frac{\partial \overline{w}^*}{\partial \underline{T}}\overline{V}^{*\prime} + (1-\theta\underline{\pi})\frac{\partial \underline{w}^*}{\partial \underline{T}}\underline{V}^{*\prime}\} \tag{11.11}$$

Substituting conditions (11.4) and (11.5) from Proposition 11.1 into (11.8)-(11.11) we get

$$\eta = \frac{1}{(1 + \frac{\partial w^*}{\partial F})\overline{V}^{*\prime}} = \frac{\overline{\sigma}_U + \overline{\sigma}_V}{\overline{\sigma}_U} \frac{1}{\overline{V}^{*\prime}} > 0 \qquad (11.12)$$

$$\frac{\overline{V}^{*\prime}}{\underline{V}^{*\prime}} = \frac{\frac{\partial w^*}{\partial \overline{T}}}{\frac{\partial w^*}{\partial \underline{T}}} = \frac{\underline{\sigma}_U}{\overline{\sigma}_U} \frac{\overline{\sigma}_U + \overline{\sigma}_V}{\underline{\sigma}_U + \underline{\sigma}_V} \qquad (11.13)$$

indicating that the principal's participation constraint (11.7) is binding because $\eta > 0$.

Condition (11.13) enables us to derive several important results. First, in equilibrium the principal's ex post utility levels are insensitive to the choice of regulatory instruments selected by the EPA that is

$$\frac{d\overline{V}^*}{d\overline{T}} = \frac{d\underline{V}^*}{d\underline{T}} = \frac{d\overline{V}^*}{d\overline{F}} = \frac{d\underline{V}^*}{d\underline{F}} = \frac{d\overline{V}^*}{d\overline{R}} = \frac{d\underline{V}^*}{d\underline{R}}$$

Second, (11.13) also implies that

$$\frac{d\mathbb{E}V^*}{d\overline{T}} = \frac{d\mathbb{E}V^*}{d\overline{F}} = \frac{d\mathbb{E}V^*}{d\overline{R}}$$

$$\frac{d\mathbb{E}V^*}{d\underline{T}} = \frac{d\mathbb{E}V^*}{d\underline{F}} = \frac{d\mathbb{E}V^*}{d\underline{R}}$$

meaning that as far as the principal's expected utility is concerned, the choice of tax instruments in both states of nature is irrelevant. From the perspective of the regulator, taxing either P or A contingently on the state of nature is equivalent. Finally, the ratio of marginal utilities of the principal in high and low states depends on the ratio of absolute risk aversions of the agent and the principal in high and low states.

If the principal's utility function V exhibits increasing absolute risk aversion which includes in particular the CARA (Constant Absolute Risk Aversion) case, regardless of utility function U, we have $\frac{\overline{V}^{*\prime}}{\underline{V}^{*\prime}} \geq 1$ ($= 1$ if CARA). Actually, if $\frac{\overline{V}^{*\prime}}{\underline{V}^{*\prime}} < 1$, then with increasing absolute risk aversion $\overline{\sigma}_V \geq \underline{\sigma}_V$ implying that $\frac{\overline{\sigma}_U + \overline{\sigma}_V}{\underline{\sigma}_U + \underline{\sigma}_V} \geq \frac{\underline{\sigma}_U + \underline{\sigma}_V}{\underline{\sigma}_U + \underline{\sigma}_V} > \frac{\overline{\sigma}_U}{\underline{\sigma}_U}$, because $\frac{y+a}{x+a} > \frac{y}{x} \; \forall x, y, a > 0$ which would imply $\frac{\overline{V}^{*\prime}}{\underline{V}^{*\prime}} > 1$. Moreover, because of the conflict of interest between EPA and P about effort, we have $\frac{\overline{V}^{*\prime}}{\underline{V}^{*\prime}} < \frac{\overline{U}^{*\prime}}{\underline{U}^{*\prime}}$. Thus $1 < \frac{\overline{V}^{*\prime}}{\underline{V}^{*\prime}} < \frac{\overline{U}^{*\prime}}{\underline{U}^{*\prime}}$ implies that $\overline{q} - \overline{F}^* - \overline{w}^*(\overline{F}^*, \overline{T}^*) - c < \underline{q} - \underline{F}^* - \underline{w}^*(\underline{F}^*, \underline{T}^*) - c$ and $\overline{w}^*(\overline{F}^*, \overline{T}^*) - \overline{T}^* < \underline{w}^*(\underline{F}^*, \underline{T}^*) - \underline{T}^*$ which implies $\overline{F}^* + \overline{w}^*(\overline{F}^*, \overline{T}^*) - \underline{F}^* - \underline{w}^*(\underline{F}^*, \underline{T}^*) > \overline{q} - \underline{q} > 0$ and therefore $\overline{R}^* = \overline{F}^* + \overline{T}^* > \underline{R}^* = \underline{F}^* + \underline{T}^*$. This implies that total taxes in the high state (higher production and higher pollution) are greater than that in the low state.

Doing a change of variable, we can show the following result even without characterizing the optimal contract between the principal and the agent.

Proposition 11.2 (Equivalence Principle - I) *Given some total tax revenue $(\overline{R}^*, \underline{R}^*)$ that the EPA wants to raise, taxing P or A or both is equivalent. Any taxation scheme satisfying $\overline{T} + \overline{F} = \overline{R}^*$ and $\underline{T} + \underline{F} = \underline{R}^*$ results in the same outcome and generates the same utility levels to all parties and hence the same welfare level. EPA's regulation only determines the optimal total tax revenue in each state and any sharing of total optimal taxes between P and A results in the same optimal solution.*

Proof 11.2 *See Appendix 11.7.*

This equivalence principle implies, for instance, that the optimal taxation scheme $(\overline{R}^*, \underline{R}^*)$ can be implemented by taxing only P $(\overline{F}^* = \overline{R}^*, \underline{F}^* = \underline{R}^*)$, or A $(\overline{T}^* = \overline{R}^*, \underline{T}^* = \underline{R}^*)$. Also, the EPA could subsidize A and tax P (for example, $\overline{T}^* = -S, \underline{T}^* = -S, \overline{F}^* = \overline{R}^*+S, \underline{F}^* = \underline{R}^*+S)$. Consequently, only total tax revenue matters. When the total tax burden increases, it is shared between P and A according to their relative risk aversion (see Proposition 11.1). It is to be noted that this equivalence principle is quite strong and robust. In particular, it can be straightforwardly extended to a version of the model with many or continuous states of nature and more or continuous levels of effort.

Of course, if EPA values the tax revenue collected from the principal and the agent differently (for example because of different administrative costs) then the equivalence principle would disappear and designing the optimal regulatory scheme would require placing the full tax burden on the party for which tax collection is the least costly. The equivalence principle is also derived under the assumption that the principal and the agent agreed upon an optimal contract after the regulatory scheme was announced by the EPA. Any rigidity or impediment in the implementation of this optimal contract would break the equivalence principle.

In the CARA case, (11.13) implies that $V^{*\prime}(\overline{q} - \overline{F} - \overline{w}^*(\overline{F}, \overline{T}) - c) = V^{*\prime}(\underline{q} - \underline{F} - \underline{w}^*(\underline{F}, \underline{T}) - c)$ that is

$$\overline{q} - \underline{q} = (\overline{F} - \underline{F}) + (\overline{w}^*(\overline{F}, \overline{T}) - \underline{w}^*(\underline{F}, \underline{T}))$$

$$\overline{q} - \underline{q} = \frac{\sigma_U}{\sigma_U + \sigma_V}[(\overline{F} + \overline{T}) - (\underline{F} + \underline{T})] + \overline{w}^*(0,0) - \underline{w}^*(0,0)$$

Moreover, (11.6) implies $\frac{\overline{U}'}{\underline{U}'} > 1$, that is, $\exp -\sigma_U[\overline{w}^*(\overline{F}, \overline{T}) - \underline{w}^*(\underline{F}, \underline{T}) - (\overline{T} - \underline{T})] > 1$ implying $\overline{w}^*(\overline{F}, \overline{T}) - \overline{T} < \underline{w}^*(\underline{F}, \underline{T}) - \underline{T}$. Therefore, net wage is lower in the high state case than in the low state case.

At this point, it is worth studying the case where the principal is risk neutral. It is frequently (but not always) the case that the principal offering production contracts to independent farmers is a large publicly traded company. Public companies are known to spread risk among their shareholders, so the assumption of risk neutrality is plausible in many situations. To carry out the analysis for the risk neutral case, we can simply set $V' \equiv 1$ (and $V'' \equiv 0$) which implies that

$$\frac{\partial \overline{w}^*}{\partial \overline{F}} = \frac{\partial \underline{w}^*}{\partial \underline{F}} = 0$$

$$\frac{\partial \overline{w}^*}{\partial \overline{T}} = \frac{\partial \underline{w}^*}{\partial \underline{T}} = 1.$$

These results show that the wages do not depend on taxes F that the principal pays but rather only on taxes that the agent pays T. Agent's net wages are constant with respect to all taxes. In contrast to the risk averse principal case where total taxes affect net wages, the EPA's taxation policy has no bearing on A's behavior and on his wage. The principal insures agent's revenue from any tax change although A's payment varies across states of nature.

11.5 Optimal Regulation under Double-Sided Moral Hazard

So far in the paper, we have assumed that x was observable by all parties and it was easy for the EPA to mandate the use of good input. However, if x is unobservable, mandating the use of environmentally friendly input $\underline{x}$ is not possible, the principal will always choose bad input $\overline{x}$. If the regulatory agency wants to stimulate the use of good input it has to design a scheme such that the principal will voluntarily choose the good input.[7]

As before, we are going to analyze only the case with conflict of interest on effort. When there is a conflict of interest between P and EPA on both effort e and input x, the EPA chooses taxes to induce the principal to implement $\underline{e}$ and $\underline{x}$. The principal's program ends up being the same as in (11.1) where x was observable. Therefore, Proposition 11.1 still holds. However with unobservable x, the EPA's problem is augmented by an additional incentive

---

[7]Of course, it will not always make sense for the EPA to give incentives to the principal to choose the good input because this can be too costly compared to the environmental benefits. Here we assume that it is always valuable to the EPA to elicit the use of environmentally friendly input.

constraint that guarantees the correct response of the principal regarding the utilization of good input

$$\max_{\underline{F},\overline{F},\underline{T},\overline{T}} \theta\pi(\overline{F}+\overline{T}-\overline{d}) + (1-\theta\pi)(\underline{F}+\underline{T}-\underline{d})$$

$$\text{s.t.} \qquad \theta\pi\overline{V}^{**} + (1-\theta\pi)\underline{V}^{**} \geq V_0 \qquad\qquad (\eta) \quad (11.14)$$

$$\theta\pi\overline{V}^{**} + (1-\theta\pi)\underline{V}^{**} \geq \pi\overline{V}^{*} + (1-\pi)\underline{V}^{*} \qquad (\gamma)$$

For notational convenience, we use ** to indicate solutions when x is not observable whereas we keep * to label solutions of Section 11.4 where x is observable. Following this convention, $\overline{V}^{**} = V(\overline{q} - \overline{F} - \overline{w}^{**} - c)$, $\underline{V}^{**} = V(\underline{q} - \underline{F} - \underline{w}^{**} - c)$, where $\overline{w}^{**}$ and $\underline{w}^{**}$ are the solutions to (11.1) and $\overline{V}^{*} = V(\overline{q} - \overline{F} - \overline{w}^{*})$, $\underline{V}^{*} = V(\underline{q} - \underline{F} - \underline{w}^{*})$ and wages $\overline{w}^{*}$ and $\underline{w}^{*}$ are the solutions to the principal's problem when he is not required to use the good input. These are the same wage functions as in Proposition 11.1 with $c = 0$ and $\theta = 1$.

Using Proposition 11.1, we can extend the equivalence principle to the unobservable input case:

Corollary 11.1 (Equivalence Principle - II) *Given some total tax revenue $\overline{R}^{**}$, $\underline{R}^{**}$ that the EPA wants to raise, taxing P or A or both is equivalent, that is, any taxation scheme satisfying $\overline{T} + \overline{F} = \overline{R}^{**}$ and $\underline{T} + \underline{F} = \underline{R}^{**}$ results in the same outcome and generates the same utility levels to all parties. The EPA's regulation only determines the optimal total tax revenue in each state and any sharing of total optimal taxes between P and A results in the same optimal solution.*

Proof 11.3 *See Appendix 11.7.*

However, in spite of the validity of the equivalence principle in the unobservable input case, the optimal regulation will change. Using the first order conditions of the problem stated above and the results of Proposition 11.1, we state the following proposition:

Proposition 11.3 *The ratio of marginal effects of taxes on the principal's utility in high and low states satisfies*

$$\frac{\frac{d\overline{V}^{**}}{d\overline{T}}}{\frac{d\underline{V}^{**}}{d\underline{T}}} = \frac{\frac{d\overline{V}^{**}}{d\overline{F}}}{\frac{d\underline{V}^{**}}{d\underline{F}}} = \left(\frac{\theta(1-\pi)}{1-\theta\pi} + \frac{\theta(1-\theta)}{(1-\theta\pi)(\theta+\gamma\overline{V}^{*\prime}\frac{\partial\overline{w}^{*}}{\partial\overline{T}})}\right)^{-1} \qquad (11.15)$$

$$1 \leq \frac{\frac{d\overline{V}^{**}}{d\overline{T}}}{\frac{d\underline{V}^{**}}{d\underline{T}}} = \frac{\frac{\partial\overline{w}^{**}}{\partial\overline{T}}\overline{V}^{**\prime}}{\frac{\partial\underline{w}^{**}}{\partial\underline{T}}\underline{V}^{**\prime}} < \frac{1-\theta\pi}{\theta(1-\pi)}$$

and the ratio of marginal utilities satisfies

$$\frac{\underline{\sigma}_U^{**}}{\overline{\sigma}_U^{**}} \frac{\overline{\sigma}_U^{**} + \overline{\sigma}_V^{**}}{\underline{\sigma}_U^{**} + \underline{\sigma}_V^{**}} \leq \frac{\overline{V}^{**\prime}}{\underline{V}^{**\prime}} < \frac{1 - \theta\pi}{\theta(1-\pi)} \frac{\underline{\sigma}_U^{**}}{\overline{\sigma}_U^{**}} \frac{\overline{\sigma}_U^{**} + \overline{\sigma}_V^{**}}{\underline{\sigma}_U^{**} + \underline{\sigma}_V^{**}}$$

Proof 11.4 *See Appendix 11.7.*

Proposition 11.3 indicates that the ratio of marginal effects of taxes on P's utility in high and low states is bounded below by 1 and bounded above by $\frac{1-\theta\pi}{\theta(1-\pi)} = \frac{P(\overline{q}|\underline{e},x)}{P(q|\underline{e},x)} \frac{P(q|\underline{e},\overline{x})}{P(\overline{q}|\underline{e},\overline{x})} > 1$. Compared to the case where x was observable, the additional incentive constraint introduces a distortion in the optimal taxation. Note that if $\gamma = 0$ (incentive constraint not binding) or $\theta = 1$ (no difference between inputs $\overline{x}$ and $\underline{x}$ in the production process) in (11.15), then we obtain the same ratio as in the observable input case

$$\frac{\overline{V}^{**\prime}}{\underline{V}^{**\prime}} = \frac{\frac{\partial \overline{w}^{**}}{\partial T}}{\frac{\partial \underline{w}^{**}}{\partial T}}$$

and the distortion disappears.

Notice also that the result about the inequality of total tax revenue in the two states of nature $(\overline{R}^* > \underline{R}^*)$ with non-decreasing absolute risk aversion of the principal and conflict of interest on effort still holds because $\frac{\underline{\sigma}_U^{**}}{\overline{\sigma}_U^{**}} \frac{\overline{\sigma}_U^{**} + \overline{\sigma}_V^{**}}{\underline{\sigma}_U^{**} + \underline{\sigma}_V^{**}} < \frac{\overline{V}^{**\prime}}{\underline{V}^{**\prime}}$ and $\frac{\overline{U}^{**\prime}}{\underline{U}^{**\prime}} > \frac{\overline{V}^{**\prime}}{\underline{V}^{**\prime}}$. In the CARA case, equation (11.15) implies $\frac{\overline{V}^{**\prime}}{\underline{V}^{**\prime}} = \exp{-\sigma_V(\overline{q} - \underline{q} - (\overline{w}^{**} - \underline{w}^{**}) - (\overline{F}^{**} - \underline{F}^{**}))} > 1 = \frac{\overline{V}^{*\prime}}{\underline{V}^{*\prime}} = \exp{-\sigma_V(\overline{q} - \underline{q} - (\overline{w}^* - \underline{w}^*) - (\overline{F}^* - \underline{F}^*))}$, that is, $\overline{q} - (\overline{w}^{**} + \overline{F}^{**}) - (\underline{q} - (\underline{w}^{**} + \underline{F}^{**})) < \overline{q} - (\overline{w}^* + \overline{F}^*) - (\underline{q} - (\underline{w}^* + \underline{F}^*)) = 0$, indicating that the utility of P in high state is lower than in bad state $(\overline{V}^{**} < \underline{V}^{**})$ whereas they are equal when x is observable $(\overline{V}^* = \underline{V}^*)$. The difference in net utility of P between low and high state is increased in absolute value compared to the case where x is observable. It means that in case where x is not observable by the EPA, the incentives to induce the implementation of the low effort are more powerful.

If the principal is risk neutral, then $\eta = 1$ and $\gamma = 0$, meaning that the principal's participation constraint is binding but not the incentive constraint. This is to say that in the equilibrium the risk neutral principal strictly prefers to use the good input rather than the bad input. With risk neutrality of the principal the distortion introduced by the fact that the regulator cannot observe x disappears.

11.6 Endogenous Contractual Organization and Regulation

It is usually reasonable to assume that the regulator is the leader of the game in the sense of first proposing a regulatory scheme to which the principal and the agent optimally respond by agreeing on a contract. In the analysis above, we have implicitly assumed that P and A would always sign a contract to jointly produce the output regardless of the regulation that the EPA imposed provided they get at least their exogenous reservation utility. The EPA took this optimal response into account but could not ex-post adjust the regulatory scheme it had committed to implement. As the contract signed between P and A is endogenous, the equivalence principle turns out to be a robust property of the optimal taxation scheme.

However, we have neglected the possibility that after observing the regulation, the parties to the contract may decide to go their separate ways instead of contracting and prefer to produce by themselves. When the regulatory agency is able to discriminate contract producers from independent producers, the optimal regulatory scheme would tax the parties contingently on whether they contract or independently produce. In this case, obviously the equivalence principle still holds. Nevertheless, when the contract producers cannot be distinguished from the independents (or if the output produced under contract cannot be disentangled from the output produced outside the contract), or if the law does not allow tax discrimination between contract and independent producers, then the regulator has to take into account that agents, after observing the regulatory scheme, may prefer to exit the contract with the principal and continue producing independently.

11.6.1 A Regulation that induces Contract Participation

We consider first the interesting situation where the regulator may prefer contracts over independent production in the targeted industry. For example, it is conceivable that due to economies of scale in feed mixing, the marginal cost of supplying environmentally friendly feed for the integrator may be smaller than for the small independent producer. In this case the EPA would design a regulatory scheme which is incentive compatible with the endogenous choice to contract given that any agent has always the opportunity to produce independently and pay only taxes T. Hence, the reservation utility of the contracting agent is endogenous as it depends on taxes T.

Let us define a "contracting compatible" regulatory scheme as follows: When facing the regulation, agents should always prefer to produce under a contract with an integrator rather than independently. If the agent produces independently his expected utility $U_a(\underline{T}, \overline{T})$ is equal to

$$U_a(\underline{T}, \overline{T}) = \max_{\substack{e \in \{\underline{e}, \overline{e}\}, \\ x \in \{\underline{x}, \overline{x}\}}} \left\{ \theta(x) \pi(e) U(\overline{q} - \overline{T} - c.\mathbf{1}_{x=\underline{x}}) \right.$$

$$\left. + (1 - \theta(x) \pi(e)) U(\underline{q} - \underline{T} - c.\mathbf{1}_{x=\underline{x}}) - e \right\}$$

which is clearly decreasing in $\underline{T}, \overline{T}$. The agent's reservation utility in the optimal wage contract between P and A writes now $U_0(\underline{T}, \overline{T}) = \max(U_0, U_a(\underline{T}, \overline{T}))$ but this does not change the properties of the optimal contract as described in Proposition 11.1. According to the equivalence principle, the optimal regulation under exogenous reservation utility is always implementable whatever the taxes $\underline{T}, \overline{T}$ because only total taxes matter and increasing taxes on the agent can be compensated by the reduced taxes on the principal. Since $U_a(\underline{T}, \overline{T})$ is decreasing in $\underline{T}$ and $\overline{T}$, it is always possible to choose taxes $(\underline{T}, \overline{T})$ such that the agent's endogenous contract participation is trivially satisfied $(U_0 \geq U_a(\underline{T}, \overline{T}))$. Then, one simply needs to choose taxes $(\underline{F}, \overline{F})$ such that the sum of taxes in each state is equal to the optimal taxes required by optimal regulation.

Proposition 11.4 *In the optimal "contracting compatible" regulatory scheme, the agent has to pay a minimum tax. Formally, given the optimal total tax revenues $(\overline{R}^*, \underline{R}^*)$, there exist two levels of taxation for the agent $\underline{T}^*_{\min}, \overline{T}^*_{\min}$ such that any taxation scheme $(\underline{F}, \overline{F}, \underline{T}, \overline{T})$ satisfying $\underline{T} \geq \underline{T}^*_{\min}, \overline{T} \geq \overline{T}^*_{\min}$ and $\overline{F} = \overline{R}^* - \overline{T}$ and $\underline{F} = \underline{R}^* - \underline{T}$ is optimal.*

This result indicates that here all shares of the total taxation scheme $(\overline{R}^*, \underline{R}^*)$ between the principal and the agent are no longer optimal. Instead, the optimal scheme is described by a minimal share that the agent has to pay and consequently a maximal share that the principal has to pay.

11.6.2   Nondiscrimination between Contract and Independent Producers

Let us consider now the situation where the EPA has to simultaneously regulate independent producers and principal-agent contracts without being able to discriminate. Consider first the regulation of independent producers only. Given the optimal contract between the principal and the agent, the *EPA's* problem is now to maximize the expected tax revenue net of environmental damage under the incentive and participation constraints of the agent:

$$\max_{\underline{T}, \overline{T}} \theta \underline{\pi}(\overline{T} - \overline{d}) + (1 - \theta \underline{\pi})(\underline{T} - \underline{d})$$

s.t.

$$\theta\underline{\pi}U(\bar{q}-\bar{T}-c)+(1-\theta\underline{\pi})U(\underline{q}-\underline{T}-c)-\underline{e}$$
$$\geq\theta\bar{\pi}U(\bar{q}-\bar{T}-c)+(1-\theta\bar{\pi})U(\underline{q}-\underline{T}-c)-\bar{e}$$
$$\theta\underline{\pi}U(\bar{q}-\bar{T}-c)+(1-\theta\underline{\pi})U(\underline{q}-\underline{T}-c)-\underline{e}$$
$$\geq U_0$$

with x observable. The incentive and participation constraints are binding and therefore the optimal taxes are uniquely determined as:

$$\underline{T}_a^* = \underline{T}^*\left(\bar{e},\underline{e},\underline{q},c,U_0,\bar{\pi},\underline{\pi}\right) = \underline{q}-c-U^{-1}\left(U_0+\underline{e}+\frac{\underline{\pi}\left(\bar{e}-\underline{e}\right)}{\bar{\pi}-\underline{\pi}}\right)$$

$$\bar{T}_a^* = \bar{T}^*\left(\bar{e},\underline{e},\bar{q},c,U_0,\bar{\pi},\underline{\pi},\theta\right) = \bar{q}-c-U^{-1}\left(U_0+\underline{e}+\frac{\underline{\pi}\left(\bar{e}-\underline{e}\right)}{\bar{\pi}-\underline{\pi}}+\frac{\bar{e}-\underline{e}}{\theta\left(\bar{\pi}-\underline{\pi}\right)}\right)$$

The equivalence principle implies that optimal regulation can now be implemented in the absence of discrimination but in a unique way as follows:

Proposition 11.5 *The optimal regulation is uniquely determined such that taxes imposed on contracting agents are also the optimal taxes to be imposed on independent producers: $\underline{T}_a^*$, $\bar{T}_a^*$. The optimal tax imposed on the principal is the difference between the optimal total tax revenue in each state and the optimal tax imposed on the agents, that is $\bar{F}=\bar{R}^*-\bar{T}_a^*$ and $\underline{F}=\underline{R}^*-\underline{T}_a^*$.*

Like in the previous case, all shares of the total taxation scheme between the principal and the agent are no longer optimal, causing the equivalence principle to break down. Instead, there exists an unique optimal share of the aggregate tax burden $(\bar{R}^*,\underline{R}^*)$ between the principal and the contract producer. Finally, notice that here we have implicitly assumed that the optimal regulation scheme should preserve the industry structure intact because the regulator is not able to discriminate among producers. Consequently the taxes imposed by the EPA are such that producers will get the same expected utility regardless of whether they are contract operators or independent producers, so there is no incentive for any of them to switch to a different mode of organization.

11.7   Conclusion

In this paper we have studied the optimal regulation of a vertically integrated polluting industry characterized by private production contracts between firms and independent agents (producers). These contractual arrangements are typical in animal agriculture, notably in poultry and swine industries. The main result shows that in a three-tier hierarchy (regulator-firm-agent) involving either a one-sided or a double-sided moral hazard problem, a principle of equivalence across regulatory schemes generally obtains.

The analysis was carried out in two steps, first by looking at the situation where there is only a moral hazard problem on the agent's side and second where there is also a moral hazard problem on the firm's side. In both situations, regardless of the tax legal incidence, for a given amount of tax revenue, the regulator can obtain the same provision of inputs and effort. Once the EPA commits to a regulatory scheme, the private production contract between the firm (principal) and the producer (agent) is such that the ex-post utility levels of both parties do not depend on the particular structure of the taxation scheme. Hence, taxing only the principal or only the agent generates the same outcome from all parties' viewpoint. The only task of the EPA is to determine the optimal total tax revenue in each state because any sharing of the tax burden between the principal and the agent would result in the same optimal solution. The way the optimal wage changes with respect to taxes is intimately related to the relative risk aversion degree between the principal and the agent.

In the CARA case, we derive some comparative statics results. We show that the optimal taxation provides full insurance to the principal because he gets the same utility levels in the high state and in the low state of nature when his provision of input is observable. However, when the regulator can not observe the principal's provision of input, the full insurance situation is not attainable. The principal receives the higher utility level in the low state due to incentives required to induce him to use the good input and therefore increase the probability of obtaining the low pollution. This result is specific to the CARA case, but is nevertheless interesting. It shows that the cost of moral hazard with respect to effort is fully loaded on the agent when the principal's provision of input is observable, whereas the cost of moral hazard is shared (i.e., the principal is not fully insured) when the case of double-sided moral hazard is considered.

The policy implications of the equivalence principle are important. It means that the EPA can implement the optimal regulation in different ways. Indeed, the optimal regulation is attainable with subsidies for one party and taxes for the other. What really matters is the total tax revenue and not the particular levels of taxes or subsidies levied on each party. However, the optimal total tax revenue that must be imposed on the contractual organization depends itself on the preferences of both parties, on their reservation utilities, and the parameters of the cost and production functions.

How can these results be interpreted in light of the new CAFO regulation? First notice that the new CAFO regulation did not fundamentally change the responsibilities of contracting parties for the provision of production inputs. Contract growers still have full responsibility for compliance with federal, state and local environmental laws regarding disposal of dead animals and animal waste. Consequently, the legal incidence of the increased costs of environmental compliance with the new CAFO rules falls entirely on contract growers. However, the economic incidence of this regulatory burden will be almost certainly different. How much different is impossible to say without analyzing individual contracts that govern particular relationships between growers and integrators. However, one can claim with certainty that, for a given total compliance cost increase, the welfare consequences for the integrator, contract growers, and the society (from perspective of achieved environmental quality improvements) will be the same had the legal incidence of the compliance cost fallen entirely on the integrator.

Finally, there are instances where the equivalence result breaks down and the design of the optimal regulatory scheme becomes substantially more subtle. One of such examples is the case where the effects of regulation on the endogenous organizational choices of the industry are explicitly taken into account. In case where the regulator may be interested in preserving contracts as a dominant mode of organizing livestock production, the taxation scheme needs to be modified such that it becomes incentive compatible with the agent's endogenous choice to contract in the presence of the alternative opportunity to produce independently. Contrary to the equivalence result obtained previously where all shares of the total taxation scheme between the principal and the agent were optimal, in this case the optimal scheme is described by the minimal and maximal shares that the agent and the principal have to pay respectively.

What are the situations where the equivalence principle of regulatory schemes would fail in the context of new CAFO regulation? Our results show that this will always happen in cases where there are some rigidities in the implementation of the optimal integrator-grower contracts, such the contract between EPA and the grower is based on a richer set of contracting variables than the contract between the integrator and the grower.

It is important to mention that prior to the passage of the Final Rule most people in the industry and in environmental circles anticipated that some form of shared responsibility for the removal and disposal of manure between the integrators and the growers will be implemented. To a big dismay of environmental groups, this did not happen. This issue of co-permitting may be important keeping in mind that most of the contract growers are in fact judgment proof firms. Facing increasingly stringent environmental regulation, growers are exposed to substantial risks of large penalties for environmentally hazardous disposal practices and especially catastrophic waste spills. Because growers generally have limited assets, the likelihood of bankruptcy is much larger for them than for the integrators who are large, sometimes publicly owned, companies. The potential insolvency can cause a reduction in care levels under strict liability because the contract operators would care only about the costs that they might actually have to pay. Also, wealthier growers may take greater care than poorer ones because they have more to lose and are less likely to escape paying damages through bankruptcy.

Ignoring externalities associated with animal waste, the observed contracts should be efficient in the sense of maximizing joint integrator plus grower surplus. Also, absent any rigidities in contracting, the legal incidence of regulation (i.e. whose name is on the permit) should be irrelevant because any form of new regulation will be endogenized via a new (redefined) optimal contract. However, with the simultaneous presence of environmental externalities and grower's bankruptcy constraint, the legal incidence of regulation is no longer irrelevant but rather matters for efficiency. For the internalization of animal waste externalities where contract operators are judgment proof entities, co-permitting may in fact be required.

Appendix

Proof of Proposition 11.1

To get the partial derivatives of $\overline{w}^*$ and $\underline{w}^*$ with respect to $\underline{F}, \overline{F}, \underline{T}, \overline{T}$, we differentiate the first order conditions with respect to $\underline{F}, \overline{F}, \underline{T}, \overline{T}$ and use them to replace λ and μ.

$$\frac{\partial \overline{w}^*}{\partial \overline{F}} = \frac{-\pi \overline{V}''}{\pi \overline{V}'' + [\lambda(\pi - \overline{\pi}) + \mu \overline{\pi}]\overline{U}''} = -\frac{-\overline{V}''/\overline{V}'}{-\overline{V}''/\overline{V}' - \overline{U}''/\overline{U}'}$$

$$\frac{\partial \overline{w}^*}{\partial \overline{T}} = \frac{[\lambda(\pi - \overline{\pi}) + \mu \overline{\pi}]\overline{U}''}{\pi \overline{V}'' + [\lambda(\pi - \overline{\pi}) + \mu \overline{\pi}]\overline{U}''} = \frac{-\overline{U}''/\overline{U}'}{-\overline{V}''/\overline{V}' - \overline{U}''/\overline{U}'}$$

$$\frac{\partial \overline{w}^*}{\partial \underline{F}} = \frac{\partial \overline{w}^*}{\partial \underline{T}} = 0$$

$$\frac{\partial \underline{w}^*}{\partial \underline{F}} = \frac{-(1 - \theta \underline{\pi})\underline{V}''}{[\lambda \theta(\overline{\pi} - \underline{\pi}) + \mu(1 - \theta \underline{\pi})]\underline{U}'' + (1 - \theta \underline{\pi})\underline{V}''} = -\frac{-\underline{V}''/\underline{V}'}{-\underline{U}''/\underline{U}' - \underline{V}''/\underline{V}'}$$

$$\frac{\partial \underline{w}^*}{\partial \underline{T}} = \frac{[\lambda \theta(\overline{\pi} - \underline{\pi}) + \mu(1 - \theta \underline{\pi})]\underline{U}''}{(1 - \theta \underline{\pi})\underline{V}'' + [\lambda \theta(\overline{\pi} - \underline{\pi}) + \mu(1 - \theta \underline{\pi})]\underline{U}''} = \frac{-\underline{U}''/\underline{U}'}{-\underline{V}''/\underline{V}' - \underline{U}''/\underline{U}'}$$

$$\frac{\partial \underline{w}^*}{\partial \overline{F}} = \frac{\partial \underline{w}^*}{\partial \overline{T}} = 0$$

Proof of Proposition 11.2

Given a taxation scheme $\overline{T}, \overline{F}, \underline{T}, \underline{F}$, denote $\overline{w}^*$, $\underline{w}^*$ solution of (11.1) and assume that the EPA can choose $\overline{T}', \overline{F}', \underline{T}', \underline{F}'$ such that $\overline{T}' + \overline{F}' = \overline{T} + \overline{F}$ and $\underline{T}' + \underline{F}' = \underline{T} + \underline{F}$. Then, the principal can set wages $\overline{w}'^*$, $\underline{w}'^*$ such that $\overline{w}'^* - \overline{T}' = \overline{w}^* - \overline{T}$ and $\overline{F}' + \overline{w}'^* = \overline{F} + \overline{w}^*$, which is possible because $\overline{T}' + \overline{F}' = \overline{T} + \overline{F}$. Then, it is clear from (11.1) that the participation and incentive constraints are unchanged (because ex-post utility levels are unchanged) and the principal's objective is the same. Therefore, $(\overline{w}'^*, \underline{w}'^*, \overline{T}', \overline{F}', \underline{T}', \underline{F}')$

implements the same outcome as $(\overline{w}^*, \underline{w}^*, \overline{T}, \overline{F}, \underline{T}, \underline{F})$. Actually, taxes F and T are perfect substitutes in the EPA's objective (only total taxes matter), wages w and taxes F are also perfect substitutes in the principal's objective, and optimal wages chosen by P are such that net wages $(w - T)$ of the agent are constant (do not depend on taxes F and T). Therefore, the same outcome can be implemented with $\overline{T} = \underline{T} = 0$ or $\underline{F} = \overline{F} = 0$ (in fact, there is an infinity of solutions in $(\underline{F}, \overline{F}, \underline{T}, \overline{T})$ including the cases where $\underline{F} = \overline{F} = 0$ or $\underline{T} = \overline{T} = 0$).

Proof of Corollary 11.1

The proof is similar as in section 11.7 except that now the EPA's program has an additional incentive constraint. However, it is clear, with the same arguments on the optimal wage contract between P and A which keeps the same properties, that given a taxation scheme $\overline{T}$, $\overline{F}$, $\underline{T}$, $\underline{F}$, the EPA can choose $\overline{T}'$, $\overline{F}'$, $\underline{T}'$, $\underline{F}'$ such that $\overline{T}' + \overline{F}' = \overline{T} + \overline{F}$ and $\underline{T}' + \underline{F}' = \underline{T} + \underline{F}$ without changing any of the net utility levels of P and A.

Proof of Proposition 11.3

The program being concave, first order conditions are sufficient

$$\theta\underline{\pi} = (\eta + \gamma)\{\theta\underline{\pi}(1 + \frac{\partial\overline{w}^{**}}{\partial\overline{F}})\overline{V}^{**\prime} + (1 - \theta\underline{\pi})\frac{\partial\underline{w}^{**}}{\partial\overline{F}}\underline{V}^{**\prime}\} \quad \text{(A.11.1)}$$

$$-\gamma\{\underline{\pi}(1 + \frac{\partial\overline{w}^*}{\partial\overline{F}})\overline{V}^{*\prime} + (1 - \underline{\pi})\frac{\partial\underline{w}^*}{\partial\overline{F}}\underline{V}^{*\prime}\}$$

$$1 - \theta\underline{\pi} = (\eta + \gamma)\{\theta\underline{\pi}\frac{\partial\overline{w}^{**}}{\partial\underline{F}}\overline{V}^{**\prime} + (1 - \theta\underline{\pi})(1 + \frac{\partial\underline{w}^{**}}{\partial\underline{F}})\underline{V}^{**\prime}\} \quad \text{(A.11.2)}$$

$$-\gamma\{\underline{\pi}\frac{\partial\overline{w}^*}{\partial\underline{F}}\overline{V}^{*\prime} + (1 - \underline{\pi})(1 + \frac{\partial\underline{w}^*}{\partial\underline{F}})\underline{V}^{*\prime}\}$$

$$\theta\underline{\pi} = (\eta + \gamma)\{\theta\underline{\pi}\frac{\partial\overline{w}^{**}}{\partial\overline{T}}\overline{V}^{**\prime} + (1 - \theta\underline{\pi})\frac{\partial\underline{w}^{**}}{\partial\overline{T}}\underline{V}^{**\prime}\} \quad \text{(A.11.3)}$$

$$-\gamma\{\underline{\pi}\frac{\partial\overline{w}^*}{\partial\overline{T}}\overline{V}^{*\prime} + (1 - \underline{\pi})\frac{\partial\underline{w}^*}{\partial\overline{T}}\underline{V}^{*\prime}\}$$

$$1 - \theta\underline{\pi} = (\eta + \gamma)\{\theta\underline{\pi}\frac{\partial\overline{w}^{**}}{\partial\underline{T}}\overline{V}^{**\prime} + (1 - \theta\underline{\pi})\frac{\partial\underline{w}^{**}}{\partial\underline{T}}\underline{V}^{**\prime}\} \quad \text{(A.11.4)}$$

$$-\gamma\{\underline{\pi}\frac{\partial\overline{w}^*}{\partial\underline{T}}\overline{V}^{*\prime} + (1 - \underline{\pi})\frac{\partial\underline{w}^*}{\partial\underline{T}}\underline{V}^{*\prime}\}$$

From proposition 11.1 we know that $\frac{\partial\underline{w}^*}{\partial\overline{F}} = \frac{\partial\underline{w}^*}{\partial\overline{T}} = \frac{\partial\underline{w}^*}{\partial\underline{F}} = \frac{\partial\underline{w}^*}{\partial\underline{T}} = 0$, $\frac{\partial\overline{w}^{**}}{\partial\overline{F}} = \frac{\partial\overline{w}^{**}}{\partial\underline{T}} = \frac{\partial\underline{w}^{**}}{\partial\overline{F}} = \frac{\partial\underline{w}^{**}}{\partial\overline{T}} = 0$, $\frac{\partial\overline{w}^{**}}{\partial\overline{F}} - \frac{\partial\overline{w}^{**}}{\partial\overline{T}} = \frac{\partial\underline{w}^{**}}{\partial\underline{F}} - \frac{\partial\underline{w}^{**}}{\partial\underline{T}} = -1$ and $\frac{\partial\overline{w}^*}{\partial\overline{F}} - \frac{\partial\overline{w}^*}{\partial\overline{T}} = \frac{\partial\underline{w}^*}{\partial\underline{F}} - \frac{\partial\underline{w}^*}{\partial\underline{T}} = -1$ therefore constraints (A.11.1) and (A.11.3) are the same and

(A.11.2) and (A.11.4) are the same. Simplifying, we get

$$\theta = (\eta + \gamma)\theta \frac{\partial \overline{w}^{**}}{\partial \overline{T}} \overline{V}^{**\prime} - \gamma \frac{\partial \overline{w}^{*}}{\partial \overline{T}} \overline{V}^{*\prime}$$

$$1 - \theta\underline{\pi} = (\eta + \gamma)(1 - \theta\underline{\pi}) \frac{\partial \underline{w}^{**}}{\partial \underline{T}} \underline{V}^{**\prime} - \gamma(1 - \underline{\pi}) \frac{\partial \underline{w}^{*}}{\partial \underline{T}} \underline{V}^{*\prime}$$

Using also $\frac{\partial \overline{w}^{*}}{\partial \overline{T}} \overline{V}^{*\prime} = \frac{\partial \underline{w}^{*}}{\partial \underline{T}} \underline{V}^{*\prime}$, we get

$$\gamma = \frac{\frac{\partial \underline{w}^{**}}{\partial \underline{T}} \underline{V}^{**\prime} - \frac{\partial \overline{w}^{**}}{\partial \overline{T}} \overline{V}^{**\prime}}{\theta(1 - \pi)\frac{\partial \underline{w}^{**}}{\partial \underline{T}} \underline{V}^{**\prime} - (1 - \theta\underline{\pi})\frac{\partial \overline{w}^{**}}{\partial \overline{T}} \overline{V}^{**\prime}} \frac{\theta(1 - \theta\pi)}{\frac{\partial \overline{w}^{*}}{\partial \overline{T}} \overline{V}^{*\prime}}$$

$$\eta = \frac{\theta(1 - \theta\underline{\pi})(\frac{\partial \underline{w}^{**}}{\partial \underline{T}} \underline{V}^{**\prime} - \frac{\partial \overline{w}^{**}}{\partial \overline{T}} \overline{V}^{**\prime}) + (1 - \theta)\frac{\partial \overline{w}^{*}}{\partial \overline{T}} \overline{V}^{*\prime}}{[(1 - \theta\underline{\pi})\frac{\partial \underline{w}^{**}}{\partial \underline{T}} \underline{V}^{**\prime} - \theta(1 - \pi)\frac{\partial \overline{w}^{**}}{\partial \overline{T}} \overline{V}^{**\prime}]\frac{\partial \overline{w}^{*}}{\partial \overline{T}} \overline{V}^{*\prime}}$$

As $\theta(1 - \pi) < (1 - \theta\underline{\pi})$, if $\frac{\partial \underline{w}^{**}}{\partial \underline{T}}\underline{V}^{**\prime} > \frac{\partial \overline{w}^{**}}{\partial \overline{T}}\overline{V}^{**\prime}$ then $(1 - \pi)\theta\frac{\partial \underline{w}^{**}}{\partial \underline{T}}\overline{V}^{**\prime} < (1 - \theta\underline{\pi})\frac{\partial \underline{w}^{**}}{\partial \underline{T}}\underline{V}^{**\prime}$ implying $\gamma < 0$ which is not possible. Therefore $\frac{\partial \underline{w}^{**}}{\partial \underline{T}}\underline{V}^{**\prime} < \frac{\partial \overline{w}^{**}}{\partial \overline{T}}\overline{V}^{**\prime}$. In addition, it must be that $(1 - \pi)\theta\frac{\partial \overline{w}^{**}}{\partial \overline{T}}\overline{V}^{**\prime} < (1 - \theta\underline{\pi})\frac{\partial \underline{w}^{**}}{\partial \underline{T}}\underline{V}^{**\prime}$ to get $\gamma \geq 0$.

We have

$$\theta = (\eta + \gamma)\theta \frac{\partial \overline{w}^{**}}{\partial \overline{T}} \overline{V}^{**\prime} - \gamma \overline{V}^{*\prime} \frac{\partial \overline{w}^{*}}{\partial \overline{T}}$$

and

$$\frac{\frac{\partial \underline{w}^{**}}{\partial \underline{T}} \underline{V}^{**\prime}}{\frac{\partial \overline{w}^{**}}{\partial \overline{T}} \overline{V}^{**\prime}} = \frac{\theta(1 - \pi)}{1 - \theta\underline{\pi}} - \frac{\theta(\theta - 1)}{(1 - \theta\underline{\pi})(\theta + \gamma\overline{V}^{*\prime}\frac{\partial \overline{w}^{*}}{\partial \overline{T}})}$$

then

$$\frac{\overline{V}^{**\prime}}{\underline{V}^{**\prime}} = \frac{\frac{\partial \underline{w}^{**}}{\partial \underline{T}}}{\frac{\partial \overline{w}^{**}}{\partial \overline{T}}} \left(\frac{\theta(1 - \pi)}{(1 - \theta\underline{\pi})} + \frac{\theta(1 - \theta)}{(1 - \theta\underline{\pi})(\theta + \gamma\overline{V}^{*\prime}\frac{\partial \overline{w}^{*}}{\partial \overline{T}})} \right)^{-1}$$

$$\geq \frac{\frac{\partial \underline{w}^{**}}{\partial \underline{T}}}{\frac{\partial \overline{w}^{**}}{\partial \overline{T}}} = \frac{\underline{\sigma}_U^{**} \, \overline{\sigma}_U^{**} + \overline{\sigma}_V^{**}}{\overline{\sigma}_U^{**} \, \underline{\sigma}_U^{**} + \underline{\sigma}_V^{**}}$$

References

Balkenborg, D., 'How Liable Should a Lender Be? The Case of Judgment-Proof Firms and Environmental Risk: Comment', *American Economic Review*, 91 (2001): 731-738.

Bontems, P., Dubois, P. and Vukina, T., 'Optimal Regulation of Private Production Contracts with Environmental Externalities', *Journal of Regulatory Economics*, 26(3) (2004), 284-298.

Boyer, M. and Laffont, J.J., 'Environmental Risks and Bank Liability', *European Economic Review*, 41 (1997): 1427-59.

Chambers, B.G. and Quiggin, J., 'Non-Point-Source Pollution Regulation as a Multi-Task Principal-Agent Problem', *Journal of Public Economics*, 59 (1996): 95-116.

EPA, 'National Pollutant Discharge Elimination System Permit Regulation and Effluent Limitations Guidelines and Standards for Concentrated Animal Feeding, (2003).

Faure-Grimaud A. and Martimort, D., 'On Some Agency Costs of Intermediated Contracting', *Economics Letters*, 71(1) (2001): 75-82.

Faure-Grimaud A., J.J., Laffont and Martimort, D., 'Collusion,Delegation and Supervision with Soft Information', *Review of Economic Studies*, 70 (2003):253-280.

Grimes, G., R. Plain and Meyer, S., 'U.S. Hog Marketing Contract Study', National Pork Board, January (2003).

Innes, R., 'The Economics of Livestock Waste and Its Regulation', *American Journal of Agricultural Economics*, 82(1) (2000): 97-117.

Macho-Stadler, I. and Pérez-Castrillo, J.D., 'Centralized and Decentralized Contracts in Moral Hazard Environments', *Journal of Industrial Economics*, 46(4) (1998): 489-510.

Pitchford, R., 'How Liable Should a Lender Be? The Case of Judgment-Proof Firms and Environmental Risk', *American Economic Review*, 85 (1995): 1171-86.

Ringleb, A. H. and Wiggins, S.N., 'Liability and Large-Scale, Long-Term Hazards', *Journal of Political Economy*, 98 (1990): 574-595.

Segerson, K. and Tietenberg, T., 'The Structure of Penalties in Environmental Enforcement: An Economic Analysis', *Journal of Environmental Economics and Management*, 23(2) (1992): 179-200.

U.S. Department of Agriculture (USDA), Economic Research Service (ERS). 'Commodity Costs and Returns: U.S. and Regional Cost and Return Data', (2001).
(www.ers.usda.gov/Data/CostsAndReturns/testpick.html)

Vukina, T., 'The Relationship between Contracting and Livestock Waste Pollution', *Review of Agricultural Economics*, 25 (2003): 66-88.

Chapter 12

Environmental Risks: Should Banks Be Liable?*

Karine Gobert
Université de Sherbrooke

Michel Poitevin
Université de Montréal

12.1 Introduction

Bank liability has been introduced by the legislator in the United States to circumvent the judgment-proof problem encountered in most environmental litigation cases.[1] The judgment-proof problem refers to a situation where a firm does not have sufficient assets to compensate victims for the harm caused by its activity. Because of limited liability, the firm must then go bankrupt, and victims are only partially compensated. Extending liability to banks has the advantage of allowing courts to have access to a deep-pocket stakeholder that can pay for environmental damages when liable firms go bankrupt. This is justified on the grounds that the bank has a privileged relationship with the firm and can therefore be considered as an "operator" of the firm.

Another effect of limited liability is that it may reduce the firm's incentives to prevent accidents by limiting its loss to the value of its assets, which may be less than total damages. Making the bank liable helps internalize the cost of environmental damages. Financial provisions would ensure an appropriate level of environmental care. The extension of environmental liability to bank therefore eliminates the adverse effects of the firm's limited liability.[2]

---

*We thank Bernard Sinclair-Desgagné, Patrick González, Pascale Viala, Andreas Richter and Nicolas Treich for helpful comments. We also acknowledge financial support from C.R.S.H. and CIRANO.

[1]The CERCLA (Comprehensive Environmental Response, Compensation and Liability Act) was introduced in the United States in the eighties and allowed for bank liability in many trials. For more details on these judgments, see Goble (1992), and Boyer and Laffont (1995) for Canadian cases. For more details on the CERCLA, see Olexa (1991) and Henderson (1994).

[2]This is shown formally in Segerson and Tietenberg (1992).

However, as shown by Segerson and Tietenberg (1992), internalization of environmental cost is achieved only if liability is given to the party that controls prevention. Most of the literature analyzes the role of bank liability when moral hazard on the level of environmental care reduces the effectiveness of financial contracts to implement an appropriate level of environmental care. In a one-period model of moral hazard and firm's limited liability, Pitchford (1995) shows that the bank's liability must be limited to the value of the firm.[3] More extensive bank liability would reduce the bank's expected payoff, and therefore increase the ex ante cost of financing for the firm. This would reduce the firm's stakes in the project, and hence reduce the firm's incentives to provide environmental care. When the repair of environmental damages has priority over financial obligations in the case of bankruptcy, Pitchford's result can be interpreted as having no bank liability since the optimal compensation to victims does not exceed the firm's assets.

Boyer and Laffont (1997) analyze a two-period model of moral hazard and firm's limited liability where the bank must decide whether to refinance the firm after an unobservable investment in prevention has been done but before an accident takes place. The usual tradeoff between efficiency and rent extraction leads to sub-optimal investment and refinancing decisions. They show that partial bank liability is better than full liability because a fully liable bank wants to induce a high level of environmental care and, to do so, must give high rents to the firm. This makes lending costly, and some projects with positive social value must be foregone after period 1.

Under the assumption that the investment in prevention is not observable, these models show that bank liability can have adverse effects on the level of environmental care provided by a firm, and therefore that the bank must be only partially liable.

Our model differs from these studies by introducing two important features: dynamics and risk aversion of the firm. Bank financing then serves two purposes: providing insurance and smoothing across periods. We also assume that the investment in prevention is observable. Investment in prevention is made at the start of the first period and an environmental accident may take place in this period. The firm's income in the period is too small to compensate for the accident so the firm would be in financial distress should an accident occur. In some circumstances, however, it may be profitable for the bank to refinance the firm after the accident to benefit from future incomes. Bank liability affects the bank's decision to refinance the firm following an environmental accident because if the bank is liable when the firm goes bankrupt, refinancing the firm would eliminate the possibility for bankruptcy, and therefore reduce its liability.

The investment in prevention by the risk averse firm must tradeoff two opposite effects. First, the firm wants to overinvest in environmental preven-

---

[3]See also Balkenborg (1997) and Dionne and Spaeter (2003).

tion in order to decrease the likelihood of bankruptcy. Second, incomplete internalization of the cost of accident induces underinvestment in prevention. With bank liability, the firm expects to be refinanced with high probability, and this reduces the incentives to overinvest. Bank liability, however, also reduces incentives to underinvest in that it allows internalization of environmental costs. In many cases, we find that less than full bank liability may be optimal. An interior value for bank liability provides a delicate balance between overinvesting because of incomplete insurance and underinvesting because of limited liability.

This result is different from what is found in Balkenborg (1997),Boyer and Laffont (1997), Dionne and Spaeter (2003) and Segerson and Tietenberg (1992) who would advocate full bank liability when the investment in prevention is observable. All these papers rest their analysis on the effect of limited liability. We show that if financial imperfections have an effect on the risk a risk-averse firm supports, then it may be optimal to reduce bank liability to prevent the firm from overinvesting in prevention in order to reduce its risk even if the investment in prevention is observable.

As in the previous literature, our analysis rests on financial-market imperfections. Such imperfections are not based on asymmetric information or moral hazard, but rather on the bank's inability to commit to future refinancing. This implies that the bank accepts refinancing the firm in states of nature of low income only if this refinancing is marginally profitable. The bank's refinancing decision depends on its outside opportunities that are, in turn, affected by the bank's liability rule. A high liability facilitates refinancing since the bank wants to avoid this liability which comes about if the firm goes bankrupt.

We expand on the existing literature in that the firm is not necessarily bankrupt following an environmental accident. In courts, bank liability applies after liable firms have gone bankrupt. In bankruptcy, courts seize the firm's assets and income. When these are insufficient to cover the cost of damages, courts typically transfer the burden to financially-secured partners. The possibility to be liable ex post is taken into account ex ante by banks. This is why the actual bankruptcy is not necessary for bank liability to have an effect on financial contracts.

In the following section, we present the model. Section 12.3 characterizes the optimal contract. Section 12.4 studies the optimal level of prevention as a function of bank liability. We solve for the case where refinancing is not constrained when there is no accident in Subsection 12.4.1, and then, in Subsection 12.4.2, we analyze the case where refinancing is constrained in non-accident states. In the last section, we argue that the legislator may want to impose less than full bank liability. The conclusion follows. All proofs are relegated to the appendix.

12.2   The Model

A firm lives for two periods. In period 1, the firm's stochastic operating income π^1 is distributed over the interval $[0, \pi_{\max}]$ with distribution function F. We note $\bar{\pi}^1 = E\pi^1$ the expected value of income. In period 2, we can assume without loss of generality that the firm earns a non-stochastic income $\bar{\pi}^2$. At the beginning of each period, the firm must finance a fixed cost K before starting production and realizing its operating income. We assume that $\bar{\pi}^t - K \geq 0$ for $t = 1, 2$. For simplicity, we assume that there is no discounting across periods.

The firm is risk averse with preferences represented by a separable utility function u, strictly increasing and strictly concave: $u' > 0$ and $u'' < 0$. We normalize $u(0) = 0$. A firm can be assumed to be risk averse if it has a motive for reducing the variability of its income. For example, it could be that its main shareholder is a risk-averse entrepreneur; or that its employees and managers must be paid a constant wage, and that this is easier to achieve the less variable is the firm's payoff.

The firm's activities represent a potential danger for the environment in that they can cause an environmental accident at the end of the first period. If there is an environmental accident, it costs an amount X for repair of damages (decontamination, compensation of victims, etc.). We assume that, in the case of an accident, the responsibility is rightly attributed to the firm which is then liable to pay the environmental damage X. Thus, the responsibility for an environmental accident is not disputed, and is imposed on the firm. This is akin to the strict liability rule.[4] This implies that the firm must pay for damages X. The firm has, however, limited liability in that it is liable up to its realized income level π^1.

The probability of an environmental accident is $p(I)$ where I is an investment in prevention made by the firm at the beginning of period 1. The probability p is decreasing and convex: $p' < 0$ and $p'' > 0$. We assume that the accident is never certain and cannot be avoided with certainty, that is, $p(0) < 1$ and $p(I) > 0$ for all I. The probability of an accident is small enough that

$$\bar{\pi}^1 - p(I)X - K - I \geq 0$$

for all relevant levels of I. An implication of this assumption is that it is always profitable to initially finance the firm.

Environmental damages X are large relative to current operating income, that is, $X > \pi_{\max}$. Since the firm is responsible for the accident and has a limited liability, this assumption implies that an environmental accident entails bankruptcy if the firm does not have access to external financing following an accident.

---

[4] See, for example, Shavell (1986) and Fluet (1999) for a discussion of various liability rules.

The firm needs financing at the beginning of the first period to finance its fixed cost K and its investment in prevention I. It may also require financing at the beginning of period 2 if the first-period income is not sufficient to finance the fixed cost for the second period. For example, if there is an environmental accident, the firm's net income is $\pi^1 - X < 0$. It can only survive through the second period if it obtains additional financing for the shortfall of the first period $\pi^1 - X < 0$ and the second-period fixed cost K. We assume that the firm can finance on a competitive financial market.

If the firm does not pay damages X, it cannot continue operating, and it must therefore declare bankruptcy. In this case, society may have to bear some of the environmental damages. The firm's limited liability causes a negative externality on society since the firm can stop operating, and hence dump the environmental cost on society. One way to (partially) internalize this externality is to hold liable a stockholder of the firm that has a deep pocket. It is in this spirit that some governments have decided to shift some of the burden onto banks. A rule that holds a bank liable for environmental damages that its borrower causes can (partially) internalize this externality. Following an environmental accident, the bank may be held liable to cover some of the shortfall between damages X and income π^1. A liability rule determines a proportion $0 \leq \alpha \leq 1$ of the shortfall $X - \pi^1$ that the bank must pay even if it does not refinance the firm and allows it to go bankrupt.

The liability rule may affect the bank's incentive to refinance the firm at the beginning of period 2 following an environmental damage. In the absence of bank liability, the bank has a reservation payoff of 0, that is, if it decides not to refinance, it gets 0. When the bank is (partially) liable, the legislator can oblige it to pay a compensation or penalty $\alpha(X - \pi^1)$ if the firm goes bankrupt, that is, even if the bank decides not to refinance the firm after the accident, a court can still force it to remain liable since it was considered an "operator" of the firm at the time of the accident. This obligation modifies the bank's reservation payoff from 0 to $-\alpha(X - \pi^1) < 0$. A bank is therefore more likely to refinance the firm following an accident if it is (partially) liable for the environmental damage. It turns out that the extent of refinancing affects the firm's incentive to invest in environmental prevention.

The object of the paper is to study the role of bank liability on the level of environmental prevention the firm undertakes, and then to analyze the socially optimal liability rule. We now characterize the optimal financial contract. In the next section, we solve for the optimal financial payments, while in the next section, we characterize the optimal level of prevention.

12.3 Optimal Financial Payments

The firm and the bank sign a financial contract at the beginning of the first period. We assume that financial markets are not completely efficient in the sense that the bank cannot commit ex ante to always refinance the firm. Any additional financing provided in period 2 must be profitable in itself. This assumption translates into refinancing constraints which are now characterized.

If there is no environmental accident, it is always profitable to refinance the firm since $\pi^1 \geq 0$ and $\bar{\pi}^2 - K \geq 0$. If there is an environmental accident, it is profitable for the bank to refinance the firm if

$$\bar{\pi}^2 - K - (X - \pi^1) \geq -\alpha(X - \pi^1), \qquad (12.1)$$

that is, if second-period profits net of financing requirements are larger than the bank's liability if it does not refinance the firm. Financing requirements consist of the second-period fixed cost K and the environmental damages X net of current income π^1. We implicitly assume that if the firm is not refinanced, the full amount π^1 is seized to pay (partially) for environmental damages. This is consistent with most bankruptcy laws where environmental liabilities have priority over a firm's shareholders.

Define implicitly the breakeven income level $\hat{\pi}^1$ by

$$\bar{\pi}^2 - K - (X - \hat{\pi}^1) = -\alpha(X - \hat{\pi}^1).$$

If $\hat{\pi}^1 < 0$, then the breakeven level is taken to be zero. Since $0 \leq \alpha \leq 1$, the firm is refinanced for all $\pi^1 \geq \hat{\pi}^1$. It is useful to compute the effect of the liability rule on the breakeven income level.

$$\frac{d\hat{\pi}^1}{d\alpha} = -\frac{X - \hat{\pi}^1}{1 - \alpha} < 0$$

When the bank's liability increases, the firm is more likely to be refinanced.

In period 1, the firm and the bank agree to a long-term financial contract that specifies, for each period t, state-contingent payments to the bank $s_0^t(\pi^1)$ when there is no accident, and $s_X^t(\pi^1)$ when there is an accident. Since prevention is observable, the financial contract can include a covenant that specifies the level of prevention that the firm must undertake. A financial contract is represented by a tuple $\{s_0^1, s_0^2, s_X^1, s_X^2, I\} = \{s, I\}$.

If there is no accident, it is optimal to refinance the firm. The financial contract must then provide incentives to do so. the bank finances the second-period fixed cost K and earns in return $s_0^2(\pi^1)$. Refinancing is profitable for the bank if

$$s_0^2(\pi^1) - K \geq 0.$$

If there is an accident, we assume that environmental liabilities have priority over financial liabilities in the settlement of bankruptcy.[5] This implies that the bank cannot get its due payment $s_X^1(\pi^1)$ if it does not refinance the

---

[5] This is indeed the case, for example, in Canada.

firm. Refinancing is profitable for the bank if

$$s_X^1(\pi^1) + s_X^2(\pi^1) - K - X \geq -\alpha(X - \pi^1),$$

that is, if financial payments net of financial requirements exceed the bank's liability in the case the firm is not refinanced and goes bankrupt.

Regardless of the financial contract that is in effect, it is optimal for the bank to refinance the firm in all states $\pi^1 \geq \hat{\pi}^1$. If the financial contract is such that the bank refinances the firm in more (or less) states, it is renegotiated so that the refinancing decision is optimal. Therefore we can impose that the firm is refinanced in all states $\pi^1 \geq \hat{\pi}^1$.

Financial markets are competitive, and the optimal contract maximizes the firm's expected intertemporal utility subject to an ex ante participation constraint for the bank and the above refinancing constraints. This corresponds to maximizing the following program.

$$\max_{s,I} \quad (1 - p(I))\mathrm{E}\left\{u(\pi^1 - s_0^1(\pi^1)) + u(\bar{\pi}^2 - s_0^2(\pi^1))\right\} +$$

$$p(I)\left[\int_{\hat{\pi}^1}^{\pi_{\max}} \left(u(\pi^1 - s_X^1(\pi^1)) + u(\bar{\pi}^2 - s_X^2(\pi^1))\right) \mathrm{d}F(\pi^1)\right]$$

s.t. $\quad (1 - p(I))\mathrm{E}\left\{s_0^1(\pi^1) + s_0^2(\pi^1) - K\right\} + p(I)\int_0^{\hat{\pi}^1} -\alpha(X - \pi^1)\mathrm{d}F(\pi^1) +$

$$p(I)\int_{\hat{\pi}^1}^{\pi_{\max}} \left(s_X^1(\pi^1) + s_X^2(\pi^1) - K - X\right) \mathrm{d}F(\pi^1) \geq K + I,$$

$$s_0^2(\pi^1) - K \geq 0 \quad \forall \, \pi^1,$$

$$s_X^1(\pi^1) + s_X^2(\pi^1) - K - X \geq -\alpha(X - \pi^1) \quad \forall \, \pi^1 \geq \hat{\pi}^1.$$

Note that the firm bears ex ante the cost for the bank's liability in that it must give the bank a fair return taking into account the possibility that the bank may be held (partially) liable. Associate the multiplier λ to the bank's participation constraint, the multipliers $\mathrm{d}F(\pi^1)\theta^0(\pi^1)$ to the first set of refinancing constraints, and the multipliers $\mathrm{d}F(\pi^1)\theta^X(\pi^1)$ to the second set. Multiply the first set of constraints by $(1 - p(I))$, and the second set by $p(I)$. The first-order conditions for financial transfers are

$$\begin{aligned}
u'(\pi^1 - s_0^1(\pi^1)) &= \lambda & \forall \, \pi^1, \\
u'(\bar{\pi}^2 - s_0^2(\pi^1)) &= \lambda + \theta^0(\pi^1) & \forall \, \pi^1, \\
u'(\pi^1 - s_X^1(\pi^1)) &= \lambda + \theta^X(\pi^1) & \forall \, \pi^1 \geq \hat{\pi}^1, \\
u'(\bar{\pi}^2 - s_X^2(\pi^1)) &= \lambda + \theta^X(\pi^1) & \forall \, \pi^1 \geq \hat{\pi}^1.
\end{aligned}$$

It is immediate from these conditions that $\pi^1 - s_0^1(\pi^1) \equiv c_0^1$ for all π^1, and that $\pi^1 - s_X^1(\pi^1) = \bar{\pi}^2 - s_X^2(\pi^1) \equiv c_X(\pi^1)$. Furthermore, using the constraints, it is easy to show that

$$c_0^2 = \min\left\{c_0^1, \bar{\pi}^2 - K\right\},$$

$$c_X(\pi^1) = \min\left\{c_0^1, \frac{\bar{\pi}^2 - K - (1-\alpha)(X - \pi^1)}{2}\right\} \quad \forall \, \pi^1 \geq \hat{\pi}^1,$$

where $c_0^2 \equiv \bar{\pi}^2 - s_0^2(\pi^1)$. When refinancing constraints are not binding, the firm's payoff is equalized in all states of nature. If a refinancing constraint is binding in one state, the firm's payoff must be reduced to ensure that refinancing is profitable for the bank.

From these definitions, we see that c_0^2 is independent of π^1, that $c_X(\hat{\pi}^1) = 0$, and that c_X is weakly increasing in π^1. We can derive the following relationships: $c_X(\pi_{\max}) \leq c_0^2 \leq c_0^1$, where equalities hold if refinancing constraints are not binding.

We first show that refinancing constraints are always binding following an environmental accident.

Lemma 12.1 *The firm's payoff in accident states is*

$$c_X(\pi^1) = \frac{\bar{\pi}^2 - K - (1-\alpha)(X - \pi^1)}{2} \quad \forall \, \pi^1 \geq \hat{\pi}^1.$$

The optimal contract seeks to insure the firm's payoff against the possibility of an environmental accident and against the variability in its first-period profits. When the bank cannot commit ex ante to always refinance the firm, such insurance is constrained. First, the bank does not always refinance the firm, that is, it does not refinance when first-period profits are too low. Second, following an accident, the cost of refinancing must come out of current and future profits, that is, there is no cross-subsidization across accident and non-accident states. The consequence is that the firm's dividend in accident states is smaller than in non-accident states, that is, $c_X(\pi^1) \leq c_0^t$ for all π^1.

We can now substitute the expression for c_X in the bank's participation constraint to obtain

$$(1 - p(I))\left(\bar{\pi}^1 + \bar{\pi}^2 - K - c_0^1 - c_0^2\right) - p(I)\alpha(X - \bar{\pi}^1) \geq K + I. \qquad (12.2)$$

This constraint is always binding. Depending on whether refinancing constraints bind or not in non-accident states, the firm's payoffs in those states are either $c_0^1 > c_0^2 = \bar{\pi}^2 - K$ or $c \equiv c_0^1 = c_0^2$. In both cases, constraint (12.2) helps solving for those values.

Suppose that the refinancing constraints are not binding in non-accident states. This implies that the firm's payoff is c in all non-accident states and both periods. Solving for c in the bank's participation constraint, we obtain

$$c = \frac{(\bar{\pi}^2 - K)(1 - p(I)) + \bar{\pi}^1 - K - I - p(I)X + p(I)(1 - \alpha)(X - \bar{\pi}^1)}{2(1 - p(I))}.$$

Refinancing constraints bind in non-accident states if $c \geq \bar{\pi}^2 - K$, that is, if

$$\bar{\pi}^1 - K - I - p(I)X + p(I)(1 - \alpha)(X - \bar{\pi}^1) \geq (1 - p(I))(\bar{\pi}^2 - K).$$

This condition says that if first-period expected profits are high enough, the optimal contract should redistribute from the first to the second period to achieve some smoothing of the firm's payoff. This reduces, however, the bank's payoff in the second period to the point of making the second-period loan non-profitable. This is not sustainable if the bank cannot commit to a non-profitable loan in period 2. In this case, the refinancing constraint would be violated and therefore $c_0^2 = \bar{\pi}^2 - K < c_0^1$.[6]

The bank's participation constraint then gives c_0^1 when $c_0^2 = \bar{\pi}^2 - K$:

$$(1 - p(I))\left(\bar{\pi}^1 - c_0^1\right) - p(I)\alpha(X - \bar{\pi}^1) = K + I.$$

This implies

$$c_0^1 = \frac{(1 - p(I))\bar{\pi}^1 - K - I - \alpha p(I)(X - \bar{\pi}^1)}{1 - p(I)}.$$

It is interesting to note that in any case, $c = (c_0^1 + c_0^2)/2$, and that c_X is unaffected by refinancing constraints in non-accident states.

12.4   Optimal Prevention

With the solution for optimal firm's payoffs as described above, what remains to be found is the optimal level of prevention. We first consider the case where refinancing constraints do not bind in non-accident states. This is the case if $\bar{\pi}^1 \leq \bar{\pi}^2$.

12.4.1   Non-binding Refinancing Constraints in Non-accident States

Since the firm's payoff is c_X in both periods following an accident, we can substitute these values in the objective function before solving for the optimal

---

[6] A sufficient condition for refinancing constraints not to bind in non-accident states is $\bar{\pi}^1 \leq \bar{\pi}^2$.

level of prevention I.

$$\max_I (1 - p(I))(u(c) + u(c)) + p(I) \int_{\hat{\pi}^1}^{\pi^{\max}} \left(u(c_X(\pi^1)) + u(c_X(\pi^1)) \right) dF(\pi^1)$$

This reduces to

$$\max_I 2 \left((1 - p(I))u(c) + p(I) \int_{\hat{\pi}^1}^{\pi^{\max}} u(c_X(\pi^1)) dF(\pi^1) \right).$$

Note that c_X is independent of I. The first-order condition is

$$-2p'(I) \left\{ u(c) - \int_{\hat{\pi}^1}^{\pi^{\max}} u(c_X(\pi^1)) dF(\pi^1) \right\} + 2(1 - p(I))u'(c)\frac{dc}{dI} = 0,$$

where

$$\frac{dc}{dI} = \frac{-p'(I)\left(K + I + \alpha(X - \bar{\pi}^1)\right)}{2(1 - p(I))^2} - \frac{1}{2(1 - p(I))} > 0.$$

After some manipulations, we show in the Appendix that the first-order condition can be written as

$$-p'(I)X - 1 = 2p'(I) \left[\left(\frac{u(c)}{u'(c)} - c \right) - \int_{\hat{\pi}^1}^{\pi^{\max}} \left(\frac{u(c_X(\pi^1))}{u'(c)} - c_X(\pi^1) \right) dF(\pi^1) \right]$$

$$+ p'(I) \int_0^{\hat{\pi}^1} \left[\bar{\pi}^2 - K - (1 - \alpha)(X - \pi^1) \right] dF(\pi^1). \tag{12.3}$$

The whole analysis revolves around the two terms on the r.h.s. of equation (12.3). These two terms reflect the two imperfections that are present in the model, namely non-commitment and limited liability.

Without the two imperfections, the firm would choose the investment level that would simply minimize the expected cost of an accident. This investment level is characterized by

$$-p'(I^1)X - 1 = 0,$$

where I^1 is the first-best level of investment.

In the r.h.s. of equation (12.3), the first term represents the non-commitment effect. If the bank can commit to a long-term contract, this term vanishes since the firm's payoff is equalized in all states, that is, $c = c_X$. When the bank cannot commit to a long-term contract, the firm's payoff is variable in accident states. Define this non-commitment effect by $NC(\alpha)$:

$$NC(\alpha) = 2p'(I) \left[\left(\frac{u(c)}{u'(c)} - c \right) - \int_{\hat{\pi}^1}^{\pi^{\max}} \left(\frac{u(c_X(\pi^1))}{u'(c)} - c_X(\pi^1) \right) dF(\pi^1) \right] < 0.$$

This effect depends on risk aversion. Because of the concavity of u, it is easy to show that $NC(\alpha)$ is negative. This implies that variability in the firm's payoff incites the firm in investing more than it would otherwise. The investment in environmental prevention becomes a remedy for the bank's inability to eliminate the variability in the firm's payoff caused by the risk of accident. Since the firm suffers in the case of an accident, it is willing to "overinvest" in environmental prevention to reduce the probability of an accident. Note that if the firm has linear preferences, the non-commitment effect vanishes since $u(x) = x$.

Bank liability affects the size of the non-commitment effect. We can compute the effect of the bank liability parameter α on this expression.

$$
\begin{aligned}
NC'(\alpha) \;=\; & 2p'(I)\left(\frac{u'(c)^2 - u''(c)u(c)}{u'(c)^2} - 1\right)\frac{dc}{d\alpha} \\
& - 2p'(I)\int_{\hat{\pi}^1}^{\pi^{\max}} \left(\frac{u'(c_X(\pi^1))u'(c)dc_X(\pi^1)/d\alpha - u''(c)u(c_X(\pi^1))dc/d\alpha}{u'(c)^2}\right. \\
& \left. \qquad\qquad\qquad\qquad -\frac{dc_X(\pi^1)}{d\alpha}\right)dF(\pi^1).
\end{aligned}
$$

Using the fact that

$$
\frac{dc}{d\alpha} = \frac{-p(I)(X - \bar{\pi}^1)}{2(1 - p(I))} < 0 \quad \text{and} \quad \frac{dc_X(\pi^1)}{d\alpha} = \frac{X - \pi^1}{2} > 0,
$$

we have

$$
\begin{aligned}
NC'(\alpha) \;=\; & 2p'(I)\int_{\hat{\pi}^1}^{\pi^{\max}} \left[\left(1 - \frac{u'(c_X(\pi^1))}{u'(c)}\right)\frac{dc_X(\pi^1)}{d\alpha}\right.\\
& \left. \qquad\qquad - \frac{u''(c)\left(u(c) - u(c_X(\pi^1))\right)}{u'(c)^2}\frac{dc}{d\alpha}\right]dF(\pi^1) \\
& - 2p'(I)\int_0^{\hat{\pi}^1}\frac{u''(c)u(c)}{u'(c)^2}\frac{dc}{d\alpha}dF(\pi^1) > 0.
\end{aligned}
$$

When the liability of the bank increases, the bank is more likely to refinance the firm. This reduces the variability of the firm's payoff and hence its incentive to overinvest.

The second term in the r.h.s. of equation (12.3) represents the usual limited-liability effect. Define this effect by $LL(\alpha)$:

$$
LL(\alpha) = p'(I)\int_0^{\hat{\pi}^1} \left[\bar{\pi}^2 - K - (1 - \alpha)(X - \pi^1)\right] dF(\pi^1) > 0.
$$

This effect is independent of the degree of risk aversion. Because the firm may go bankrupt in the case of an accident, it may avoid some of the cost for the accident. This is the externality that we mentioned before. The firm therefore invests less in environmental prevention than it would if it had full liability. This effect has been well documented in the literature.[7]

Bank liability affects the size of the limited-liability effect. We can compute the effect of the bank liability parameter α on this expression.

$$LL'(\alpha) = p'(I) \int_0^{\hat{\pi}^1} (X - \pi^1) dF(\pi^1) < 0.$$

When the liability of the bank increases, the bank is more likely to refinance the firm. This reduces the size of the externality and hence its incentive to underinvest.

The two effects work in opposite directions. We now characterize the effect of bank liability on the chosen level of investment, and how this level relates to the first-best level.

Suppose first that the bank's liability is such that it always refinances following an accident, that is, $\hat{\pi}^1 = 0$. This is the case if $(1 - \alpha)X \leq \bar{\pi}^2 - K$. From the definition of $\hat{\pi}^1$, we see that the firm refinances for all liability levels $\alpha \geq \hat{\alpha}$ where $\hat{\alpha} = 1 - (\bar{\pi}^2 - K)/X$. Note that $\hat{\alpha} < 1$. In this case, the limited-liability effect vanishes, and the level of investment in prevention is larger than the first-best level I^1 because of the non-commitment effect $NC(\alpha)$. Furthermore, as α varies from $\hat{\alpha}$ to 1, the level of investment decreases even though it remains above I^1.

Now suppose that the bank has no liability in the case of an accident. The firm is then refinanced if $\pi^1 \geq \hat{\pi}^1 = X - (\bar{\pi}^2 - K)$. To keep the problem interesting, assume that $X - (\bar{\pi}^2 - K) > 0$, that is, there is a positive probability that the firm is not refinanced. Otherwise, the firm is always refinanced, the limited-liability effect vanishes and the investment is larger than I^1. If there is a positive probability that the firm is not refinanced, then both imperfections are present. The net effect depends on the relative strength of each effect. If the non-commitment effect dominates, then the investment is larger than I^1. If the limited-liability effect dominates, the investment is smaller than I^1.

We now provide an example that shows that the limited-liability effect can dominate for low values of α. Consider a HARA utility function

$$u(c) = \frac{1 - \gamma}{\gamma} \left(\frac{c}{1 - \gamma} \right)^\gamma$$

with $\gamma = 0.7$. Suppose that the probability of accident is

$$p(I) = \frac{a}{1 + I}$$

---

[7] For example, see Segerson and Tietenberg (1992) or Van't Veld, Rausser and Simon (1997).

with $a = 0.2$. Suppose that F is uniformly distributed over $[0, \pi_{\max} = 40]$, that $\bar{\pi}^2 = 20$, that $K = 10$, and that $X = 50$. Computations reveal that, for $\alpha \geq 0.48$, the non-commitment effect dominates, while the limited-liability effect dominates for $\alpha < 0.48$. This implies that the firm overinvests in accident prevention if the bank liability is sufficiently high, while it underinvests if it is low.

When the firm is less risk averse, the limited-liability effect becomes relatively more important. Suppose that $\gamma = 0.9$. The non-commitment effect dominates for $\alpha \geq 0.69$. The extent of overinvestment diminishes. In the limit, when $\gamma = 1$, the non-commitment effect vanishes, and only the limited-liability effect is present. This is the effect that has been well documented in the literature. The firm underinvests for all values $\alpha < 0.8$. When $\alpha \geq 0.8$, the firm is always refinanced, and hence investment is at its first-best level.

12.4.2 Binding Refinancing Constraints in Non-accident States

Refinancing constraints bind in non-accident states if second-period net profits are low enough, that is, if $\bar{\pi}^2 - K$ is sufficiently close to 0.

The effect of investment on first-period payoff c_0^1 when refinancing constraints bind in non-accident states is:

$$\frac{dc_0^1}{dI} = \frac{-p'(I)\left\{\alpha(X - \bar{\pi}^1) + K + I\right\} - (1 - p(I))}{(1 - p(I))^2}.$$

It is easy to show that the first-order condition for the choice of an optimal investment level is then:

$$-p'(I)X - 1 = p'(I)\left\{\left[\sum_{t=1}^{2}\left(\frac{u(c_0^t)}{u'(c_0^t)} - c_0^t\right)\right]\right.$$
$$\left. -2\int_{\hat{\pi}^1}^{\pi^{\max}}\left(\frac{u(c_X(\pi^1))}{u'(c)} - c_X(\pi^1)\right)dF(\pi^1)\right\}$$
$$+ p'(I)\int_0^{\hat{\pi}^1}\left[\bar{\pi}^2 - K - (1 - \alpha)(X - \pi^1)\right]dF(\pi^1). \quad (12.4)$$

This condition is similar to that of the preceding section with the exception that in non-accident states consumption is not equalized across periods 1 and 2.

The condition for the choice of investment can also be decomposed into a limited-liability effect and a non-commitment effect. The first term in the r.h.s. of equation (12.4) represents the non-commitment effect. Define this effect by $\widetilde{NC}(\alpha)$:

$$\widetilde{NC}(\alpha) = p'(I)\left\{\left[\sum_{t=1}^{2}\left(\frac{u(c_0^t)}{u'(c_0^t)} - c_0^t\right)\right]\right.$$
$$\left. -2\int_{\hat{\pi}^1}^{\pi^{\max}}\left(\frac{u(c_X(\pi^1))}{u'(c)} - c_X(\pi^1)\right)dF(\pi^1)\right\} < 0.$$

It is easy to show that $\widetilde{NC}'(\alpha) > 0$. Since we know that $c = (c_0^1 + c_0^2)/2$, and that u is concave, $\widetilde{NC}(\alpha)$ is less negative than $NC(\alpha)$, so that the non-commitment effect induces less overinvestment when refinancing constraints bind in non-accident states than when they do not. The non-commitment effect is now alleviated by the fact that perfect insurance is not achievable in non-accident states, that is, even when there is no accident, the financial contract does not provide a high dividend to the firm in the second period. This makes the contrast between accident and non-accident states less pronounced, and thus reduces the incentives to overinvest in prevention in order to avoid accident states. Nevertheless, the effect of bank liability α on the level of investment is qualitatively the same as in the preceding case.

The second term in the r.h.s. of equation (12.4) is the same limited-liability effect $LL(\alpha)$, which is positive and decreasing in α.

The results are qualitatively similar to those of the preceding section. Consider our preceding example with $\gamma = 0.7$, $\bar{\pi}^2 = 10.5$, $K = 10$, and $X = 41$. Computations reveal that the non-commitment effect dominates for $\alpha \geq 0.81$. When $\gamma = 0.9$, the non-commitment effect dominates for $\alpha \geq 0.94$. Again, a reduction in risk aversion reduces the strength of the non-commitment effect.

12.5 Optimal Bank Liability

Suppose that the social objective of the legislator is to implement the first-best level of investment. This is a reasonable assumption given that we have adopted a partial-equilibrium framework. It amounts to say that the extent of bank liability cannot be used to alleviate financial imperfections but only to induce an optimal level of prevention. Suppose also that the legislator cannot verify the level of prevention. It does not have "insider" privileges such as a bank has, and therefore cannot produce information about the level of prevention that would make it verifiable by a court. The legislator is then to pick an optimal level of liability for banks knowing that the extent of liability has an effect on the level of prevention that the firm takes. The objective is then to choose α such that $NC(\alpha) + LL(\alpha) = 0$ when evaluated at I^1, if at all possible.

This is a hard problem to solve and the solution will generally depend on parameter values and functional forms. It may be that the solution is at $\alpha = 1$ if $NC(\alpha) + LL(\alpha) < NC(1) + LL(1) < 0$ for all $\alpha < 1$. In this case, imposing full liability implements a level of prevention that is larger than I^1 because of incomplete insurance provided by the financial contract.

In many cases, however, an interior level of liability can implement the optimal level of prevention I^1. In all examples cited in the previous section, the limited-liability effect is strong enough to offset the non-commitment effect for small values of α. This is a different result than what is found in the literature which is based on asymmetric information. Here, the investment is observable but financial markets are imperfect. When introducing financial imperfections and concave preferences, the firm has a tendency to overinvest in prevention. Reducing bank liability can then restore prevention to its first-best level.

12.6 Conclusion

Laws such as CERCLA seek to apply the principle that polluters should pay for the pollution they generate. These laws then reduce the social burden of environmental risks in two ways: compensation payments do not have to be supported by tax-payers' money and they provide better incentives for prevention. When the search for a payer entails the legislator to turn against banks, however, the financial system can suffer distortions whose consequences on the pollution level are not easily quantifiable.

Court judgments that followed the introduction of CERCLA in the United States justified the imposition of bank liability by the fact that banks had close and long relationships with their borrowers and, hence, possibly an influence on their decisions. Firms and banks entering in contractual relationships take into account bank liability when writing contracts. Therefore, the consequences of past court decisions have an impact on future environmental prevention.

Since banks offer insurance and smoothing to their borrowers, the usual limited-liability investment reducing effect is mitigated by the efficiency of the financial contract in providing such insurance and smoothing. Financial imperfections induce firms in overinvesting in prevention. The total effect on firms' incentives in prevention is ambiguous. In many cases, it is optimal for the legislator to impose less than full liability to banks. Bank liability then becomes a useful tool to implement an optimal level of prevention, and this even if investment in prevention is observable.

Appendix

Proof of Lemma 12.1 Suppose the refinancing constraint does not bind in some accident state $\tilde{\pi}$. This implies $c_X(\tilde{\pi}^1) = c_0^1$. If refinancing constraints are binding in non-accident states ($\theta^0(\pi^1) > 0$), we have:

$$\tilde{\pi}^2 - K = c_0^2 < c_0^1 = c_X(\tilde{\pi}^1) \le \left(\tilde{\pi}^2 - K - (1-\alpha)(X - \tilde{\pi}^1)\right)/2.$$

Since this is impossible, it has to be that if $\theta^X(\pi^1) = 0$ for some π^1, then $\theta^0(\pi^1) = 0$ for all π^1.

This implies $c_0^1 = c_0^2 = c_X(\pi^1)$ for all $\pi^1 \ge \tilde{\pi}^1$ and

$$c_X(\pi^1) = \left(\tilde{\pi}^2 - K - (1-\alpha)(X - \pi^1)\right)/2 < c_0^1$$

for all $\hat{\pi}^1 \le \pi^1 \le \tilde{\pi}^1$. By continuity, it must be that

$$c_0^1 = c_X(\tilde{\pi}^1) = \left(\tilde{\pi}^2 - K - (1-\alpha)(X - \tilde{\pi}^1)\right)/2.$$

The bank's participation constraint gives the expression for consumption c_0^1 that also depends on $\tilde{\pi}^1$:

$$c_0^1 = \frac{1}{2\left(1 - p(I)F(\tilde{\pi}^1)\right)} \Bigg((1 - p(I))(\tilde{\pi}^1 + (\tilde{\pi}^2 - K)) + p(I) \int_0^{\tilde{\pi}^1} -\alpha(X - \pi^1)dF(\pi^1)$$

$$+ p(I) \int_{\tilde{\pi}^1}^{\pi^{\max}} (\pi^1 - X)dF(\pi^1) + p(I)(1 - F(\tilde{\pi}^1))(\tilde{\pi}^2 - K) - K - I \Bigg).$$

Since we have two expressions for c_0^1, it must be that they are equal:

$$\frac{\tilde{\pi}^2 - K - (1-\alpha)(X - \tilde{\pi}^1)}{2} = \frac{1}{2\left(1 - p(I)F(\tilde{\pi}^1)\right)}\Big((1 - p(I))(\tilde{\pi}^1 + (\tilde{\pi}^2 - K))$$

$$+ p(I) \int_0^{\tilde{\pi}^1} -\alpha(X - \pi^1)dF(\pi^1)$$

$$+ p(I) \int_{\tilde{\pi}^1}^{\pi^{\max}} (\pi^1 - X)dF(\pi^1)$$

$$+ p(I)(1 - F(\tilde{\pi}^1))(\tilde{\pi}^2 - K)\Big)$$

$$-(1-\alpha)\left(1 - p(I)F(\tilde{\pi}^1)\right)(X - \tilde{\pi}^1) = (1 - p(I))\tilde{\pi}^1 - K - I$$

$$+ p(I) \int_0^{\tilde{\pi}^1} -\alpha(X - \pi^1)dF(\pi^1)$$

$$+ p(I) \int_{\tilde{\pi}^1}^{\pi^{\max}} (\pi^1 - X)dF(\pi^1)$$

$$(1-\alpha)\left(1-p(I)F(\tilde{\pi}^1)\right)\tilde{\pi}^1 - (1-\alpha)X \;=\; \tilde{\pi}^1 - p(I)X - K - I$$
$$-\,(1-\alpha)p(I)\int_0^{\tilde{\pi}^1}\pi^1 dF(\pi^1)$$

$$(1-\alpha)(\tilde{\pi}^1 - X) + (1-\alpha)p(I)\int_0^{\tilde{\pi}^1}(\pi^1 - \tilde{\pi}^1)dF(\pi^1) = \tilde{\pi}^1 - p(I)X - K - I$$

But this is impossible since the left-hand side of this condition is negative and the right-hand side is the positive ex-ante value of the firm. This implies that our initial supposition that the refinancing constraint did not bind in some accident state is false. Hence, refinancing constraints bind in all accident states, and c_X is given by the expression in Lemma 12.1. ∎

First-order condition for the choice of investment. Suppose no refinancing constraint is binding in non-accident states. Then, the firm's payment is the same in all states of nature and in both periods : $c_1^0 = c_2^0 = c$. The first-order condition for investment is then:

$$-2p'(I)\left[u(c) - \int_{\tilde{\pi}^1}^{\pi^{\max}}u(c_X(\pi^1))dF(\pi^1)\right]$$
$$-2(1-p(I))u'(c)\left[\frac{1}{2(1-p(I))} + \frac{p'(I)}{2(1-p(I))^2}\left(\alpha(X-\tilde{\pi}^1)+K+I\right)\right]=0,$$

that is,

$$-2p'(I)\left[\frac{u(c)}{u'(c)} - \int_{\tilde{\pi}^1}^{\pi^{\max}}\frac{u(c_X(\pi^1))}{u'(c)}dF(\pi^1)\right] - \frac{p'(I)}{1-p(I)}\left(\alpha(X-\tilde{\pi}^1)+K+I\right)=1.$$

Hence,

$$-2p'(I)\left[\left(\frac{u(c)}{u'(c)}-c\right) - \int_{\tilde{\pi}^1}^{\pi^{\max}}\left(\frac{u(c_X(\pi^1))}{u'(c)} - c_X(\pi^1)\right)dF(\pi^1)\right]$$
$$=\; 1+2p'(I)\left[c - \int_{\tilde{\pi}^1}^{\pi^{\max}}c_X(\pi^1)dF(\pi^1)\right] + \frac{p'(I)}{1-p(I)}\left(\alpha(X-\tilde{\pi}^1)+K+I\right)$$
$$=\; 1 - p'(I)\int_{\tilde{\pi}^1}^{\pi^{\max}}(\tilde{\pi}^2 - K - (1-\alpha)(X-\pi^1))dF(\pi^1)$$
$$+\frac{p'(I)}{1-p(I)}\left(\alpha(X-\tilde{\pi}^1)+K+I\right)$$
$$+\frac{p'(I)}{1-p(I)}((\tilde{\pi}^2 - K)(1-p(I)) + \tilde{\pi}^1 - K - I - p(I)X$$
$$+p(I)(1-\alpha)(X-\tilde{\pi}^1)$$

$$= 1 - p'(I) \int_{\hat{\pi}^1}^{\pi^{\max}} (\bar{\pi}^2 - K - (1 - \alpha)(X - \pi^1)) \mathrm{d}F(\pi^1) +$$

$$\frac{p'(I)}{1 - p(I)} \left((\bar{\pi}^2 - K)(1 - p(I)) - (1 - p(I))(1 - \alpha)(X - \bar{\pi}^1) + p'(I)X \right).$$

This simplifies to

$$-p'(I)X - 1 = 2p'(I) \left[\left(\frac{u(c)}{u'(c)} - c \right) - \int_{\hat{\pi}^1}^{\pi^{\max}} \left(\frac{u(c_X(\pi^1))}{u'(c)} - c_X(\pi^1) \right) \mathrm{d}F(\pi^1) \right]$$

$$+ p'(I) \int_0^{\hat{\pi}^1} \left[\bar{\pi}^2 - K - (1 - \alpha)(X - \pi^1) \right] \mathrm{d}F(\pi^1).$$

References

Balkenborg, D., 'Bargaining Power and the Impact of Lender Liability for Environmental Damages', University of Southampton D.P.97-09 (1997).

Boyer, M. and Laffont, J.J., 'Environmental Protection, Producer Insolvency and Lender Liability', CIRANO scientific series, (1995) 95s-50, Montréal.

Boyer, M. and Laffont, J.J., 'Environmental Risk and Bank Liability', *European Economic Review*, 41 (1997): 1427-1459.

Dionne, G. and Spaeter, S., 'Environmental Risks and Extended Liability: The Case of Green Technologies', *Journal of Public Economics*, 87 (2003): 1025-1060.

Fluet, C., 'Régulation des risques et insolvabilité : le rôle de la responsabilité pour faute en information imparfaite', *L'Actualité Economique*, 75 (1999): 379-399.

Goble, R.A., 'EPA's CERCLA Lender Liability Proposal: Secured Creditors 'Hit the Jackpot'. Comment', *Natural Resources Journal*, 32 (1992): 653-679.

Henderson, D.A., 'Environmental Liability and the Law of Contracts', *The Business Lawyer*, 50 (1994): 183-266.

Olexa, M.T., 'Contaminated Collateral and Lender Liability: CERCLA and the New Age Banker', *American Journal of Agricultural Economics*, 73 (1991): 1388-1393.

Pitchford, R., 'How Liable Should the Lender be? The Case of Judgement-Proof Firms and Environmental Risks', *American Economic Review*, 85 (1995): 1171-1186.

Segerson, K. and Tietenberg, T., 'The Structure of Penalties in Environmental Enforcement: An Economic Analysis', *Journal of Environmental Economics and Management*, 23 (1992): 179-200.

Shavell, S., 'The Judgement-Proof Problem', *International Review of Law and Economics*, 6 (1986): 45-58.

Van't Veld, K., G.C. Rausser and Simon, L.K., 'The Judgement Proof Opportunity', Fondazione Eni Enrico Mattei Note di Lavoro, (1997): 83/97.

Chapter 13

Sharing Liability between Banks and Firms: The Case of Industrial Safety Risk[*]

Marcel Boyer
Université de Montréal

Donatella Porrini
Universita di Lecce

13.1 Introduction

The determinants of industrial or environmental accidents are numerous and their interactions are complex. To design and implement proper control policies to reduce the probability and severity of those accidents, those determinants and their interactions must be better understood. We consider in this paper that those determinants or factors can be regrouped into three different sets: technological, eventual, and organizational. The first set regroups all factors that are directly linked to the characteristics of the products being handled, both inputs and outputs, and their current production and distribution technologies. Clearly, handling relatively unstable, harmful or explosive chemicals is intrinsically more dangerous and producing or distributing those products by using technologies that are relatively labor intensive or through urban areas or on public transportation networks is also intrinsically more dangerous. The second set of determinants or factors includes exogenous and in part uncontrollable purely random natural events such as lightning, flooding and even residual unavoidable human errors, which may ignite a process leading more or less inexorably to an accident. Finally, the third set contains the organizational characteristics which may contribute to the occurrence of a severe accident by allowing more of less control, coordination or timely information transmission by and among agents who may be called "stakeholders" in the occurrence of industrial or environmental accidents. This paper deals

[*]We are grateful to Eric Gravel for research assistance.

with this third set, namely the organizational factors which include among others the institutional characteristics and informational asymmetries which constitutes significant background features.

More precisely, we study the impact of different institutional and informational constraints in the implementation of a public policy aimed at preventing environmental or industrial accidents. The most important constraints to be taken into account are: limited liability (judgement proofness) of firms; limited capacity of governments to intervene in business decisions and transactions, which requires that governments use policy instruments that are either of the ex-ante regulation type, or of the ex-post liability type, including extended liability and financial responsibility, or of both types; limited power of the court system to search and find all the facts relevant to a judgement; asymmetric information between the agents, namely governments, firms, banks, and courts, whose decisions and behavior may have an impact on the observed probability and severity of environmental or industrial accidents.

We discuss in the following sections the institutional frameworks of the American and European liability systems, the specific roles of the four main actors in the determination of industrial/environmental accidents and the relevant academic literature, before proceeding in section 4 with the presentation of our model followed by a section where the main results are discussed. We conclude in section 7 with some policy recommendations.

13.2 The Internalization of Environmental Damage in the U.S. and Europe through Extended Liability and Financial Responsibility

The term 'financial responsibility' refers to the set of instruments with which potential polluters can demonstrate ex ante that their financial resources are adequate for the restoration of environmental damage they may cause. In one practical application, financial responsibility requires that the authorization to carry out production activities in risky sectors be restricted to firms who can demonstrate an appropriate financial or insurance coverage for future obligations resulting from the assignment of environmental liability.[1]

---

[1]Financial responsibility includes various kinds of instruments: letters of credit and surety bonds; cash accounts and certificates of deposit; self-insurance and corporate guarantee (Boyd, 2002). Letters of credit and surety bonds are purchased from banks or insurance companies; they require the latter to pay a third party beneficiary (often the government) under specific circumstances, such as the failure of the purchaser to perform certain obligations. Cash accounts and certificates of deposit place cash or some other form of interest-bearing security into accounts that are assigned or made payable to a regulatory authority. Companies with relatively deep pockets may self-insure to satisfy coverage requirements by demonstrating sufficient financial strength. Finally, a corporate guarantee allows another

Financial responsibility has been widely applied in the United States since the 1980s within the framework of the liability assignment system for environmental damage, particularly in the U.S. Comprehensive Environmental Response, Compensation and Liability Act (CERCLA) that states that the owners and operators of a facility causing an accident are strictly liable for the costs involved in cleaning up the contaminated sites and in compensating the victims.[2]

The U.S. experience shows that financial responsibility may be a (sometimes necessary) complement to legislation on liability assignment of environmental damage. It is usually needed to ensure that the damaged natural resources are reclaimed. Its different applications all stem from a common motivation, which is to ensure the proper internalization of the financial burden engendered by polluters in order to indemnify victims and discourage various forms of environmental deterioration.

As in the U.S. system, the proposed EC system considers the need for instruments that foster the internalizing of environmental damage in cases of insolvency and judgment-proofness of the liable parties. In the wording of the Proposal:[3] "The insolvency of operators is one factor that may hinder cost recovery in line with the 'polluter pays' principle by competent authorities, but the impact of this may be limited by adequate financial insurance of potential damage" (page 4, point 2). Thus the EC legislator recognizes an important role for financial institutions, although it does not specifically define any compulsory financial security in order to meet the flexibility requirements of the new liability system. In fact, the Proposal states that "Financial assurance of environmental liability is beneficial for all stakeholders: for public authorities and the public in general, it is one of the most effective, if not the only, way of ensuring that restoration actually takes place in line with the 'polluter pays' principle; for industry operators, it provides a way of spreading risks and managing uncertainties; for the insurance industry, it is a sizeable market" (page 7, point 4).

But while in the U.S., due to the legal provisions, financial responsibility has evolved to provide a wide array of financial instruments tailored to in-

firm, such as a parent company, to satisfy the coverage requirement; financial guarantors must then explicitly agree to cover the liabilities of the firm.

[2]Financial responsibility is also provided for by CERCLA, by the Safe Drinking Water Act (SDWA), by the Outer Continental Shelf Lands Act (OCSLA), and by the Surface Mining Control and Reclamation Act (SMCRA). Moreover, it is part of the Resource Conservation and Recovery Act (RCRA) and of the Oil Pollution Act (33 U.S.C. §2716 of 1990). Under CERCLA, the Environmental Protection Agency (EPA) can proceed promptly with the clean up of contaminated sites with funding from the Hazardous Substances Response Trust Fund, commonly known as the Superfund. The fund is financed through a combination of federal appropriations, industry taxes, and penalties imposed under judgments entered against responsible parties.

[3]*Proposal for a Directive of the European Parliament and of the Council on Environmental Liability with regards to the Prevention and Remedying of Environmental Damage*, COM (2002), 17 final, Brussels, January 23, 2002.

dividual firms and regulatory needs, in the EC this kind of instrument has a corresponding importance but relatively little diffusion. In fact, the White Paper[4] (§ 4.9 "Financial security") contains the following statement: "When looking at the insurance market – insurance being one of the possible ways of having financial security, alongside, among others, bank guarantees, internal reserves or sector-wise pooling systems – it appears that coverage of environmental damage risks is still relatively undeveloped, but there is clear progress being made in parts of the financial markets specializing in this area." The imposition of such instrument also seems to be delayed in time, as indicated by the statement that "[...] the EC regime should not impose an obligation to have financial security, in order to allow the necessary flexibility as long as experience with the new regime still has to be gathered. The provision of financial security by the insurance and banking sectors for the risks resulting from the regime should take place on a voluntary basis." This, however, disregards the fact that financial responsibility instruments have already been made mandatory within individual Member States.[5]

From an economic point of view, the enforcement of financial responsibility ensures that the expected costs related to environmental risks are recorded in the firm's balance sheet and accounts. Since financial guarantees are purchased from banks or insurance companies, a contract relation is established that makes the latter keen on protecting their investment, for instance by monitoring the production activity of their corporate customers. As in the economic models of lenders' liability found in the literature, the bank, acting as principal in a financial responsibility regime, is encouraged to monitor the environmental risk prevention activity of its corporate customers. The agent firm, in turn, pays to the principal the cost associated with its risk level (through possibly the cost of the loan), and is therefore encouraged to adopt preventive measures to reduce its risk and, as a result, its total borrowing costs.

In this respect, the "financial guarantors" may provide a remedy to information asymmetry issues – in particular those involving moral hazard – through monitoring. They may also help addressing issues related to adverse selection, through contract design and by setting an appropriate loan cost, thereby offering lower cost guarantees to firms that make less risky choices

---

[4] *White Paper on Environmental Liability*, COM (2000), 66 final, Brussels, February 9, 2000.

[5] For example in Italy, the Ministero dell'Ambiente, in a decree of October 8, 1996, defined the method for granting financial guarantees in favor of the State by companies that carry out waste transportation activities related to reclaiming, restoration of site conditions, waste transportation and disposal, as well as the reimbursement of any further damage caused to the environment. Another example is the Flemish experience - in particular the proposals of the Interuniversity Commission for the revision of environmental law in the Flemish region, which has provided for elaborate provisions concerning financial guarantees (Faure and Grimeaud 2000).

from an environmental standpoint and to those that implement prevention schemes (Feess and Hege 2000, 2003).

The efficiency of extending liability to the firm's major partners, as part of the broader policy issue of preventing environmental and/or industrial accidents, has been recently addressed in the literature, as a principal-agent problem between a bank or financier and a firm.[6] Extending environmental liability becomes a crucial factor once a firm has reached the stage of judgment-proofness (Shavell 1986) a problem which arises when identified polluters prove unable to pay for the damages they caused.[7] Given that they may fail to pay the full cost associated with the environmental damage, firms face weaker incentives to implement preventive measures at an efficient level.

A particular case of extended lenders' liability stems from significant jurisprudence in the United States. Under the CERCLA-based system of strict liability of the polluting firms, the notion of owner and operator used in the wording of the law was gradually extended to include lenders who were particularly active in supervising the firm's operations. In so doing, the court have been instrumental in finding deep-pocket parties to compensate victims and pay for clean-up costs caused by insolvent firms.[8] In the case United States v. Mirabile (Environmental Law Reports 20, 994 (E.D. Pa 1985)), the bank was found liable for the damage because the Court found that it had been significantly involved in the supervision of the firm's operations. The Court stated that in this case, the firm's lenders could be considered as operators and, as such, the so-called "secured creditor exemption clause" was not applicable, and therefore that the bank was liable for the damage under CERCLA regulation. In another case, United States v. Maryland Bank and Trust (632 F. Supp. 573 (d. Md. 1986)), the Court stated that the bank was liable for the decontamination costs based on the fact that it became the owner of the plant after the closing of the firm before the damage was identified. In contrast with the previous case, no consideration was given to the fact that the bank had been involved or not in the decisions made with respect to the polluting substances. Rather, the lender's liability was established on the sole ground that the bank had become the owner of the property. Finally, in the case United States v. Fleet Factors Corp. (901 F. 2d 1550 (11th Cir. 1990), cert. Denied, 498 U.S. 1046 (1991)), the bank was deemed liable for the decontamination

---

[6]Numerous contributions have appeared recently in the literature addressing the question of lenders' liability: see for instance Beard (1990), Pitchford (1995, 2001), Heyes (1996), Boyer and Laffont (1996, 1997), Boyd and Ingberman (1997), Balkenborg (2001), and Lewis and Sappington (2001), Boyer and Porrini (2004), Gobert and Poitevin (2004).

[7]Judgment-proofness is not necessarily connected with firm size: "History has unfortunately shown that even companies with limited financial means may cause huge environmental damage and may thereafter be judgment-proof. Moreover, the insolvency risk may even arise with larger companies since almost all (larger) companies are organized as legal entities and therefore enjoy the benefits of the limited liability of the corporation" (Faure 2001, p. 192).

[8]See Boyer and Porrini (2002) for a discussion of that jurisprudence.

costs based on the fact that its financing of the firm had conferred it the ability to influence management, even if it had not been directly involved in the firm's operations. This rule would provide for lenders to be considered parties to the management of the firm simply because, in granting the loan, they could if they wished force their clients to carry out preventive actions against environmental abuses and/or may review their preventive practices.

In this vein, it is possible to define a form of "extended principal-agent liability," under which liability is shared between the firm and its financier, thereby giving financial institutions an incentive to properly identify and monitor risky firms. This is the modeling strategy we use in this paper. It implies an active preventive role for financial institutions through financial responsibility in the spirit of the proposed common liability system mentioned in a European Community Directive.[9] Imposing liability on financial institutions is thus not seen any more simply as a way to relax the firms' resources constraint. Rather, in the new system, the legislator will delegate to financial institutions part of its control upon the firms' preventive behavior by inducing them to internalize environmental damage.

Rather than an extended liability system, we will consider a liability sharing system under which the bank is responsible for a fixed part (percentage) of the damage caused - even in the case where the firm actually has the financial resources to cover more than its complementary part of the damage. With the implementation of a liability sharing system the banks are called upon to influence the firms' environmental prevention activities through different forms of monitoring or incentive financial contracts when the banks suffer from an asymmetric information structure regarding the firms' preventive activities.

However the principal-agent relationship between the bank and the firm is only part of the complex network of relationships that characterizes the business environment of a firm whose activities give rise to a probability of industrial/environmental accident. Thus, we will also consider explicitly the behavior and decisions of other parties, such as governments and courts. In a sense, the role of providing incentives for preventive environmental care by monitoring firms' activities is shared between regulatory bodies, firms and courts.[10]

---

[9]The European Parliament and the Council, *Directive of the European Parliament and of the Council on Environmental Liability with regard to the Prevention and Remedying of Environmental Damage*, Joint text approved by the Conciliation Committee, Brussels, 10 March 2004.

[10]An important point is then the comparison of ex ante and ex post instruments; see Boyer and Porrini (2001, 2004).

13.3   Environmental Protection Policy: The Roles of Government, Financial Institutions, Firms, and Courts

The environmental economics literature on lenders' liability mainly applies the principal-agent framework to the relationship between firms and financial institutions. In this paper we extend the concept of principal-agent liability to include other relationships that are important determinants of the industrial/environmental safety decisions aimed at preventing environmental accidents.

The first kind of relationship is the one between firms and financial institutions. Our model will take into account the recent trend described above with respect to the instrument of financial responsibility. Thus insurance companies and banks can be called upon to provide financial guarantees for firms that operate in risky sectors. Informational asymmetries are present in this relationship and want to model the monitoring role that can be played by the financial institutions regarding the prevention care activities implemented by firms.

The second kind of relationship is the one between firms and the government/legislator. The government/legislator determines a recommended level of safety and sets the rule to share liability between the firm and the financial institution. Unlike other contributions, such as Balkenborg's (2004), which consider a punitive liability that can exceed the damage costs, our model is built on a full liability that covers all but only the damage costs. In the model, the sharing rule takes into account the incentives towards the optimal level of prevention in a framework that is again characterized by the presence of informational asymmetries about the prevention decisions of the firms.

Finally, the third kind of relationship is the one between firms and courts. The courts are called upon to decide if and when firms are negligent in choosing a level of prevention that is lower than the level recommended by government. Also, financial institutions may have an interest in suing firms that are (or appear to be) guilty of insufficient preventive action, with the intent of recovering from them their part of the shared burden.

This model developed here includes the three types of relationships described above between the four actors involved: firms, financial institutions, government/legislator, and courts. In the literature addressing the choice of instruments to implement an environmental policy, the focus is generally on one of these relationships only. But in a realistic framework meant to design an optimal policy, it is clearly necessary to consider the roles of all four actors, in particular if one wishes to account for the optimal delegation to financial institutions of the monitoring of prevention activities realized by the firms.

Thus although this paper will not explicitly address the issue of the comparison between *ex ante* and *ex post* regulations, the role of the courts in evaluating (*ex post*) the firm's behavior is connected to the command-and-control standards set by *ex ante* regulation. In this sense, the role of providing incentives for preventive care by the means of monitoring activities is shared between regulatory agencies and the courts (Boyer and Porrini 2001). Working in this direction, Boyer and Porrini (2004) compare *ex ante* and *ex post* environmental regulation by modeling a liability system, where liability is extended to the financial partners of the firm when the firm goes bankrupt following an environmental accident, and an incentive regulation system, where the environmental protection agency may be "captured" by the regulated firms. Their model integrates the following variables and features, whose effects on the choice of instrument for environmental protection is characterized: limited liability; the cost of low and high levels of preventive care; the non observability of care; the social cost of public funds; the (net) profitability of the firm; the level of damages if an accident occurs; and finally a regulatory capture factor. They show that a relatively large cost of care favors the extended liability regime because the captured regulator regime would imply too much care, or too few environmental accidents, and too much financing of risky business, that is, an over-development of environmentally risky industries. This is due to the fact that the social value of the informational rents so allowed is not large enough to compensate for the social cost of the extra care activities. They also show that a relatively low cost of public funds, i.e. a relatively efficient (nondiscretionary) taxation system, favors the captured regulator regime because the extended lender liability regime would imply too little care and too little financing. This is because the benefits of a reduced expected cost of environmental accidents are not large enough to compensate for the loss of profits (informational rents), whose social cost is small when the cost of public funds is low. Finally, they show that a larger regulatory capture factor favors the extended liability regime: the captured regulator induces too much preventive care and hence too few accidents, allowing the firm to reap a costly informational rent. Hence, the best instrument in terms of social welfare may not be the one which ensures the better environmental protection as measured by the lower probability of environmental accidents. In a similar vein, Hutchison and van't Veld (2004) consider a context where some care activities (unobservable) reduce the probability of accident while others (observable) reduce the level of damage if an accident occurs. In such a context, extended liability improves welfare but does not induce the first-best levels of care. With free entry and exit in the industry, extended liability generates too much exit (although second-best optimal given the levels of care). The authors show that if the regulator is constrained to one instrument only, then direct regulation of the observable type of care strictly welfare dominates extended liability.

Referring to the law and economics literature, principal-agent liability relates to cases where instead of a single actor as injurer, harm is caused by an individual or a firm that is under the control or supervision of someone else. Principal-agent schemes are applied to analyze cases in which liability is extended from the person that directly caused the damage to a party that is in some sense related to that person. This is the so-called "secondary" or "vicarious" liability (Sykes 1984; Kornhauser 1982), such as that of the parents who are deemed liable for damage caused by their children, or the employer who is deemed liable for the damage caused by the activities of its employees. Some aspects remain to be developed in modeling this form of delegated control, including the presence and possible effects of monitoring costs (Dari Mattiacci and Parisi 2004) as well as the definition of the optimal sharing rule between the principal and the agent in terms of *ex ante* incentives to undertake the optimal level of prevention activities.

In Polinsky (2003) the assignment of liability to the principal is crucial to achieve the optimal level of prevention in three cases. The first case is when the agent's risk aversion lowers the optimal level of liability that the agent himself should bear as for instance, in product liability analysis when the harm is accidental in character. The second case relates to the principal's formal monitoring role when it is difficult to determine whether the agent caused harm; in such circumstances, a strictly liable principal has a strong incentive to control her agents' preventive behavior. The third case is the judgment-proofness of the agent, which induces him to carry out inadequate prevention activities and, again, renders important the monitoring role of a liable principal that will be called upon to compensate part of the damage.

Among other contributions, Newman and Wright (1990) address the problem of defining a principal-agent liability framework where moral hazard is also being considered. In particular, the authors examine the socially optimal level of prevention compared with the level achieved under strict liability in the case where an environmental/industrial accident is caused by employees' action. Given that the preferences of the employee/agent differ from those of the employer/principal and the agent's prevention activities are unobservable, a moral hazard problem arises and the principal's strict liability will influence the choice of employment contract offered.

The law and economics literature represents principal-agent liability as a framework where maximizing agents choose their preventive care level under the monitoring activities of some principal (Shavell 2004). In these models there is no room for considering how the agent's prevention decision may be influenced by factors other than the maximization hypothesis and the principal's influence. Daughety and Reinganum (2003) proposed to widen the standard law and economics literature framework by assuming that market and the tort system interact to affect the decision about prevention care level with reference to product liability. In particular, their analysis includes an endogenously-determined fixed-cost component to prevention that comes from the characteristics of the output market, the relevant litigation costs, a representation of victims as a consumer, a group of consumers or third parties, and a variety of imperfectly competitive market structures.[11]

In this paper we analyze extended liability and consider as liable parties the financial institutions that provide financial resources for production activities that may be environmentally risky. Thus with respect to the law and economics literature described above, we widen the traditional framework by adopting the legislator's point of view which, in these cases, is to determine what mix of liability to impose on the principal and the agent in order to maximize a social welfare criteria. However this maximum welfare depends upon the incentives the firm as the agent faces given its financial contract with the bank. When the firm's preventive care behavior is observable by the bank, the financial contract can impose a specific level of preventive measures as a condition for financing the firm. In general, social optimum is achieved in a full information setting by having one party or the other bear full responsibility for the damages. However, if the firm is subject to limited liability, the damage may not be fully internalized when holding the firm solely liable for the accidents it causes and, therefore, some liability should be assigned to the bank as principal in order to achieve socially optimal care.

The case of incomplete information is considered in Segerson and Tietenberg (1992). They assume the existence of moral hazard related to the contract with standard debt and risk sharing. In this case, the financial contract cannot directly depend on the firm's prevention effort. It is thus socially optimal to assign liability to the firm, i.e. to the informed party, until the damage cost is fully internalized. However, within a system of limited liability this may be unenforceable and may impede full damage internalization, leading to a sub-optimal prevention effort. The authors suggest making up for the limited liability problem by defining non-monetary penalties, including criminal prosecution for decision-makers and managers, in case of accident.

Pitchford (1995) develops a model of a judgment-proof firm, a lender and a victim in which a full lender liability policy can increase accident frequency and reduce efficiency. Under moral hazard and limited liability, the prevention effort is sub-optimal, hence the search for some liability sharing between the firm and its lender as a mechanism to bridge in part the gap between actual and optimal unobservable efforts to reduce the probability of accident, whose damage is known. The loan contract considered by Pitchford provides for payments conditional upon the occurrence or not of an accident. The prevention effort level chosen by the firm depends then on the difference between the loan reimbursement by the firm in the two states, accident and no-accident. If the creditor or lender "cannot observe the precautionary choice undertaken by the firm, and if that firm is potentially judgment-proof, then increasing the liability of the creditor can lead to an increase in the probability of acci-

---

[11]See Woodfield (1004) for a discussion of liability Internet Service Providers may incur for transporting content material which violates copyrights.

dent" (page 1182), due to the fact that the creditor, requiring that a proper insurance-like premium be paid by the firm in the no-accident case reduces the difference between the firm's profits in the two states and therefore reduces incentives for care: the firm gets no profit in the accident case and a reduced profit in the no-accident case due to the higher insurance premium. Pitchford concludes that the optimal lenders' liability level from an accident prevention standpoint is equal to the amount of the firm's available capital, that is, the project value added plus the equity invested. If lender liability is lower or higher than this level, the contract ends up providing an incentive for the firm to lower its prevention effort.

Boyer and Laffont (1997) studied the lenders' liability issue with a two-period model, including a refinancing decision in the second period. This decision is made in an incomplete information setting, which may alternatively take the form of moral hazard on the prevention effort and of adverse selection on the profitability of the firm. The firm is liable for the accident damage up to the value of its assets and the bank is liable for the residual damage. In the first period, the bank offers a loan contract. If the firm accepts, it makes a prevention effort and achieves a first period profit. At the beginning of the second period, the bank must decide whether to refinance the firm or not, a decision which is used in the adverse selection case, to elicit truthful revelation of the first period profit level. If the firm is refinanced, another profit is realized and an environmental accident may or may not occur. In the moral hazard case, Boyer and Laffont conclude by advising against full lender liability, i.e. a liability rule by which, following a major accident which makes the firm bankrupt, the bank pays the total amount of damage and recovers whatever it can from the firm's assets. The private contract solution approaches the socially optimal solution if the bank's liability is defined as a share of such damage. The authors derive a formula for the optimal bank extended liability share, which depends on the model's parameters, and particularly upon the cost of public funds.

Hiriart and Martimort (2006) consider the optimal regulation of a risky project where a buyer (principal) has a contract with a seller (agent). The information on the level of safety care exerted by the agent is private to the agent (moral hazard). The authors derive conditions under which extending liability to the principal (buyer) improves social welfare. They show that if the principal has all the bargaining power, then extended liability favors the internalization of environmental damage and so improves welfare. However, when principals are competitive, extending liability has no value under complete contracting. But if the buyer-seller relation is plagued by adverse selection problems, then extending liability can again contribute to raising welfare.

13.4   The Model

We consider that safety activity levels and therefore environmental or industrial accidents are the result of interactions between, and choices made by four actors: the government, the firm, the financier, and the court. We model the role of each actor as follows.

The government decides on the liability system, that is the liability rules and the standards for safety. The liability system may take different forms combining strict and/or negligence based liability, joint and several and/or distributed liability between the different parties. The safety standards are also determined by the government and they stipulate the proper or socially desirable behavior recommended to or imposed on firms. As discussed above, the jurisprudence of extended lender liability and financial responsibility requirements in the US under CERCLA, in Canada and in Europe, can be considered as different forms of shared liability. Building on such jurisprudence, we model the government role as maximizing a social welfare function in setting a (strict) liability sharing formula for accident damage between the bank and the firm, as well as setting safety care standards for the firms. In exchange for its involvement in covering the cost of accidents as a deep pocket provider, the bank will be given a right to sue the firm for negligence in order to recuperate some or all its accident liability costs.

The firm decides on care activities, that is, the design and implementation of a safety program of self-protection or care activities, which can reduce the probability of an accident.[12] In most cases, the safety plan itself can be considered as observable by all parties but its implementation is much more difficult to observe and must therefore, at least to some extent, be considered as unobservable by external parties. We therefore consider a moral hazard context where the firm chooses the level of care, which is a private information of the firm. Under the government chosen strict liability sharing formula, the firm is strictly responsible for its share of the cost of an accident and for reimbursing the bank's share if it is found guilty of negligence.

The financial institutions (we will refer to those institutions as banks although in many cases they could be insurers or other financial syndicates) make possible the operations of the firm and, through their involvement in the firms' affairs, undertaken to protect their financial interest (loan, equity, and guarantees), can be considered as potentially efficient supervisors of the firm's safety care program. In the Fleet Factor case,[13] the judge wrote: "Although similar, the phrase 'participating in management' and the term 'operator' are not congruent. Under the standard we adopt today, a secured creditor may incur [...] liability without being an operator, by participating in the finan-

---

[12] We do not consider here self-insurance activities which could reduce the cost or severity of an accident if and when such an accident occurs.

[13] United States v. Fleet Factors Corp. (901 F. 2d 1550 (11th Cir. 1990), cert. Denied, 498 U.S. 1046 (1991)). For a discussion of that case, see Boyer and Laffont (1996).

cial management of a facility to a degree indicating a capacity to influence the corporation's treatment of hazardous wastes. It is not necessary for the secured creditor to actually involve itself in the day-to-day operations of the facility in order to be liable — although such conduct will certainly lead to the loss of the protection of the statutory [secured creditor] exemption. Nor is it necessary for the secured creditor to participate in management decisions relating to hazardous waste. Rather, a secured creditor will be liable, if its involvement with the management of the facility is sufficiently broad to support the inference that it could affect hazardous waste disposal decisions if it so chose. We therefore specifically reject the formulation of the secured creditor exemption suggested by the district court in *Mirabile*."

We model the role of the bank as financing the firm through a loan contract. Under the government chosen strict liability sharing formula, the firm is responsible for its share of the cost of an accident. However, to represent the bank monitoring capacity, we assume that the liability rule allows it to sue its client firm for negligence in order to recover its share of the accident cost, if it thinks that the firm has been careless in preventing accidents, that is, has been negligent by exerting a level of care lower than the level required by the government determined standards. If the bank decides to sue the firm following an accident, then the case is argued in court.

The court is responsible for deciding on liability or breach of contract in litigation cases. The court is therefore seen as in part responsible for implementing the government policy regarding the liability sharing formula and the safety care standards. In so doing, the court imperfectly assesses the level of care exerted by the firm and finds the firm guilty of insufficient care or negligence if this assessed level is lower than the level required by the safety care standards set by the government. If found guilty, the firm either reimburses the bank for the latter's share of the cost of the accident if it can do so or goes bankrupt if it cannot.[14]

The game unfolds as follows. The government first decides on the liability sharing rule and the level of safety care standard: in so doing, it makes the firm strictly responsible for a share of $\alpha\%$ of the cost of an industrial accident while it makes the bank strictly responsible for a share of $(1 - \alpha)\%$ of that cost, including both clean-up costs and compensation for victims. We will assume that the cost of an accident is known to be L. In spite of the fact that liability is strict in the sense that payments for damages are made in the above proportions to affected parties, the government also sets a minimum standard s for the level of safety a firm should undertake. The standard s

---

[14]There is a fifth group of stakeholders in the determination of the probability and severity of accidents, namely the customers of the firm. Insofar as these customers internalize or not in their buying behavior the industrial or environmental risk that the products they buy represent or cause, they in a sense contribute to the occurrence of such accidents. Boyer, Mahenc and Moreaux (2004) develop an analysis of the impact of the environmental awareness of buyers. We do not consider such an impact in this paper.

is legally enforceable by courts in the following sense. Given the occurrence of an accident generating damages L, the bank can sue the firm in court to recuperate its part $(1-\alpha)L$ of the cost of an accident, which, under the strict liability rule, it has already paid in lieu of clean-up costs or compensation to the victims.

Although the level of safety care exerted by the firm is a private information of the firm, the bank is allowed to sue the firm in case of an accident to recuperate its share of the cost. In order to induce the firm to exert a proper level of care, the bank chooses the probability ν with which it will sue the firm in case of accident and in so doing incurs a cost $C(\nu)$. This cost corresponds to maintaining within the bank a legal apparatus capable of launching court cases and arguing negligent behavior by client firms responsible for accidents; the expected number of such cases depends directly on the probability ν. Litigation costs supported by the firm are simply C_F and are assumed to be incurred only if the firm is sued. If the court finds the firm guilty of an insufficient level of safety care, the firm must compensate the bank for the latter's share of the cost of the accident if it can do it (if the firm is solvent) and goes bankrupt if it cannot, in which case the bank seizes the firm's assets but cannot recuperate the residual part of the cost it has incurred.

To avoid the daunting task of explicitly modeling the process by which the firm is found guilty or not, we adopt a simple reduced form description of the court's complex and imperfect process. We assume that given s and q there is a probability denoted by $P(q,s) \in [0,1]$ that the firm will be found guilty for violating the recommended standard. For example, the court cannot perfectly observe the level of care but rather can observe a signal $\tilde{q}$ of the firm's care level with $E(\tilde{q}) = q$ (imperfect but unbiased verifiability). We assume that the court's complex and imperfect process is such that the probability $P(q,s)$ is decreasing at a decreasing rate as the firm exert more safety care $(P_q < 0,\ P_{qq} > 0)$. We also assume that a tightening of the government determined standard s increases *ceteris paribus* the probability that the firm will be found guilty $(P_s > 0)$.

Hence, the firms are subject to both a strict liability rule for $\alpha\%$ of the cost of an accident and a liability for negligence rule. It is important to notice though that the liability for negligence, which may reach $(1 - \alpha)L$ for fully solvent firms and less for judgement proof firms, is incurred only if the bank decides to sue the firm and the court finds the firm guilty, which occurs under imperfect observation of the level of care exerted by the firm.[15]

---

[15]This modeling strategy differs from the more usual negligence rule under which non-negligent firms are exempt from liability. See for instance Innes (2004) for a model where the government chooses in an integrated way the liability design, enforcement, and regulatory policy when the probability of accident depends on care activities. He shows that the ex-ante regulation of care may be more efficient than ex-post liability even if monitoring care is relatively expensive. See Boyer, Lewis and Liu (2000) for a model where the cost of monitoring affects the setting of standards for care. See also Shavell (1993), Viscusi

Firms and banks negotiate a standard Townsend-Gale-Hellwig loan contract. We assume that the firm needs a loan of K to operate. The firm invests the amount borrowed K in a risky project. The project generates profits of π_1 with probability μ and π_2 with probability $1 - \mu$, with $\pi_2 > \pi_1$. An industrial/environmental accident with loss L occurs with probability $p(q) \in (0, 1)$ with $p(q)$ decreasing and convex ($p'(q) < 0$ and $p''(q) > 0$). The amount to be repaid by the firm to the bank is simply $(1 + r)K$ where r is the competitive interest rate for the risk class the firm and the industry belong to.[16]

We solve the above three stage game by backward induction as we are looking for subgame perfect equilibria. In stage 1, the government sets the values of α and s to maximize social welfare. In stage 2, the firm and the bank negotiate a loan contract (r, K) and the bank announces its choice of ν, the probability with which it will sue the firm in case of an accident. This probability is set before observing whether an accident occurs or not, and the bank is committed to apply this probability of suing the firm. The firm, knowing ν, chooses the safety care level q at cost $Q(q)$ (we assume $Q'(q) > 0$ and $Q''(q) > 0$), implying a probability of accident of $p(q)$. The total amount to be paid by the firm to the bank will therefore depend on both ν and q, as well as on α and s. In stage 3, all actors observe whether an accident occurs or not. If an accident occurs, the bank sues the firm with probability ν; if the bank sues the firm, the court is called into play and finds the firm guilty of negligence with probability $P(q, s)$, in which case the firm makes the additional payment $(1 - \alpha)L$ if it can do so and otherwise go bankrupt while its assets, valued for the bank at $\max\{0,\ \pi_1 - \alpha L - (1 + r)K - Q(q) - C_F\}$, are seized by the bank.

The third stage
In the third stage, the court is called into action if an accident occurred and the bank sued the firm for breach of contract, that is, for (alleged) insufficient care. If a suit is brought against the firm, the court finds the firm liable with probability $P(q, s)$. Therefore, in stage 3, the bank realizes an expected net cash flow $E\Pi_B$ which depends on whether an accident occurred or not, on whether a suit is brought or not, on the findings of the court, and on the profits realized by the firm which determines the firm's capacity to avoid bankruptcy. Given our assumption of a standard Townsend-Gale-Hellwig loan contract, those profits are strictly observed by the bank and the court only if, following the court's decision, the firm is not fully reimbursing the bank as instructed by the court, a situation which will indeed occur only if profits are low.

(1989), and Boyer and Porrini (2001) for comparisons between ex-ante regulatory policies and ex-post liability schemes.

[16]Typically, banks do not charge firm specific interest rates but rather interest rates based on defined risk classes. Once the risk class of the firm is determined and the rate of interest fixed, the bank nevertheless will typically monitor carefully its financial interest in the firm.

The total expected profit of the bank $E\Pi_B$ is therefore:

$$E\Pi_B(\nu, q, \alpha, s; \; K, r, \pi_1, \pi_2, \mu) = (1 + r)K - C(\nu)$$

$$-p(q)[(1 - \nu) + \nu(1 - P(q, s))](1 - \alpha)L$$

$$-p(q)\nu P(q, s)(1 - \mu)[0]$$

$$-p(q)\nu P(q, s)\mu\Big[(1 - \alpha)L - \max\{0, \; \pi_1 - \alpha L - (1 + r)K - Q(q) - C_F\}\Big].$$

$$(13.1)$$

By assumption, the bank always receives a repayment of $(1 + r)K$ and the cost of the bank's strategy (the probability ν with which it sues the firm) is borne whether an accident occurs or not. In the accident state, occurring with probability $p(q)$, the bank suffers a loss which may take three different values: (i) a loss of $(1 - \alpha)L$ if it does not sue the firm for negligence, or if it sues the firm but the firm is found not guilty by the court, (ii) a loss of 0 if it sues the firm and the firm, when found guilty of negligence by the court, can reimburse the bank for the latter's share $(1 - \alpha)L$ of the accident cost, or (iii) a loss of $[(1 - \alpha)L - \max\{0, \; \pi_1 - \alpha L - (1 + r)K - Q(q) - C_F\}]$ if it sues the firm and the firm is found guilty of negligence by the court while profits are too low to allow the firm to fulfill its obligations, in which case the firm goes bankrupt and the bank gets only partial reimbursement of its share of the accident cost. The third term above indicates that the firm's expenses $(1 + r)K - Q(q) - C_F$ as well as the liability expenses αL are covered in priority before the bank can get some of its liability expenses reimbursed. We further assume (see below) that when the firm goes bankrupt, the (operating) expenses $(1 + r)K - Q(q) - C_F$ are covered before the firm's liability expenses αL. We will concentrate our analysis to the more interesting case where, if the firm is sued and found guilty of negligence, situation (ii) would occur when profits are high ($\pi_t = \pi_2$) while situation (iii) would occur when profits are low ($\pi_t = \pi_1$).

Similarly, the total expected profit of the firm $E\Pi_F$ can be written as follows where $E\pi = \mu\pi_1 + (1 - \mu)\pi_2$:

$$E\Pi_F(q, \nu, \alpha, s; \; K, r, \pi_1, \pi_2, \mu) = E\pi - Q(q) - (1 + r)K$$

$$-p(q)(1 - \nu)\Big[(1 - \mu)\alpha L + \mu\min\{\alpha L, \; \pi_1 - Q(q) - (1 + r)K\}\Big]$$

$$-p(q)\nu(1 - P(q, s))\Big[(1 - \mu)(\alpha L + C_F)$$

$$+\mu \min\{\alpha L + C_F, \ \pi_1 - Q(q) - C_F - (1+r)K\}\Big]$$

$$-p(q)\nu P(q,s)\Big[(1-\mu)[L + C_F] + \mu \max\{0, \ \pi_1 - Q(q) - C_F - (1+r)K\}\Big].$$
$$(13.2)$$

The firm's extra payment in case of an accident can take three different forms: (i) if it is not sued for negligence by the bank, the firm pays its liability expenses αL if it can do so, that is if it can avoid bankruptcy (recall that bankruptcy can occur if profit is low, in which case the firm, being protected by limited liability, will not pay more than $\min\{\alpha L, \ \pi_1 - Q(q) - (1+r)K\}$); (ii) if sued by the bank but found not guilty by the court, the firm is in a situation similar to the previous one except for the additional litigation cost C_F; (iii) if sued and found guilty, the firm reimburses the bank if it can (when profit is π_2), in which case it covers the full cost of the accident, or goes bankrupt and pays at most its assets value given by $\max\{0, \ \pi_1 - Q(q) - C_F - (1+r)K\}$ to cover its liability expense L.

No decision is made in stage 3. It is just recording the unfolding of the game. If no accident occurs, the firm repays $(1+r)K$. If an accident occurs but the bank does not sue the firm, then the latter repays $(1+r)K$ while the bank and the firm support their respective share of the cost of the accident. If an accident occurs and the bank sues the firm but the court finds the firm not guilty, then again the latter repays $(1+r)K$ while the bank and the firm support their respective share of the cost of the accident and the firm supports its litigation costs. Finally, if an accident occurs, the bank sues the firm, and the court finds the firm guilty of negligence, then the latter repays the loan $(1+r)K$ and covers its share of the accident costs αL plus the bank's share of the accident costs $(1-\alpha)L$ if profits are π_2, but repays only its net profits or assets value and goes bankrupt otherwise.

The second stage
In the second stage, the bank and the firm, observing the government determined values of α and s, choose respectively ν and q with the bank acting as first mover.[17] Let us first consider the choice of safety care q made by the firm as a function of the predetermined and observed values of α, s and ν. The firm chooses the level of q which maximizes its expected profit (13.2). Safety care activities affects the firm's expected profits through three channels: first, through the probability $p(q)$ of an accident; second, through the probability $P(q,s)$ of being found guilty of insufficient safety care or negligence if an accident occurs and the bank sues the firm; and third, through the cost $Q(q)$ the firm incurs for exerting safety care level q.

---

[17]This is the natural assumption to make here. Indeed, banks typically determine their suing policy well in advance of determining a given firm risk class and therefore the terms of its loan. This policy must be assumed to be well known by firms.

The firm chooses q satisfying

$$\frac{\partial E\Pi_F}{\partial q} = 0, \tag{13.3}$$

giving rise to the best reply function $q(\nu \mid \alpha, s; \ K, r, \pi_1, \pi_2, \mu, L, C_F, C_B)$ to the choice of ν made by the bank and of α and s set by the government.

The expected return of the bank on its loan to the firm depends on the probability that it will sue the firm, on the probability that the firm will be found guilty if sued, and on the probability that the firm makes high profits. Since profits will not be observed unless the firm is not meeting its obligations to the bank, the probability that the bank will sue the firm in the accident state will not depend on the level of those profits. However, the expected benefits of suing depend on the expected capacity of the firm to avoid bankruptcy if found guilty by the court, that is, on the probability that realized profits are high. Hence, the bank chooses ν satisfying

$$\frac{dE\Pi_B}{d\nu} = 0, \tag{13.4}$$

considering the best reply care level function $q(\nu \mid \alpha, s; \ K, r, \pi_1, \pi_2, \mu, L, C_F, C_B)$ of the firm. The solution to conditions (13.4) and (13.3) give us the second stage *equilibrium values* of variables ν and q as functions of the government-determined variables α and s, namely: $\nu^*(\alpha, s; \ K, r, \pi_1, \pi_2, \mu, L, C_F, C_B)$ and $q^*(\alpha, s; \ K, r, \pi_1, \pi_2, \mu, \ L, C_F, C_B)$.

The first stage
We are now ready to define the government objective function (social welfare function) and characterize the optimal liability sharing value α^* and the optimal safety standard s^*. We will assume that the determination of the liability sharing formula involves a "political economy" cost $A(\alpha)$ if the government wants to implement a formula which moves away from the most acceptable formula from a social or political standpoint (assumed below to correspond to an equal liability sharing). The political correctness of the sharing formula is a social constraint that is quite realistic in the present context. The cost itself corresponds to the efforts necessary to convince the population of firms and banks of the desirability of the intended liability sharing formula.

The social welfare function SWF is composed of the total gross benefits W from allowing the firm to operate minus the expected cost of an accident, the cost of the firm's safety effort q, the cost of the bank's strategy ν, the political cost of diverging from the most socially acceptable liability sharing, the firm's expected costs of litigation, and the cost of public funds necessary to cover that part of the cost of an accident which is covered neither by the firm nor by the bank. This last cost is proportional to the amount disbursed by the government in case of an accident under the assumption that the government will one way or another cover the cost of an accident (clean up costs and compensation of victims) not covered by the total payments made by the firm and the bank. This cost of public funds therefore depends on whether the bank sues the firm or not and in the former case on whether the firm is found guilty of negligence or not, and on whether the firm goes bankrupt or not following the occurrence of an accident and the decision of the court. The $SWF(\alpha, s)$ is given by:

$$SWF(\alpha, s) = W - p(q^*)L - Q(q^*) - C(\nu^*) - A(\alpha)$$

$$-\lambda p(q^*)\Big[(1 - \nu^*)\mu \max\{0, \ \alpha L - (\pi_1 - (1+r)K - Q(q^*))\}$$

$$+\nu^*(1 - P(q^*, s)) \max\{0, \ \alpha L - (\pi_1 - (1+r)K - C_F - Q(q^*))\}\Big]$$

$$-\lambda p(q^*)\nu^* P(q^*, s)\mu \max\Big\{0, \ \alpha L - \max\{0, \ \pi_1 - (1+r)K - C_F - Q(q^*)\}\Big\}$$

$$(13.5)$$

where $\nu^* = \nu^*(\alpha, s; \ K, r, \pi_1, \pi_2, \mu, L, C_F, C_B)$ and $q^* = q^*(\alpha, s; \ K, r, \pi_1, \pi_2, \mu, L, C_F, C_B)$.

The government chooses the liability sharing factor α and the standard of care s to maximize the social welfare function (13.5):

$$\frac{\partial SWF(\alpha, s)}{\partial \alpha} = 0$$

$$(13.6)$$

$$\frac{\partial SWF(\alpha, s)}{\partial s} = 0$$

The first-best solution
The above modelization clearly includes many institutional and informational constraints. We want to compare the solution obtained in the different cases (parameter values) not only between themselves but also with the first-best solution. The first-best solution is obtained under full information when the government chooses directly the level of safety effort q and the value of α which maximizes social welfare. From the above, the first best least cost value of α, namely α_{FB}, is the one minimizing $A(\alpha)$. As for the optimal value of q, it corresponds simply to the value which minimizes the expected cost of an accident plus the cost of safety itself, that is the value q_{FB} satisfies $p'(q_{FB})L + Q'(q_{FB}) = 0$. The cost of public funds λ plays no role here since under full information, one can assume that $\lambda = 0$.

13.5   A Simplified Realistic Case

Characterizing the best government strategy under the above institutional and informational constraints is a daunting, possibly impossible endeavor. Rather than pursue such a goal here, we will rather tackle a more limited but more tractable task, namely obtaining numerical best values of α in a simplified but realistic case. Clearly, this is a first step in a long journey. But we like to believe that it is a useful step if one aims at characterizing the complex set of forces which govern the intricate relationships underlying the determination of industrial/environmental accidents as seen for a social point of view.

We first present the numerical function we will use in the simulations and then show their properties in an attempt to convince the reader that the functions are sufficiently realistic and flexible to account for many significant real situations. Let us consider the following functions:

$$p(q) \; \equiv \; p_0 + (p_M - p_0)(1 - e^{-\eta q}) \begin{cases} = p_0, & \text{if } q = 0 \\ \\ \to p_M, & \text{as } q \text{ becomes very large} \end{cases} \tag{13.7}$$

$$P(q,s) \; \equiv \; e^{-\delta(q/s)} \begin{cases} = 1, & \text{if } q = 0 \\ \\ = e^{-\delta}, & \text{if } q = s (P(s,s) = 0.5 \text{ if } \delta = ln2) \\ \\ \to 0, & \text{as } q \text{ becomes very large} \end{cases} \tag{13.8}$$

$$Q(q) \; \equiv \; zq^b \text{ where } b > 1 \text{ and } z \text{ is some positive parameter.} \tag{13.9}$$

$$C(\nu) \; \equiv \; B\nu^n. \tag{13.10}$$

$$A(\alpha) \; \equiv \; A(\alpha - 0,5)^a \tag{13.11}$$

The probability $p(q)$ of an accident is assumed to be between p_M and p_0, where p_M is the minimum unavoidable probability level representing the effect of all causal factors still at work when care is as large as it can be, and p_0 is the probability of an accident when the firm exerts no care ($p_0 > p_M$) with $p'(q) = \eta(p_M - p_0)e^{-\eta q} < 0$ and $p''(q) = -\eta^2(p_M - p_0)e^{-\eta q} > 0$. Parameter η measures the efficiency of care in reducing the probability of an accident.

The probability $P(q,s)$ that the court will find the firm guilty of insufficient care satisfies the following desirable and/or realistic properties, namely:
(i) it decreases at a decreasing rate with care q (from 1, when the firm exert no care, toward 0, when the level of care becomes very large), so that both errors of type 1 (finding a firm with $q > s$ guilty of insufficient care) and errors of type 2 (finding a firm with $q < s$ not guilty) always remain possible, with

$$P(s,s) \; = \; e^{-\delta},$$

$$P_q(q,s) \; = \; -(\delta/s)e^{-\delta(q/s)} < 0,$$

$$P_{qq}(q,s) \; = \; (\delta/s)^2 e^{-\delta(q/s)} > 0;$$

(*ii*) it increases with the safety standard s with

$$P_s(q,s) = (\delta q/s^2)e^{-\delta(q/s)} > 0,$$

$$P_{ss}(q,s) = [(\delta q/s^2)^2 - (2s\delta q)/s^4]e^{-\delta(q/s)},$$

the latter being positive if $s < \frac{1}{2}\delta q$ (that is, the increase in the standard increases the probability of conviction, given the level of care q, at an increasing rate if s is relatively small) and negative if $s > \frac{1}{2}\delta q$ (that is, the increase in the standard increases the probability of conviction, given the level of care q, at a decreasing rate if s is relatively large);

(*iii*) when $q = s$, it is constant at $e^{-\delta}$ for all s (it is equal to 0.5 if $\delta = ln2$);

(*iv*) moreover, it satisfies

$$P_{qs}(q,s) = [(\delta/s^2)(1 - \delta q/s)]e^{-\delta(q/s)},$$

which is positive if $\delta q < s$ (that is, the increase in the standard, given the level of care q, decreases the marginal efficiency of care in reducing the probability of conviction if s is relatively large) and negative if $\delta q > s$ (that is, the increase in the standard, given the level of care q, increases the marginal efficiency of care in reducing the probability of conviction if s is relatively small). Hence, for $\delta = ln2$, an increase in the standard s increases the efficiency of safety care in reducing the probability of conviction iff the level of safety care q is above $(ln2)s \approx 0.6931s$, in which case an increase in the government determined standard would induce the firm to increase *ceteris paribus* its level of safety care. If the level of safety care q is below that critical level, an increase in the government determined standard would induce the firm to *decrease* its safety care level, the reason being that it becomes more costly at the margin to avoid being convicted of negligence.

The cost of care function $Q(q)$ is increasing in q at an increasing [decreasing] rate if $b > [<]1$. So, parameter b measures the economies/diseconomies of scale which may be present in care programs and activities. As to parameter z, it simply measures the linear cost efficiency of care.

The bank's litigation service cost function $C(\nu)$ increases in the probability that the bank will indeed sue a client firm responsible for an accident. To be able to do so efficiently, we assume that the bank has at all times an internal legal service whose size depends on ν and whose cost increases with ν at an increasing [decreasing] rate if $n > [<]1$. Hence, parameter n measures the economies/diseconomies of scale which may be present in the bank's litigation capacity. Parameter B simply measures the linear cost efficiency of litigation capacity.

Finally, the political economy cost function of a given value of liability sharing is assumed to be minimized when liability is shared half and half between the parties, namely the bank and the firm here. Clearly, other "politically correct liability sharing" reference points could be envisaged.

The socially optimal liability sharing value α^* and the optimal safety standard s^* can then be related to the parameters. We can run some comparative statics and illustrate the results in tables. That is what we do next. Among all possibilities, we will concentrate on the impact of the profitability of the firm (represented by parameter μ, the probability that the project/firm generate a low level of profit or benefit), the cost of care activities (represented by parameter z appearing in $Q(q)$, the cost of care function), the efficiency of care in reducing the probability of accident (represented by parameter η appearing in $p(q)$, the probability of accident), the bank's cost of monitoring the firm's care activities (measured indirectly here by parameter B appearing in the cost $C(\nu)$ incurred by the bank for maintaining a legal service, that is, a court litigation capacity as well as the expected costs of litigation and court procedures themselves), and finally the social cost of public funds (represented by parameter λ).

We first illustrate a base case scenario (Case 1) in Table 13.1.

Case 1. Parameters: $\pi_1 = 1000$, $\pi_2 = 5000$, $\mu = 0.2$, $K = 75$, $r = 0.1$, $p_0 = 0.4$, $p_M = 0.05$, $\delta = ln(2)$, $z = 10$, $b = 1.2$, $L = 4000$, $\eta = 0.2$, $C_F = 0$, $B = 1$, $n = 2$, $\lambda = 0.3$, $A = 25$, $a = 2$; four values of s, namely 6, 10, 18, 28. We obtain the following: $\alpha_{FB} = 0.5$, $q_{FB} = 13.17$, $p(q_{FB}) = 0.0751$, and the values given in Table 13.1.

This base case scenario shows among other things that if the liability share of the firm increases, then as expected the bank chooses a lower probability of suing ν, while the firm chooses a higher level of safety care, implying a lower probability of accident and a lower probability, for a given level of recommended care s, of being found guilty by the court in case the firm is taken to court by the bank. This translates into a level of social welfare which first increases and then decreases, under the combined opposite effects of a lower probability of accident and a higher cost of care activities. Table 1 shows also the effect on the equilibrium values of variables α, ν and q of different levels of care recommended by the government. The maximal social welfare is obtained when the government chooses $s = 18$ and $\alpha = 35\%$, thereby inducing the bank to choose a relatively high probability of litigation of 91%; those values of α, s and ν induce the firm to exert a relatively low level of care at 11.987 which implies a probability of accident of 8.2% and a probability of conviction for negligence of 63%. It is interesting to note here that the levels α and s are both lower than their first best values, which in this base case scenario are respectively 50% and 13.17 implying a first best probability of accident of 7.5%. For low [high] values of s, the government sets the firm's liability share above [below] its full information first best level.

Case 2. Parameters: same as Case 1 except for μ which varies between 0.1 and 0.3 and s which is fixed at $s = 10$; we obtain the following: $\alpha_{FB} = 0.5$, $q_{FB} = 13.17$, $p(q_{FB}) = 0.0751$, and the values given in Table 13.2.

We look in this case at the effect of a lower profitability of the firm on the social welfare maximizing liability sharing (considering the level of recommended care fixed exogenously at $s = 10$) and equilibrium levels of litigation probability ν and care q. As the profitability of the firm/project decreases (as μ increases), liability is transferred from the firm to the bank, inducing the latter to increase the litigation probability. The net effect of these conflicting forces on the firm's incentives for care is to lower the level of care, thereby increasing the probability of accident and the probability of conviction for negligence. The latter probability remains lower than 50% though since the level of care chosen by the firm remains higher than the recommended level $s = 10$. The incomplete information and partial control social welfare maximizing liability sharing between the firm and the bank falls mainly on the firm when the firm/project is very profitable (μ small) but falls mainly on the bank when it is not (μ large).

Case 3. Parameters: same as Case 1 except for z which varies between 5 and 20 and s which is fixed at $s = 10$; we obtain the following: $\alpha_{FB} = 0.5$, q_{FB} variable, $p(q_{FB})$ variable, and the value given in Table 13.3.

We consider in case 3 the effect of increasing parameter z which translates into higher cost of care activities. Although the first best level of liability sharing remains at 50%, the first best level of care decreases with z and therefore the probability of accident increases with z. The liability share of the bank increases with z, inducing it to increase the probability of litigation. The net effect of these three factors (higher z, lower α, higher ν) is to induce the firm to reduce its care level, thereby increasing the probability of accident and the probability of conviction for negligence which goes from under to over 50% as the level of care goes from higher to lower than the recommended level $s = 10$.

Case 4. Parameters: same as Case 1 except for η which varies between 0.1 and 0.3; we obtain the following: $\alpha_{FB} = 0.5$, q_{FB} variable, $p(q_{FB})$ variable, and the values given in Table 13.4.

We report in table 4 the results of a greater efficiency of care in reducing the probability of accident (higher η). The first best level of care, therefore the probability of accident, is sensitive to increases in η as the higher efficiency of care could allow for a reduction in care and therefore the cost of care, with no cost in terms of probability of accident. We observe that the first best level of care indeed decreases with η. The combined effects of the increase in η and the reduction in first best care level q_{FB} is to decrease the first best probability of accident. As the efficiency of care increases, the incomplete information and partial control social welfare maximizing liability share of the firm decreases while the probability of litigation first increases and then decreases. The net effects on the equilibrium level of care is to reduce it while reducing also the probability of accident, which nevertheless remains higher than the decreasing the first best levels. As the level of care falls due to the increase in the efficiency of care, the probability of conviction in litigated cases increases.

Case 5. Parameters: same as Case 1 except for B which varies between 0.5 and 2.5; we obtain the following: $\alpha_{FB} = 0.5$, $q_{FB} = 13.17$, $p(q_{FB}) = 0.0751$, and the values given in Table 13.5.

We illustrate in this case the effects of increases in the cost of the bank's litigation capacity (including court costs). As the cost parameter B increases, we observe that the incomplete information and partial control social welfare maximizing liability share of the firm increases while the probability of litigation by the bank drops, inducing a non-monotonic effect on the level of care exerted by the firm, as well as on the probability of accident, over the 0.5 to 1.5 range of B values. The non-monotonicity in the social welfare function implies that there is a discontinuity in the optimal liability share of the firm which jumps suddenly to 1 somewhere between $B = 2$ and $B = 2.5$.

Case 6. Parameters: same as Case 1 except for λ which varies between 0.1 and 0.5; we obtain the following: $\alpha_{FB} = 0.5$, $q_{FB} = 13.17$, $p(q_{FB}) = 0.0751$, and the values given in Table 13.6.

Finally, we show in table 6 the results of increases in λ, the parameter representing the social cost of public funds. In the full information first best solution, the cost of public funds is irrelevant since it can be assumed to be 0. Hence the first best level of q is not sensitive to changes in λ. But under the informational and regulatory (control) constraints considered here, the efficiency of the taxation system is relevant insofar as the government may be called to cover part of both the clean up costs and the compensation of victims if an accident occurs. As Table 6 illustrates, increases in the social opportunity cost of public funds parameter λ reduces the welfare maximizing value of the liability share of the firm, prompting the bank to increases its probability of litigation in case of accident. The lower liability share of the firm combined with the higher probability of litigation induces the firm to reduce its care activities while maintaining them at a level above the exogenously fixed recommended level $s = 10$ in order to avoid too large an increase in the probability of conviction for negligence. Nevertheless, the probability of accident increases as well as the probability of conviction. A higher λ induces the government to increase the liability share of the bank in order to avoid having to cover, at a higher social cost, the accident damages that the judgment-proof firm may impose on the government.

13.6 Conclusion

The importance of the debate about the choice of instruments in the environmental realm stems from the accelerating diffusion of environmentally risky production activities in modern industrial societies and the ensuing necessity to properly compensate the victims of accidents. To those elements, must be added the need to induce an efficient level of care by the potential injurers in contexts characterized by asymmetric information and partial control. Finally, and perhaps equally important, these issues and their analysis are important, in particular in the light of the challenging attempt by EC countries to design and implement a common environmental protection system.

The game being played by the four actors, namely governments, firms, banks, and courts, involves a set of complex interactions between them in the determination of industrial accidents. Although the modeling strategy we followed here is admittedly not the only possible way to make explicit the interactions between the four players, we believe that our modeling strategy captures many if not all the important characteristics, stemming from institutional and informational constraints, of a large number of real situations. The level of complexity in the interactions between the four players is significant, in particular in terms of the firm's limited liability, the asymmetric information on safety care levels chosen by firms and the profits made by firms, and the difficulty for courts to ascertain through the litigation process the level of care a firm has effectively exerted.

Nevertheless, we were able, through realistic numerical cases, to derive statements regarding the socially efficient government policies for industrial safety and the way those policies interact with the private interests of banks and firms as well as the imperfection of the court system. We characterized in particular the socially efficient liability sharing formula and standard of safety care, based on the interactions between the banks and the firms and on the efficiency of the courts in assessing the level of care exerted by the firm.

Our main results is to characterize, in a series of comparative statics exercises on a simplified but realistic functional example, the determination of the liability sharing between the firm and the bank. We showed that the incomplete information and partial control social welfare maximizing liability or financial responsibility share of the bank *increases* as the recommended level of care s increases, as the profitability of the firm decreases, as the cost of care increases, as the efficiency of care increases, and as the social cost of public funds increases.

Hence, our results indicate or suggest that the government would be justified to implement an extended lender liability regime,[18] or more generally a financial responsibility regime, or more specifically a reduced and more restrictive firm liability in favor of an increased and more extensive lender liability for environmental accidents. The higher is the recommended level of care, the more stringent and aggressive the courts are instructed to be with regards to firms' conviction for negligence, the less profitable is the firm/project (insofar as the firm is allowed to operate, that is, is globally profitable), the higher the cost of care, the higher the efficiency of care in reducing the probability of industrial/environmental accidents, the smaller the cost for the bank of maintaining a litigation capacity (including the court costs), and the higher the social cost of public funds.

References

Balkenborg, D., 'How Liable Should a Lender Be? The Case of Judgment-Proof Firms and Environmental Risk: Comment', *American Economic Review*, 91(3) (2001): 731-738.

Balkenborg, D., 'On extended Liability in a Model of Adverse Selection', (in this book), (2004).

Beard, R., 'Bankruptcy and Care Choice', *RAND Journal of Economics*, 21(4) (1990): 627-634.

Boyd, J. and Ingberman, D., 'The Search of Deep Pocket: Is 'Extended Liability' Expensive Liability?', *Journal of Law, Economics, and Organization*, 13(1) (1997): 233-258.

Boyd, J., 'Financial Responsibility for Environmental Obligations: An Analysis of Environmental Bonding and Assurance Rules', in T. Swanson, *Law and Economics of Environmental Policy*, Research in Law and Economics Series, (Elsevier, 2002).

Boyer, M. and Laffont, J.J., 'Environmental Protection, Producer Insolvency and Lender Liability', in A. Xepapadeas (ed.), *Economic Policy for the Environment and Natural Resources*, (Edward Elgar Pub. Ltd, 1996).

Boyer, M. and Laffont, J.J., 'Environmental Risk and Bank Liability', *European Economic Review*, 41 (1997): 1427-1459.

Boyer, M., Lewis, T.L. and Liu, W.L., 'Setting Standards for Credible Compliance and Law Enforcement', *Canadian Journal of Economics*, 33(2000): 319-340.

Boyer, M., Mahenc, P. and Moreaux, M., 'Environmental Protection, Consumer Awareness, Product Characteristics, and Market Power', (in this book), (2004).

---

[18]See Boyer and Porrini (2004) for an analysis of factors favoring an extended lender liability regime over a regulatory regime subject to capture by the regulatees.

Boyer, M. and Porrini, D., 'Law versus Regulation: A Political Economy Model of Instruments Choice in Environmental Policy', in A. Heyes (ed.), *Law and Economics of the Environment*, (Edward Elgar Publishing Ltd., 2001).

Boyer, M. and Porrini, D., 'The Choice of Instruments for Environmental Policy: Liability or Regulation?', in T. Swanson and R. Zerbe (eds.), *An Introduction to the Law and Economics of Environmental Policy: Issues in Institutional Design*, Research in Law and Economics Series, (Elsevier, 2002).

Boyer, M. and Porrini, D., 'Modelling the Choice Between Regulation and Liability in Terms of Social Welfare', *Canadian Journal of Economics*, 37(3) (2004): 590-612.

Dari Mattiacci, G. and Parisi, F., 'The Cost of Delegated Control: Vicarious Liability, Secondary Liability and Mandatory Insurance', *International Review of Law and Economics*, 23 (2004): 453-475.

Daughety, A.F. and Reinganum, J.F., 'Markets, Torts and Social Inefficiency', Vanderbilt University (2003).

Gobert, K. and Poitevin, M., 'Environmental Risks: Should Banks be Liable?', (in this book) (2004).

Faure, M., 'The White Paper on Environmental Liability: Efficiency and Insurability Analysis', *Environmental Liability*, 4 (2001): 188-201.

Faure, M. and Grimeaud, D., 'Financial Assurance Issues of Environmental Liability', Report for the European Commission (2000).

Feess, E. and Hege, U., 'Environmental Harm and Financial Responsibility', *Geneva Papers of Risk and Insurance: Issue and Practice*, 25 (2000): 220-234.

Feess, E. and Hege, U., 'Safety Monitoring, Capital Structure, and 'Financial Responsibility'', *International Review of Law and Economics*, 23(3) (2003): 323-339.

Heyes, A., 'Lender Penalty for Environmental Damage and the Equilibrium Cost of Capital', *Economica*, 63 (1996): 311-323.

Hiriart, Y. and Martimort, D., 'Environmental Risk Regulation and Liability under Adverse Selection and Moral Hazard', (in this book) (2006).

Hutchison, E. and van't Veld, K., 'Extended Liability for Environmental Accidents: What You See Is What You Get', *Journal of Environmental Economics and Management*, 49 (2005): 157-173.

Innes, R., 'Enforcement costs, optimal sanctions, and the choice between ex-post liability and ex-ante regulation', *International Review of Law and Economics*, 24 (2004): 29-48.

Kornhauser, L.A., 'An Economic Analysis of the Choice Between Enterprise and Personal Liability for Accidents', *California Law Review*, 70 (1982): 1345-92.

Lewis, T. and Sappington, D.M., 'How Liable Should a Lender Be? The Case of Judgment-Proof Firms and Environmental Risk: Comment', *American Economic Review*, 91(3) (2001): 724-730.

Newman, H.A. and Wright, D.W., 'Strict Liability in Principal-Agent Model', *International Review of Law and Economics*, 10 (1990): 219-231.

Pitchford, R., 'How Liable Should a Lender Be?', *American Economic Review*, 85(5) (1995): 1171-1186.

Pitchford, R., 'How Liable Should a Lender Be? The Case of Judgment-Proof Firms and Environmental Risk: Reply', *American Economic Review*, 91(3) (2001): 739-45.

Polinsky, M.A., 'Principal-Agent Liability', Stanford University Law School, *Working Paper* 258 (2003).

Segerson, K. and Tietenberg, T., 'The Structure of Penalties in Environmental Enforcement: an Economic Analysis', *Journal of Environmental Economics and Management*, 23 (1992): 179-200.

Shavell, S., 'The Judgement Proof Problem', *International Review of Law and Economics*, 6 (1986): 45-58.

Shavell, S., 'The Optimal Structure of Law Enforcement', *Journal of Law and Economics*, 36(1) (1993): 255-287.

Shavell, S., *Foundations of Economic Analysis of Law*, (Harvard University Press, Cambridge, MA, 2004).

Sykes, A.O., 'The Economics of Vicarious Liability', *The Yale Law Journal*, 93 (1984): 1231-80.

Viscusi, W.K., 'Toward a Diminished Role fro Tort Liability: Social Insurance, Government Regulation, and Contemporary Risks to Health and Safety', *Yale Journal of Regulation*, 6 (1989): 65-107.

Woodfield, A.E., 'When Should the Bell Toll? The Economics of New Zealand's Debate on Indirect Liability for Internet Copyright Infringement', *Review of Economic Research on Copyright Issues*, 1 (2004), 119-149.

Table 13.1: Base case scenario
[$\alpha_{FB} = 0.5$, $q_{FB} = 13.17$, $p(q_{FB}) = 0.0751$]

α	ν	q	SWF	$p(q)$	$P(q,s)$
			$s = 6$		
0.05	0.918	10.343	3558.29	0.094	0.303
0.25	0.898	10.830	3570.93	0.090	0.286
0.45	0.869	11.215	3575.20	0.087	0.274
0.65	*0.820*	*11.611*	*3576.53*	*0.084*	*0.261*
0.75	0.776	11.814	3576.28	0.083	0.255
0.95	0.445	12.251	3574.53	0.080	0.243
			$s = 10$		
0.05	0.926	11.187	3567.00	0.087	0.461
0.25	0.908	11.522	3576.91	0.085	0.450
0.45	*0.887*	*11.723*	*3578.30*	*0.084*	*0.444*
0.65	0.852	11.933	3577.48	0.082	0.437
0.75	0.821	12.043	3576.45	0.081	0.434
0.95	0.577	12.296	3573.94	0.080	0.426
			$**s = 18**$		
0.05	0.929	11.809	3572.97	0.083	0.635
0.25	0.917	11.943	3578.52	0.082	0.631
0.35	*0.910*	*11.987*	*3579.03*	*0.082*	*0.630*
0.55	0.889	12.081	3578.27	0.081	0.628
0.75	0.847	12.187	3575.98	0.081	0.625
0.95	0.648	12.327	3573.31	0.080	0.622
			$s = 28$		
0.05	0.932	12.001	3573.43	0.082	0.743
0.25	0.922	12.084	3578.44	0.081	0.741
0.35	*0.915*	*12.110*	*3578.81*	*0.081*	*0.741*
0.55	0.896	12.167	3577.90	0.081	0.740
0.75	0.858	12.236	3575.53	0.080	0.739
0.95	0.676	12.338	3572.94	0.797	0.737

Table 13.2: Variable μ
[$\alpha_{FB} = 0.5$, $q_{FB} = 13.17$, $p(q_{FB}) = 0.0751$]

μ	α	ν	q	SWF	$p(q)$	$P(q,s)$
0.10	0.65	0.855	12.310	3983.42	0.080	0.426
0.15	0.55	0.873	12.016	3580.83	0.082	0.435
0.20	0.45	0.887	11.723	3578.30	0.084	0.444
0.25	0.35	0.898	11.432	3375.84	0.086	0.453
0.30	0.30	0.903	11.189	3173.48	0.087	0.460

Table 13.3: Variable z

z $(\alpha_{FB}, q_{FB}, p(q_{FB}))$	α	ν	q	SWF	$p(q)$	$P(q,s)$
5.0 (0.5, 16.417, 0.063)	0.45	0.860	14.885	3707.87	0.068	0.356
7.5 (0.5, 14.513, 0.069)	0.45	0.876	13.008	3638.74	0.076	0.406
10.0 (0.5, 13.171, 0.075)	0.45	0.887	11.723	3578.30	0.084	0.444
12.5 (0.5, 12.137, 0.081)	0.45	0.895	10.752	3524.11	0.091	0.475
15.0 (0.5, 11.298, 0.087)	0.40	0.907	9.934	3474.78	0.098	0.502
17.5 (0.5, 10.591, 0.092)	0.40	0.912	9.292	3429.48	0.105	0.525
20.0 (0.5, 9.983, 0.098)	0.35	0.920	8.714	3387.53	0.111	0.547

Table 13.4: Variable η

η $(\alpha_{FB}, q_{FB}, p(q_{FB}))$	α	ν	q	SWF	$p(q)$	$P(q,s)$
0.10 (0.5, 18.709, 0.104)	0.65	0.865	16.056	3737.42	0.120	0.329
0.15 (0.5, 15.433, 0.085)	0.55	0.877	13.524	3490.49	0.096	0.392
0.20 (0.5, 13.171, 0.075)	0.45	0.887	11.723	3578.30	0.084	0.444
0.25 (0.5, 11.536, 0.070)	0.40	0.890	10.439	3635.30	0.076	0.485
0.30 (0.5, 10.297, 0.066)	0.40	0.888	9.456	3675.29	0.070	0.519

Table 13.5: Variable B

$[\ \alpha_{FB} = 0.5,\ q_{FB} = 13.17,\ p(q_{FB}) = 0.0751\ ]$

B	α	ν	q	SWF	$p(q)$	$P(q,s)$
0.5	0.40	0.938	11.746	3581.58	0.084	0.443
1.0	0.45	0.887	11.723	3578.30	0.085	0.444
1.5	0.55	0.845	11.777	3575.93	0.083	0.442
2.0	0.60	0.809	11.793	3574.10	0.082	0.441
2.5	1.00	0.000	12.413	3572.71	0.079	0.423

Table 13.6: Variable λ

$[\ \alpha_{FB} = 0.5,\ q_{FB} = 13.17,\ p(q_{FB}) = 0.0751\ ]$

λ	α	ν	q	SWF	$p(q)$	$P(q,s)$
0.10	0.75	0.821	12.043	3583.05	0.081	0.434
0.20	0.60	0.863	11.879	3580.22	0.083	0.439
0.30	0.45	0.887	11.723	3578.30	0.084	0.444
0.40	0.30	0.904	11.571	3577.26	0.085	0.448
0.50	0.30	0.904	11.571	3676.84	0.085	0.448

Index